Geographies of
Agriculture

We work with leading authors to develop the strongest
educational materials in Geography, bringing cutting-edge
thinking and best learning practice to a global market.

Under a range of well-known imprints, including
Prentice Hall, we craft high quality print and
electronic publications which help readers to understand
and apply their content, whether studying or at work.

To find out more about the complete range of our
publishing, please visit us on the World Wide Web at:
www.pearsoned.co.uk

Geographies of Agriculture: Globalisation, restructuring and sustainability

GUY M. ROBINSON

School of Geography, Kingston University, London

PEARSON
Prentice Hall

Harlow, England • London • New York • Boston • San Francisco • Toronto • Sydney • Singapore • Hong Kong
Tokyo • Seoul • Taipei • New Delhi • Cape Town • Madrid • Mexico City • Amsterdam • Munich • Paris • Milan

Pearson Education Limited
Edinburgh Gate
Harlow
Essex CM20 2JE
England

and Associated Companies throughout the world

Visit us on the World Wide Web at:
www.pearsoned.co.uk

ISBN 0582 356628

British Library Cataloguing-in-Publication Data
A catalogue record for this book is available from the British Library

Library of Congress Cataloging-in-Publication Data
Robinson, G. M. (Guy M.)
    Geographies of agriculture : globalisation, restructuring, and sustainability / Guy M. Robinson.
        p. cm.
    Includes bibliographical references and index.
    ISBN 0–582–35662–8 (pbk.)
    1. Agricultural geography.   I. Title.

    S494.5.G46R63 2003
    338.1′09—dc21

                                                                    2003049882

10 9 8 7 6 5 4 3 2 1
06   05   04

Typeset in 10/12pt Sabon by 35
Printed and bound in Malaysia, KVP

Contents

Figures

Maps

Tables

Preface

Agricultural geography has long been an important component of the geographical discipline and has been the subject of numerous textbooks. These have reflected the prevailing methodology of their time, and so have moved from a regional emphasis in the first half of the twentieth century towards more systematic approaches, drawing initially upon simple economic and behavioural theories. For example, in the early 1980s there was a concern with behavioural approaches to the subject, emphasising decision-making by individual farmers. However, the dramatic changes within human geography over the last quarter century have also transformed agricultural geography, bringing significant methodological and philosophical changes to both discipline and sub-discipline alike. These have rendered existing textbooks at best deficient or even obsolete. There has been a glaring need for a new introductory text on agricultural geography to cover the wide-ranging work of recent years, which has broadened the scope of enquiry by treating agriculture as an integral part of a wider food and fibre production system encompassing input supply, farming, food processing, wholesaling and retailing. This broader subject matter has also been tackled from a number of new perspectives so that Marxist theory, political economy and regulationist approaches have been embraced as appropriate tools for investigation of new global trends in agricultural development. These methods and ideas form a significant part of this new text that embraces agricultural geography's place in the 'new' geography.

The book is intended as an undergraduate textbook, relevant to those taking a second-year or Honours option in agricultural geography or a related area, but also suitable for some first years taking more general courses in human geography. In addition, it is intended to be of interest to a broader range of students interested in agriculture, food production and supply, and rural development, especially those from sociology, economics and development studies. It is aimed at students throughout the English-speaking world.

The book consists of ten chapters, incorporating some traditional subject matter, such as the factors of agricultural production and classification of agricultural systems, but deliberately emphasising topics reflecting globalisation processes, the integration of agriculture in the wider food system, the concern with attaining sustainable systems, and the importance of government and supra-government policy. Although including many examples from the Developed World (especially North America and Western Europe), agricultural issues affecting the Developing World are not neglected as in some previous agricultural geography texts, and there is a chapter dealing explicitly with hunger and starvation ('the world food problem'). The concluding chapter is forward looking, with references to the impacts of biotechnology, the formulation of new policies for agriculture and changing demands of agriculture upon the environmental resource base.

I would like to acknowledge the help and encouragement, much of it at key times, over my thirty years as a professional geographer that I received from a number of distinguished former colleagues, who helped shape my thinking about life as much as on matters geographical. Sadly all of the following are now deceased: Terry Coppock, Frank Emery, Jean Gottmann, Jack Hotson, John House, Kath Lacey, Mary Marshall, Walter Newey, Paul Paget, Derrick Sewell and Wreford Watson.

I am indebted to Matthew Smith, formerly of Addison Wesley Longman, for his perseverance and encouragement. At several times it must have seemed to him that I would never complete the manuscript, but important activities such as golf, gardening and support for West Bromwich Albion notwithstanding, it did get finished. That it did so owes much to inputs from various people, including my colleagues at Kingston University, the University of Otago, New Zealand, where I was in receipt of a William Evans Visiting Fellowship, and the University of Guelph, Canada, where I spent a short period of research leave as I completed this book. I received great assistance from Claire Ivison, who produced all the maps and diagrams and throughout showed great patience regarding my pathetically inadequate efforts to describe what each figure should contain. Most of all, though, I must thank my wife, Susan, who corrected my English, helped structure my arguments, criticised, cajoled and encouraged in equal amounts, and constantly reminded me that Philosophy is much harder to write than Geography!

Note

Agricultural geography abounds with instances of agro- and agri-. Hence agribusiness and agri-environmental seem to be universal, whilst agro-food network and agro-industrial are widely used. Personally, agro- always puts me in mind of crowd violence at soccer matches so I have taken a unilateral decision and opted for agri- throughout this book. I make no apologies to lovers of agro-.

Guy M. Robinson
Epsom Downs, Surrey
November 2002

Acknowledgements

We are grateful to the following for permission to reproduce copyright material:

Figures 1.2 and 1.4 from *The ecology of agricultural systems* (Bayliss-Smith, T. P. 1982), pp. 10 and 109, and Figure 7.1 from *Rural Africa* (Grove, A. T. and Klein, F. M. G. 1979), published by Cambridge University Press; Figure 1.5 from Adaptability of agricultural systems to global climate change: a Renfrew County, Ontario, Canada pilot study (Brklacich, M., McNabb, D., Bryant, C. and Dumanski, J. 1997) in Ilbery, B. W., Chiotti, Q. and Rickard, T. (eds), *Agricultural restructuring and sustainability: a geographical perspective*, p. 186, published by CAB International; Figure 2.1 from *Spatial behavior: a geographical perspective* (Golledge, R. G. and Stimson, R. J. 1997), p. 27, published by The Guildford Press; Figure 2.2 from Whatmore, S. J. (1995), From farming to agribusiness: the global agro-food system, in Johnston, R. J., Taylor, P. J. and Watts, M. (eds), *Geographies of global change: remapping the world in the late twentieth century*, pp. 57–67, and Figure 6.1 from Chul-Kyoo, K. and Curry, J. (1993), Fordism, flexible specialization and agri-industrial restructuring: the case of the US broiler industry, *Sociologia Ruralis*, Vol. 33, pp. 61–80, published by Blackwell Publishing Ltd.; Figure 2.3 reprinted from Family farmers, real regulation, and the experience of food regimes, *Journal of Rural Studies*, Vol. 12, pp. 245–58 (Moran, W., Blunden, G., Workman, M. and Bradly, A. 1996) and Figure 8.5 reprinted from Shifting global strategies of US foreign food aid, 1955–90, *Political Geography*, Vol. 12, pp. 232–46 (Kodras, J. E. 1993), with permission from Elsevier; Figure 4.1 from The Greek fresh-fruit market in the framework of the Common Agricultural Policy, unpublished PhD thesis, University of Coventry (Kaldis, P. E. 2002),

reprinted with the kind permission of the author; Map 1.2 from *The Food Resource* (Pierce, J. T. 1990) and Figure 5.1 from *Government and agriculture: a spatial perspective* (Bowler, I. R. 1979), published by Pearson Education Limited; Figure 6.4 from Farm-based recreation in England and Wales, unpublished PhD thesis, University College Worcester (Chaplin, S. P. 2000), reprinted with the kind permission of the author; Figure 7.2 from Changes within small-scale agriculture. A case study from south-western Tanzania, *Danish Journal of Geography*, Vol. 96, pp. 60–9 (Birch-Thomsen, T. and Fog, B. 1996); Figure 7.3 from An agricultural transition on the Pacific Rim: explorations towards a model, in Magee, T. and Watters, R. (eds), *Asia Pacific: new geographies of the Pacific Rim* (Hill, R. D. 1997), pp. 93–112, published by C. Hurst & Co. (Publishers) Ltd; Figure 8.1 from The space of vulnerability: the causal structure of hunger and famine, *Progress in Human Geography*, Vol. 17, pp. 43–67 (Watts, M. J. and Bohle, H. G. 1993), published by Hodder Arnold; Figure 9.1 from Land transformation: trends, prospects and challenges, *Geographical Papers, University of Reading*, no. 125, p. 25 (Mannion, A. M. 1998), reprinted with the kind permission of the author; Figure 9.3 from Land use conflict in the urban fringe, *Journal of the Scottish Association of Geography Teachers*, Vol. 18, pp. 4–11 (Pacione, M. 1989), published by the Scottish Association of Geography Teachers; Figure 9.5 from *Techniques in map analysis* (Worthington, B. and Gant, R. L. 1983), p. 94, published by Palgrave Macmillan; Figure 9.6 from Half a century of cropland change, *Geographical Review*, Vol. 91, pp. 525–43 (Hart, J. F. 2001), published by the American Geographical Society; Figure 10.1 from *The state of the nation's birds* (Gregory, R. D., Noble, D. H., Campbell, L. C. and Gibbons,

D. W. 1999), published by RSPB/BTO/Defra; Map 10.1 from Geographical aspects of the 2001 outbreak of foot and mouth disease in the UK, *Geography*, Vol. 87, pp. 142–7 (Ilbery, B. W. 2002), published by the Geographical Association.

Whilst every effort has been made to trace the owners of copyright material, in a few cases this has proved impossible and we would be grateful to hear from anyone with information which would enable us to do so.

1 Agricultural systems

1.1 Introduction

Agriculture, or farming, is the rearing of animals and the production of crop plants through cultivating the soil (Mannion, 1995a, p. 2). It is a manifestation of the interaction between people and the environment, though the nature of this interaction has evolved over a period of at least 10,000 years since the first domestications of wild plants began in the Fertile Crescent of the Near East around 10,000 years before present (BP) (MacNeish, 1992). Sheep, pigs, goats, cattle, barley and wheat were first domesticated in this area, followed by six other independent origins of agriculture: East Asia (between 8400 and 7800 BP) utilising rice, millet, pigs, chickens and buffalo; Central America (4700 BP) and South America (4600 BP) produced potato, maize, beans, squash, llama, alpaca and guinea pigs; North America (4500 BP) had goosefoot and sunflower, whilst Africa (4000 BP) had cattle, pigs, rice, millet and sorghum.

The domestication of plants and animals spread from the Near East into south-eastern Europe, where the combination of improved cultivation methods and an extensive trading network supported first the Greek and then the Roman empires. It was these that gave rise to the term 'agriculture', which is derived from the Latin word *agar*, and the Greek word *agros*, both meaning 'field', and symbolising the integral link between land-based production and accompanying modification of the natural environment (Mannion, 1995b).

This modification produces the agri-ecosystem in which an ecological system is overlain by socio-economic elements and processes. This forms 'an ecological and socio-economic system, comprising domesticated plants and/or animals and the people who husband them, intended for the purpose of producing food, fibre or other agricultural products' (Conway, 1997, p. 166). Agricultural geographers have viewed this agri-ecosystem as part of a nested hierarchy that extends from an individual plant or animal and its cultivator, tender or manager, through crop or animal populations, fields and ranges, farms, villages, watersheds, regions, countries and the world as a whole.

Agricultural geography includes work that spans a wide range of issues pertaining to the nature of this hierarchy, including the spatial distribution of crops and livestock, the systems of management employed, the nature of linkages to the broader economic, social, cultural, political and ecological systems, and the broad spectrum of food production, processing, marketing and consumption. The principal focus for research by agricultural geographers in the last four decades has been the economic, social and political characteristics of agriculture and its linkages to both the suppliers of inputs to the agri-ecosystem and to the processing, sale and consumption of food products (Munton, 1992). However, it should not be forgotten that at the heart of farming activity, underlying the chain of food supply from farmers to consumers, is a set of activities directly dependent upon the physical conditions within which farming takes place. Hence, before concentrating upon the principal foci of contemporary agricultural geography in the rest of the book, this chapter outlines the key physical aspects of agriculture that

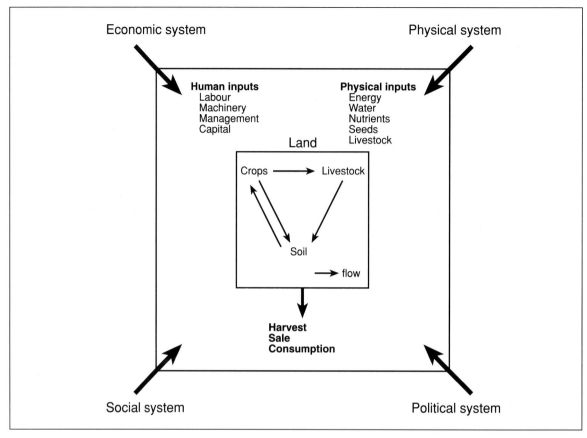

Figure 1.1 Simple conceptualisation of a farming system

form the foundations to which the multi-faceted human dimensions of farming activity are applied.

Six key factors can be recognised as influencing the distribution of farming types: biological, physical, economic, political, socio-cultural and marketing (the food trade) (Ganderton, 2000, p. 161). These factors are part of the simple conceptualisation of a farming system shown in Figure 1.1, in which a series of inputs to the land generates a series of outputs. Social, economic, political and environmental factors affect the nature of these inputs and outputs, producing tremendous variation in the pattern of the world's farming systems. This chapter examines the nature of the physical basis to agriculture and considers the role that physical and other factors play in determining the nature of farming systems.

1.2 The agri-ecosystem

Unlike many aspects of economic activity, the contributions made by the physical environment can be of fundamental significance to the nature of the farming system. In the Developed World especially, farmers may have capital at their disposal to enable purchase of inputs that can substantially modify some of the physical characteristics of the land upon which farming is based. Yet, the changeable nature of weather and hydrological regimes can inject elements of risk and uncertainty unknown to other areas of economic activity. Despite the influence of non-physical factors upon farming, farming retains strong parallels with the natural ecosystems from which agricultural systems derive, and hence farming can be portrayed as an agri-ecosystem.

There is a reciprocal relationship between environmental factors and agricultural activity. Environment affects the nature of farming, exerting a wide range of controls, but, in turn, farming affects the environment. Agricultural systems are modifications of natural ecosystems; they are artificial human creations in which productivity is increased through control of soil fertility, vegetation, fauna and microclimate. This is intended to generate a greater biomass than that of natural systems in similar environments, though this may also generate undesirable environmental consequences. In particular, farming alters the character of the soil and, through runoff, effects can be extended to neighbouring areas, e.g. nitrate pollution of the watercourses and groundwater, and effects on wildlife (Parry, 1992).

Agriculture can also be differentiated from many other economic activities by virtue of the fact that it deals with living things. The plants and animals have inherent biological characteristics that largely determine their productivity. They function best in environments to which they are well adapted, and this exerts a strong influence on the nature and location of agricultural production. Despite the diversity of agricultural systems they all have many common features, notably the human control of ecosystems, for example by varying the amounts of energy inputs. The extent and exact nature of this control varies largely in response to social and economic requirements. However, the control is also affected by environmental characteristics acting as constraints.

In an agri-ecosystem the farmer is the essential human component that influences or determines the composition, functioning and stability of the system. The system differs from natural ecosystems in that the agri-ecosystems are simpler, with less diversity of plant and animal species and with a less complex structure. In particular, the long history of plant domestication has produced agricultural crops with less genetic diversity than their wild ancestors. In agri-ecosystems the biomass of the large herbivores, such as cattle and sheep, is much greater than that of the ecologically equivalent animals normally supported by unmanaged terrestrial ecosystems. Cultivation means that a higher proportion of available light energy reaches crops and, because of crop harvesting or consumption of crops by domestic livestock, less energy is supplied to the soil from dead and decaying organic matter and humus than is usually the case in unmanaged ecosystems in similar environments. Agri-ecosystems are more open systems than their natural counterparts, with a greater number and larger volume of inputs and outputs. Additional inputs are provided in the form of direct energy from human and animal labour and fuel, and also in indirect forms from seeds, fertilisers, herbicides, pesticides, machinery and water. The dominant physical or natural resource inputs to the farming system are climate and soils.

1.3 Climate and agriculture

The greatest physical constraints upon agricultural activity are generally imposed by average temperatures, the amount of precipitation, and their annual distribution. More localised limitations are imposed by soil type, nutrient availability, topography, aspect and drainage. In particular, though, climate determines the broad geographical region in which any given crop can be cultivated. Whilst modern plant breeding has extended the moisture and temperature requirements of many plants, they still have their limits, and hence it is still legitimate to refer to a strong degree of climatic determinism in the distribution of agricultural crops. Rice and Vandermeer (1990) have combined the influence of climatic controls with edaphic factors to produce a classification of the world's agro-climatological characteristics (Table 1.1). This classification is one of several ways in which agricultural systems may be differentiated. This is discussed further below with specific reference to various types of classification of agricultural systems.

1.3.1 Temperature

Both plant and animal growth are affected by several climatic variables, notably receipt of solar energy, precipitation available for transpiration and temperature during the growing season. Relationships between these variables are rarely linear, but optimum growth conditions can be recognised

Table 1.1 A classification of agricultural systems based on climate–soil–crop inter-relationships (agri-climatological types)

Agri-climatology	Approximate distribution	Soil type	Cropping systems
1. Wet tropical	Lat. 5°N to 5°S	Oxisols, Ultisols; Nutrient poor	Shifting cultivation, plantation cropping
2. Wet–dry tropical	Lat. 5° to 25°N 5° to 25°S	Vertisols, Alfisols, Mollisols; Water content varies; high clay content; fire is important	Shifting cultivation, rice cultivation, maize production, dryland rainfed agriculture
3. Cool tropical	Mountainous zones of the tropics, elevation >1000 m, e.g. Andes and high regions of S. Asia	Soils are highly varied	Diverse agriculture, e.g. tea plantations, coffee plantations, dairy cattle
4. Moist mid-latitude	Lat. 25° to 55°, mostly northern hemisphere. Frost threat present	Utlisols, Mollisols, Alfisols	Cotton, peanuts. tobacco, soy beans, rice, maize, tomatoes, multiple cropping systems, e.g. three main crops per year
5. Dry mid-latitude	Lat. 30° to 50°, mostly northern hemisphere	Mollisols; high clay content	Small grains, e.g. wheat, maize (US Corn Belt), oil-seed crops
6. Mediterranean	Lat. 30° to 40°N, Wet winter, dry summer	Inceptisols; high clay content	Small grain production, grazing, rainfed cereals, viticulture, olives, citrus fruit production
7. Arid	23.5°N, 23.5°S, band either side of tropics	Aridisols; moisture deficit all year; irrigation	Pastoral nomadic systems, some rainfed agriculture

Note: The soil classification is the US Seventh Approximation or Comprehensive Soil Classification Scheme, which is discussed below.
(Sources: based on Rice and Vandermeer (1990) and Mannion (1995a))

where plants give the highest yields, i.e. the largest weight of the edible part of the crop per unit area. Generally it will be most economic to cultivate a crop in a physical region around the optimum and well removed from the absolute limit to the plant's growth. However, there are many other factors that can affect the economic limits to production, notably those impinging upon production costs and market demand. Increases in production costs and/or falls in price promote contraction of the margin of cultivation towards the optimum area. For example, with respect to the production of maize and wheat in the area west of Buenos Aires, Argentina, both crops can give high yields in this area. However, maize is preferred because, from the same inputs, its yields tend to be highest (Grigg, 1995, p. 23). Further south, where there is less rainfall, maize yields decline more rapidly than those for wheat as maize is less drought resistant, and hence wheat becomes the dominant crop.

With reference to crop growth, provided there is adequate water, the crucial determinants are temperature and light, which effectively enable distinctions to be drawn between tropical, sub-tropical, temperate and cool-temperate agri-climatic regions. These are broadly related to different biochemical pathways of carbon dioxide fixation in photosynthesis, which, in turn, reflect basic physiological differences. Further adaptations of crops to climate are associated with plant responses to seasonal variations in weather (termed 'phenology').

There are certain inherent differences in photosynthetic efficiency between species. In particular these relate to two different ways of using carbon

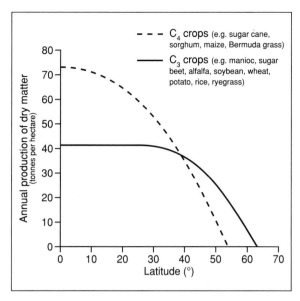

Figure 1.2 Maximum recorded yields in different latitudes of C$_3$- and C$_4$ crops (based on Bayliss-Smith, 1982)

dioxide, the C$_3$-pathway and the C$_4$-pathway, which are strongly influenced by temperature regimes. The former is common in temperate species and the latter in species of tropical and/or arid origins, and especially grasses. As shown in Figure 1.2, both types of plant excel in tropical and sub-tropical regions, although there is a sharper decline with rising latitude for C$_4$ crops like maize, sugar cane, sorghum and fodder grasses. These tend to be tall, upright plants able to cope with high light intensities during the middle of the day. C$_3$ species tend to outperform C$_4$ crops in mid to high latitudes. C$_3$ crops include sugar beet, alfalfa, soybean, wheat, potatoes, rice and ryegrass.

For each crop there is a temperature range within which growth and development can take place. The critical temperatures are:

- the minimum, below which there is insufficient heat for biological activity;
- the optimum, at which rates of metabolic processes are at their maximum;
- the maximum, beyond which growth ceases. Higher temperatures may be harmful or lethal.

For cool-season cereals Yao (1981) gives the ranges for these three critical temperatures as 0–5°C,

25–31°C and 21–37°C respectively. The corresponding temperatures for warm-season cereals are 15–18°C, 31–37°C and 44–50°C.

Some crops have other particular temperature requirements, such as needing an alternation of low night-time and higher daytime temperatures. Others need a degree of winter chilling before flowering and seed-setting can occur within the available growing period. Other crops are termed photoperiodic, if it is day-length that is the trigger necessary to initiate flowering. Four groups are normally recognised (Tivy, 1990, p. 23):

- Short-day/long-night, with a photoperiod of under ten hours, e.g. soybean, sweet potatoes, millet. These occur in low latitudes where spring or autumn seasons are warm enough to allow their harvest cycle to be completed.
- Long-day/short-night, with a photoperiod of over 14 hours, e.g. small grains, timothy, sweet clover. These occur in high latitudes.
- Intermediate day, with a photoperiod of 12 to 14 hours and an inhibition of reproduction either above or below these levels.
- Day-neutral, unaffected by variations in day-length.

Variations in crop-growing habits in response to climate, especially temperature, have played an important part in the application of scheduling techniques whereby farmers phase the planting and harvesting of annual crops in order to make the most efficient use of the time and space available. This has become especially significant in the production of fruit and vegetable crops for the chilled and frozen food market, and was recognised in the 1950s when the climatologist C. W. Thornthwaite constructed climatic calendars for the planting and harvesting of peas for freezing by the Seabrook Farms Co. in the USA (Wang, 1972).

There are various definitions of the growing season, with perhaps the most useful being the vegetative season, the period during which there is production of sufficient vegetative growth to support either continuously or subsequently the necessary yield-forming activities (Map 1.1). A common threshold of 6°C as the mean temperature has often been adopted to represent the commencement of growth in temperate cereals and

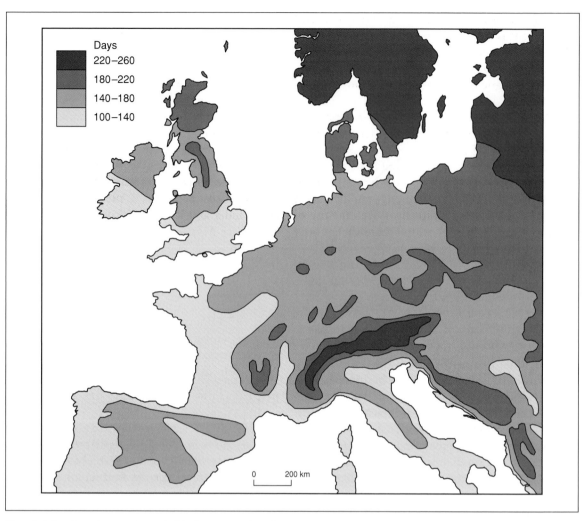

Map 1.1 Duration of vegetative period in Europe: number of days between seeding of summer grains in spring and of winter wheat in autumn (based on Seeman *et al.*, 1979)

grasses, though individual crops can deviate quite widely from this. A related concept is that of accumulated temperatures, which represents a measure of the relative warmth of a growing season of a given length. However, this notion tends to assume a linear relationship between increase in heat and crop growth, which is misleading as organic growth tends to occur at an exponential rate. Factors other than temperature also affect growth, thereby modifying the assumed linearity. For example, the length and effectiveness of the growing season also depends on availability of sufficient soil moisture, and indeed, in many parts

of the world, water is the main factor limiting crop production.

1.3.2 Water

Water supply from precipitation is fundamental to all agricultural systems, though its management by the farmer can compensate for problems in the natural supply. Water transports nutrients to and through plants and also plays a vital role in weathering, leaching and erosion. Therefore it largely controls inputs of nutrients to and losses from the system. Losses may occur through evapotranspiration, drainage to groundwater, and lateral flow as runoff and throughflow to streams. Water is stored in the soil, plant tissues and in the bodies of livestock.

In many agricultural systems, moisture deficiencies are a vital limitation on crop yields (e.g. Cooke, 1979). Hence, management of water supply by farmers can form a significant component of their activities, forming part of the substantial modifications of the natural hydrological cycle that agricultural activity creates. 'The character and extent of crop cover, tillage, land drainage and irrigation practices, and – more indirectly – even the use of fertilisers and pesticides all influence the amount of water stored in the system and the quantities lost by drainage, run-off and evapotranspiration' (Briggs and Courtney, 1989, p. 11). These impacts are of major concern in temperate areas.

In effect, agricultural practices disrupt the pattern of the annual water balance that is associated with the functioning of any ecosystem. This balance is the outcome of the input from precipitation versus losses from evapotranspiration. If a surplus of water occurs then water can accumulate in the soil until maximum storage capacity (field capacity) occurs, after which there is waterlogging or runoff. If evapotranspiration exceeds precipitation there is a moisture deficit. Plants then deplete the store of soil moisture until, eventually, the soil is said to reach wilting point, when evapotranspiration ceases and plants start to wilt. In arid areas large soil moisture deficits often develop during the growing season.

When the balance between precipitation, evaporation and runoff is considered worldwide, it is clear that certain areas are in much greater need of irrigation and very careful waste management in order to conserve water. Thus in North Africa and the Middle East the water requirement for all uses is around 97 per cent of the usable resource. A high proportion of the available resource is also used in the semi-arid areas of southern and eastern Europe, and north, central and south Asia. In these and many other parts of the world irrigation can be important, reflecting the significance of variations in climate, other physical characteristics, the intensity of demand for water by a given agricultural system and other demands on the water resource. In total, irrigated agriculture consumes 2500 km^3 of water on 18 per cent of the world's cultivated land (Pierce, 1990, p. 126). This represents a seven-fold increase in irrigated area during the twentieth century (though there are problems in defining what constitutes irrigated land). The greatest contribution of irrigation to national food output occurs in countries where padi rice is a significant crop and/or where semi-arid climates occur, e.g. Pakistan (65 per cent of the cultivated area is irrigated), China (50 per cent), Indonesia (40 per cent), Chile and Peru (35 per cent), India and Mexico (30 per cent) (Rangeley, 1987) (see Map 1.2). 'The majority of the irrigation development in Asia is for the expansion of rice cultivation, which already accounts for three-quarters of food grain consumption there' (Pierce, 1990, p. 131). Hence it is not surprising that in India, Pakistan, the Philippines (and also Mexico) much of the capital assistance for the so-called 'Green Revolution' package to revolutionise agricultural output in the 1960s and 1970s was dominated by expenditure for the extension, upgrading and maintenance of irrigation systems. In the same decades the majority of Middle Eastern countries allocated between 60 and 80 per cent of their agricultural investment to irrigation (UN Water Conference, 1977). Many of these extensions to the irrigated area, as typified by the opening of the Aswan High Dam in Egypt in 1969, have been costly, large-scale projects reflecting the fact that most of the sites with plentiful supplies of water for irrigation have been utilised already. Moreover, there is little evidence that economies of scale are present in the larger schemes and hence costs are very high, contributing in several cases to the high indebtedness of Developing Countries (Kreuger et al., 1992).

Whilst irrigation has had dramatic impacts upon crop productivity and in extensions to the cultivated area, especially in dry climate regions, there have been some negative impacts associated with alterations to the natural water-salt balance, increasing the extent and risk of saline and alkaline soils. Secondary salinisation and alkalisation occur when the natural drainage system is unable to accommodate the additional water input. This causes a rise in groundwater levels, and capillary

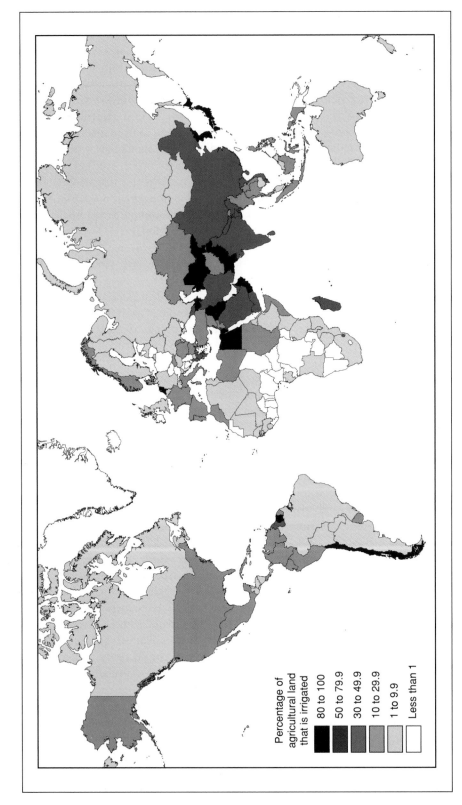

Map 1.2 Proportion of agricultural land that is irrigated (based on Pierce, 1990, p. 131)

Percentage of
agricultural land
that is irrigated

80 to 100
50 to 79.9
30 to 49.9
10 to 29.9
1 to 9.9
Less than 1

action can transport dissolved salts to the active root-zone and surface areas. The extent of this process depends on the depth of the groundwater, but generally the higher the salt content of the groundwater, the greater the depth through which this saline solution can damage crops.

Following the expansion of irrigation in the 1970s, one estimate claimed that nearly 70 per cent of the 30 million ha of irrigated land in Egypt, Iran, Iraq and Pakistan were suffering from moderate to severe salinity problems (Schaffer, 1980). A further 7 million ha in India were also being adversely affected following extensions of irrigation in the central and western portions of the Indo-Ganges plain, Gujerat and Rajasthan. Such problems also occur outside the Developing World, with major problems occurring in Australia, especially in conjunction with irrigation in the main river basin, the Murray–Darling (Robinson *et al.*, 2000, pp. 57–62). Similar problems have been recorded in several parts of former Soviet Central Asia where there has been increased extraction of river water for growing cotton. In addition to exacerbating salinity problems, the high levels of water consumption have also contributed hugely to the demise of the Aral Sea, which has dramatically shrunk in size in recent years.

1.4 Agricultural soils

The primary agricultural management practice is the cultivation of the soil, which acts as the reservoir of the water, minerals and nutrients that are needed for plant growth. Cultivation involves selecting plants likely to produce a satisfactory yield; propagation, in which tillage of the soil ensures suitable conditions for planting or sowing and for feeding the crop; and protection from competition for the primary resources by weeds and from direct or indirect reduction in yield potential by animal pests and pathogenic organisms (Tivy, 1990, pp. 1–2).

Soils can vary considerably by virtue of changes in their structure, depth, texture, plant nutrient content, and acidity. These characteristics influence not only the types of crops that can be grown but also their yields on both a macro-, world-scale and also a micro-, field-scale. It is possible to argue that there are optimum edaphic conditions for particular plants, analogous to the climatic optimum referred to above, but rarely is there sufficiently detailed information for this concept to be of practical value (Ellis and Mellor, 1995).

The edaphic optimum applies to soils in which a wide range of crops may be grown with high yields attainable without the need for extensive modifications of the soil. For many plants this optimum will refer to deep, well-drained loams that are well supplied with plant nutrients. Some, though, require highly specific soil conditions, for example padi rice, which needs impermeable subsoil so that it can grow in waterlogged conditions. However, the value of this optimum is limited in terms of interpreting crop distributions. In part, this is because of the lack of sufficiently detailed widespread information on soils. Many countries produce soil maps based on inference from underlying geology and climate rather than use of field survey. Moreover, it is difficult to make strict correlations between soil type and crop growth without reference to a range of other environmental variables. For example, soil texture is generally regarded as an important determinant of crop yields as it influences moisture availability, and in temperate climates texture is usually the main edaphic determinant of yield, but this is also dependent on rainfall and evaporation, not simply texture.

A vital characteristic of soil is its depth. In general, the deeper the soil, the greater will be its capacity to store water and minerals. Shallow soils, as found in many glaciated areas, cannot carry enough moisture to support plant growth, supply sufficient nutrients or support root development. Some thin soils, such as those frequently developed on limestones, may give good yields of shallow-rooted cereals. In contrast, certain plants only thrive in deep loams, as in the case of potatoes.

Soil texture refers to the relative importance of particles of different sizes. The large particles of sandy soils provide light, well-drained land that is readily warmed for early spring planting. In contrast, fine particles of clay soils retain water, are slower to warm in spring and are heavy to cultivate. However, clay soils release potassium only slowly so they are less likely to suffer from potash deficiency. Their water-retaining capacity can be of advantage in dry conditions, but their tendency to waterlogging has meant that underdrainage has proved particularly important in helping to improve yields (e.g. Phillips, 1975). Loams are a combination of clay and sands, which neither tend to suffer low moisture content nor excess water.

Acidity is another significant soil variable. This is usually measured on the potential hydrogen (pH) scale that runs from 0, the most acid, to 14, the most alkaline, with a pH of 7 indicating a neutral soil. The degree of soil acidity is determined by the chemical composition of the underlying parent material and the rate of leaching, which, in turn, is closely related to the amount and type of precipitation. In temperate climates soil acidity is greater in areas in receipt of heavy rainfall, e.g. in Britain acid soils occur in the west and in upland areas, where a pH of around 4.9 can ensure good crops of potatoes and a pH of around 6.2 supports good crops of lucerne, grown for cattle feed. Increasing soil acidity reduces the amount and activity of nitrogen-fixing bacteria, and it also reduces organisms that improve soil texture and structure. As a result, few crops thrive in acidic soil. Similarly, few like highly alkaline conditions, with most preferring neutral or mildly acid conditions. Some cereals, notably oats and rye, can tolerate relatively high acidity. Highly alkaline soils are common under semi-arid conditions and where irrigation produces waterlogging. High alkalinity may be tolerated by barley, cotton and the date palm.

1.4.1 Soil classification

Work on the classification of soils in the late nineteenth century by the Russian pedologist, V. V. Dokuchaev, and on soil-forming factors in the 1930s by Jenny (1941) produced both a basis for classifying soils and an understanding of the relationships between soil properties and environmental factors. These factors are climate, parent material, biotic factors (vegetation, animals and human activity), relief and time over which the factors have operated.

Dokuchaev focused on large-scale soil variations associated primarily with the relationships between soils, natural vegetation and climate in Russia. He argued that environmental factors were crucial in producing dynamic processes that formed different soil layers or horizons, but with an equilibrium that could be established eventually, along the same lines as was later suggested for vegetation and 'climax' plant communities (Tansley, 1953).

Three basic soil classes were recognised: zonal, intra-zonal and azonal, the first of which was identified as soils that had developed in particular climatic and/or vegetational regimes. As shown in Table 1.2, seven soil types were recognised in this zonal category. There were three in the intrazonal or transitional category, where local physiographic or lithological factors could override zonal factors in influencing soil development. Azonal soils occur where erosional and depositional processes dominate other soil processes.

Table 1.2 The Dokuchaev soil classification

Zone	Soil type
Zonal classes	
Boreal	Tundra (dark brown) soils
Taiga	Light grey podzolised
Forest–steppe	Grey and dark grey soils
Steppe	Chernozem
Desert–steppe	Chestnut and brown soils
Desert	Aerial soils, yellow soils, white soils
Sub-tropical	Laterite or red soils
Intrazonal classes	
	Dryland moor soils or moor–meadow soils
	Soils containing carbonate (rendzina)
	Secondary alkaline soils
Azonal classes	
	Moor soils (e.g. moorland peats)
	Alluvial soils (e.g. riverine wetland soils)
	Aeolian soils (e.g. sand dune soils)

In the first half of the twentieth century there were several examples of classification schemes based largely on Dokuchaev's ideas, notably Baldwin et al.'s (1938) in North America. However, the breadth of the categories was problematic as was its over-emphasis upon the influence of climate and vegetation (Avery, 1969). It was recognised in the 1950s and 1960s that 'many of the world's agricultural soils have been influenced for centuries by man's [sic] activity and are only in a limited sense "natural"' (Curtis et al., 1976, p. 32). This contributed to a move away from typological classifications, inferred from genetic factors, to definitional systems based on recognisable soil properties. New systems devised in individual countries were popularised, with notable developments occurring in Canada, the Netherlands (De Bakker and Schelling, 1966), the USA (NRCS, 1998) and the UK (Avery, 1973), reflecting local inputs and conditions.

The United States Department of Agriculture (USDA) produced a classification, known as the Seventh Approximation, based on soil pedons, an artificial cuboid unit with a cross-sectional area dependent on the lateral variability of properties that define classes. This recognised twelve soil orders, as shown in Table 1.3, but with a detailed set of sub-orders that has permitted preparation of tables of approximate relations, notably making comparisons with the classification developed in 1974 by the Food and Agriculture Organisation (FAO) of the United Nations. In contrast to the system used in the USA, that adopted by the Soil Survey of England and Wales was essentially a classification of soil profiles or vertical soil sections, and applied to profiles deeper than 10 cm using standard criteria (Table 1.4).

Most of the world's major soil groups (Map 1.3) are deficient in one or more of the key attributes relating to physical and/or chemical properties. For example, unproductive entisols, inceptisols, mountain soils and spodosols cover large parts of the Northern Hemisphere's cold climate zone. They tend to be young soils with little profile development; they are low in organic matter, high in acidity and offer limited depth of rooting potential. Spodosols in particular are often leached, acidic, poorly drained and may be bog-like in places. Aridisols are associated with dry savannah, steppe and desert climates. They have low humus content and are prone to high levels of salinity and alkalinity. Their potential for agriculture depends greatly on irrigation development and techniques to improve water retention abilities of the soil.

To both the north and south of the principal areas of aridisols are the oxisols and ultisols of the humid tropics, covering around two-thirds of this climatic zone. Generally these are well drained, deep and granular, though they possess poor mineral properties and are low in nutrient supply.

Table 1.3 Soil taxonomy in the USA

Soil orders	Characteristics
Gelisols	Soils with permafrost within 2 m of the surface
Histosols	Organic soils
Spodosols	Acid soils with a subsurface accumulation of metal-humus complexes
Oxisols	Intensely weathered soils of tropical and subtropical environments
Vertisols	Clayey soils with high shrink/swell capacity
Aridisols	$CaCO_3$ – containing soils of arid environments with moderate to strong development
Ultisols	Soils with a subsurface zone of silicate clay accumulation and <35% base saturation
Mollisols	Grassland soils with high base status
Alfisols	Soils with a subsurface zone of silicate clay accumulation and ≥35% base saturation
Inceptisols	Soils with weakly developed subsurface horizons
Entisols	Soils with little or no morphological development

(*Source*: www.nhq.nrcs.usda.gov/CCS/soilmnth.html)

Table 1.4 Soil classification for England and Wales

Major group	Group
Lithomorphic soils Normally well-drained soils with distinct, humose or organic topsoil and bedrock or little altered unconsolidated material at 30 cm or less	Rankers Sand-rankers Ranker-like alluvial soils Rendzinas Pararendzinas Sand-pararendzinas Rendzina-like alluvial soils
Brown soils Well-drained to imperfectly drained soils (excluding Pelosols) with an altered sub-surface horizon, usually brownish, that has soil structure rather than rock structure and extends below 30 cm depth	Brown calcareous earths Brown calcareous sands Brown calcareous alluvial soils Brown earths Brown sands Brown alluvial soils Argillic brown earths Paleo-argillic brown earths
Podzolic soils Well-drained to poorly drained soils with black, dark brown or ochreous sub-surface horizon in which aluminium and/or iron have accumulated in amorphous forms associated with organic matter. An overlying bleached horizon, a peaty topsoil, or both, may or may not be present	Brown podzolic soils Gley-podzols Podzols Stagnopodzols
Pelosols Slowly permeable non-alluvial clayey soils that crack deeply in dry seasons with brown, greyish or reddish blocky or prismatic sub-surface horizon, usually slightly mottled	Calcareous pelosols Argillic pelosols Non-calcareous pelosols
Gley soils With distinct, humose or peaty top-soil and grey or grey-and-brown mottled (gleyed) sub-surface horizon altered by reduction, or reduction and segregation, of iron caused by periodic or permanent saturation by water in the presence of organic matter. Horizons characteristic of podzolic soils are absent	Alluvial gley soils Sandy gley soils Cambic gley soils Argillic gley soils Stagnogley soils Humic-alluvial gley soils Humic-sandy gley soils Stagnohumic gley soils
Man-made soils With thick man-made topsoil or disturbed soil more than 40 cm thick	Man-made humic soils Disturbed soils
Peat soils With a dominantly organic layer at least 40 cm thick formed under wet conditions and starting at the surface or within 30 cm depth	Raw peat soils Earthy peat soils

(*Source*: based on Avery, 1973; 1980; 1990)

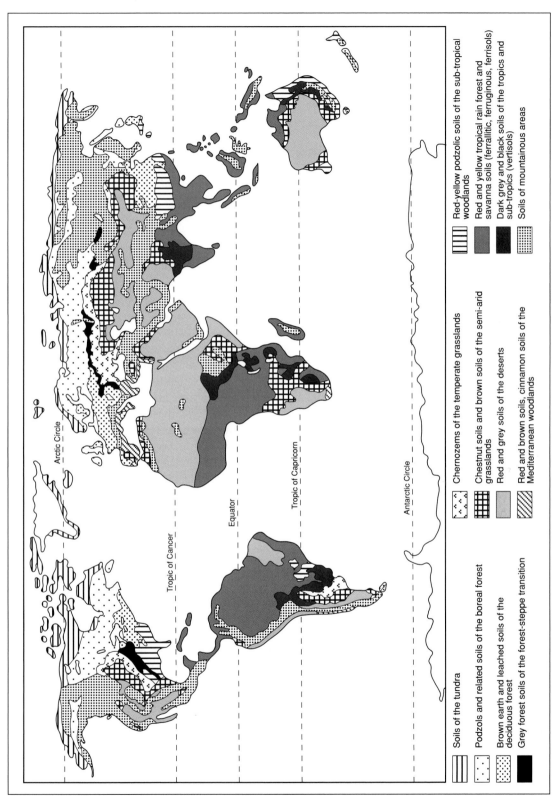

Map 1.3 Major soil types (compiled from various sources)

Legend:

- Soils of the tundra
- Podzols and related soils of the boreal forest
- Brown earth and leached soils of the deciduous forest
- Grey forest soils of the forest-steppe transition
- Chernozems of the temperate grasslands
- Chestnut soils and brown soils of the semi-arid grasslands
- Red and grey soils of the deserts
- Red and brown soils, cinnamon soils of the Mediterranean woodlands
- Red-yellow podzolic soils of the sub-tropical woodlands
- Red and yellow tropical rain forest and savanna soils (ferrallitic, ferruginous, ferrisols)
- Dark grey and black soils of the tropics and sub-tropics (vertisols)
- Soils of mountainous areas

Arctic Circle
Tropic of Cancer
Equator
Tropic of Capricorn
Antarctic Circle

Once cleared of vegetation these soils are susceptible to having their base nutrients easily leached away so that soil acidity remains high. However, applications of nitrogenous and phosphoric fertilisers, lime to reduce acidity, manures and careful land management to control erosion have provided the basis for long-term agricultural production in south-east USA and many parts of China (Chapman, 2001). Other soils found in similar climatic conditions, such as alfisols, vertisols and mollisols have better physical and chemical properties for crop production and hence have been centres for major concentrations of population (Sanchez and Buol, 1975). These have been the soils upon which successful applications of the Green Revolution have been based (see Chapter 8). The same soils occur in cooler latitudes on either side of the humid tropics and have been the basis for the world's major centres of food production, especially in Central Europe, North America, Russia and China.

This very brief overview of soil factors and the distribution of the world's major soils provides some indication of the delicate physical and biological balance that renders agriculture possible and restricts production in various ways. It helps to explain why less than 15 per cent of the earth's land mass has been cultivated and why current estimates claim that this proportion can only be extended to 25 per cent with huge investment in either or both irrigation and other technological inputs (Pierce, 1990, p. 22).

1.5 Energy

Although a range of physical factors affects the distribution of agricultural crops and animals, domestication for over 10,000 years has sought to modify or ameliorate the influence of these factors. This can be seen most readily in terms of 'artificial' alterations to nutrient availability, especially nitrate, through applications of fertilisers. The structure and tilth of soils may also be improved by mechanised means, as can availability of water through irrigation.

Domestication of plants involved modifications to the existing plant stock by genetic changes through human selection, either deliberately or unconsciously. In particular, over the past three centuries, plant and animal breeding programmes and recent applications of biotechnology have improved the inherent productivity of plants and livestock. This has been achieved in various ways, initially through reducing competition for light and nutrients between crops and 'pests'. Pesticides, fungicides and other products may be applied to reduce this competition.

Another significant modification is to increase inputs of energy to the agricultural system, usually by additions of fossil-fuel energy that supplement solar power, the prime input to the system. The amount of available energy has a major effect upon photosynthesis of plants, the process whereby organic matter is formed by plants through a chemical process sustained by sunlight. However, the rate of photosynthesis is also constrained by environmental conditions external to the plant community, including light intensity, temperature regime, water and nutrient availability, topography and soil structure (Mannion, 1995a, p. 20). Hence one key aspect of farm management is to reduce or remove these constraints in order to maximise the useable solar radiation.

To assist the process of plant growth, additional supplies of fertiliser are usually added by the farmer, in the form of animal manure, other types of organic matter or chemical additives. In most parts of the Developed World the amount of non-animal-based fertilisers has increased substantially over the last century. For example, the quantity of artificial fertiliser applied to crops in Britain between 1939 and 1975 rose sevenfold (Briggs and Courtenay, 1989, p. 33). In Developed Countries the type of fertilisers applied has also changed substantially over time, moving from simple forms, such as ammonium nitrate, ammonium phosphate and potassium chloride to compound fertilisers comprising a mixture of nitrogen, phosphorus and potassium. The largest increases have been of nitrogenous fertilisers, with research showing direct links between increased nitrogen usage and rising crop yields (Austin, 1978).

The agri-ecosystem can be regarded as a dynamic system of flows of matter and energy, including water, solutes (nutrients) and solids (e.g.

soil particles). Inputs take the form of weathering of underlying bedrock, to produce soil and nutrients; energy from solar radiation; precipitation; transfers from adjacent land surfaces; and inputs by the farmer in the form of seeds, livestock, manure, fertiliser, animal feeds and fuel energy. In addition, the farmer can control many outputs from the system. The major output occurs in the form of the crop harvest, with inputs of manures and fertilisers required to effect replenishment. Land drainage and irrigation affect water loss whilst crop husbandry practices, such as tillage, soil conservation measures and crop rotation, can control soil erosion (Boardman, 1992; Robinson and Blackman, 1990). Foster *et al.* (1997) illustrate this last point when referring to the influence of centuries-old cultivation, removal of hedgerows, new methods of seed-bed preparation and changes in the timing of seed-bed preparation upon flooding of farmland in the English Midlands (see also Evans, 1997). Relatively small amounts of the overall energy inputs to an agricultural system are actually consumed by people. For potato cultivation the proportion of energy inputs available as human food may be as high as 0.25 per cent. For cattle produced on an extensive ranching system it may be as low as 0.002 per cent (Duckham and Masefield, 1970).

In terms of the management of agricultural systems, key aspects are the ways in which certain key cycles are controlled, especially energy, water and nutrients. Solar radiation provides the fundamental energy source to support plant and animal growth. The amount of radiation received by plants depends on latitude and albedo or reflection, which varies considerably for different surfaces (Jones, 1976). This energy is the driving force for cycling nutrients through the agri-ecosystem (Figure 1.3). Crucial aspects of this cycling are the carbon (or inorganic) cycle and the nitrogen cycle. Different farming systems and their accompanying management strategies have varying effects upon nutrient cycling, thereby producing differential impacts upon the soil base upon which the systems operate. For example, soluble nitrates are vulnerable to removal by being leached and so they have to be maintained by careful management. In traditional hill sheep farming in Europe, for instance,

the soil is maintained in a more or less steady state whereas, once grazing land is improved via addition of artificial fertiliser, the total soil pool gains nitrogen and phosphorus, but may lose potassium (Frissel, 1978; Tivy, 1987). Therefore, additions from the fertiliser are not entirely balanced by losses in animal product and leaching.

The constraints imposed by solar radiation, temperature and rainfall are less readily controlled by farmers than nutrient deficiency, as nutrients can be managed to a certain extent via careful husbandry. The most vital nutrients are nitrogen, phosphorus, potassium, calcium and magnesium, which together can comprise up to 10 per cent of a plant's dry weight, the plant having derived the minerals from the soil. In a natural ecosystem, a large degree of recycling of nutrients occurs via decomposition of dead plant litter by bacteria and fungi or by manure from animals that have consumed the plants. Small nutrient losses through leaching or runoff may be balanced by weathering of bedrock or input via precipitation. However, no such balance occurs readily in an agricultural system. Tillage of the soil creates bare patches that accelerate losses through runoff and leaching, whilst crop harvesting interrupts the natural recycling of nutrients. For example, Bayliss-Smith (1982, p. 14) reports that, in the case of sweet potato production in the Solomon Islands, at least $105 \, \mathrm{g \, m^{-2}}$ of soil nutrients are removed in the leaves, stems and tubers of the crop. This means that sustained cultivation of the same piece of land is rendered impossible unless efforts are made to replace the lost nutrients. This can take the form of manuring, mulching or adding artificial fertilisers. In addition, leguminous crops such as beans, peas, clover and lucerne can be grown which have *Rhizobia* bacteria that add up to $10 \, \mathrm{g \, m^{-2}}$ nitrogen per annum. Appreciation of these nitrogen-fixing properties led to crop rotations being developed in Europe that usually included a legume. The classic example was the Norfolk four-course rotation, first developed in East Anglia in the eighteenth century, which consisted of clover, wheat, turnips and barley grown in rotation to enable farmers to use a plot of land continuously without recourse to fallowing (Orwin and Whetham, 1971) (Table 1.5). This type of husbandry was subsequently superseded in the

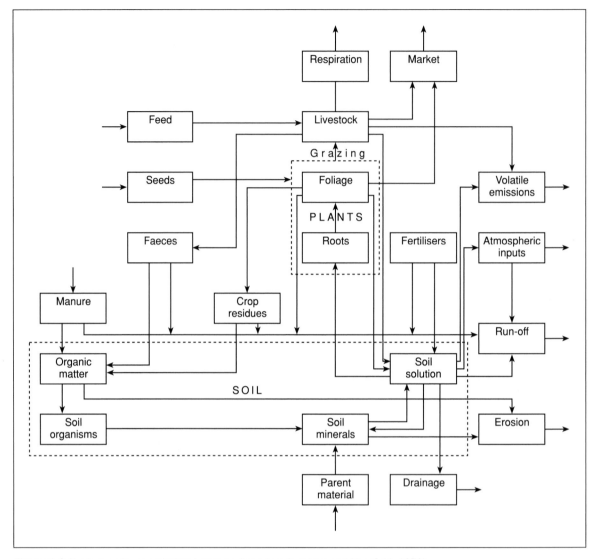

Figure 1.3 Nutrient cycles in agricultural systems (based on Duckham and Masefield, 1970)

Table 1.5 The Norfolk four-course crop rotation

Year	Crop	Use
1	Turnips or Swedes	Folded with sheep in winter
2	Spring barley	Cash crop
3	Red clover	Grazed in spring and summer
4	Winter wheat	Cash crop

(*Source*: Briggs and Courtney, 1989, p. 29)

Developed World by the practice of adding artificial fertilisers to the soil, though this has substantially increased the overall energy consumption in farming as such fertilisers are energy-intensive products.

In temperate climates the main legumes are peas, beans, lucerne and clover, whilst their counterparts in the tropics are chick peas, groundnuts and soybeans, though nitrogen-fixation is less efficient in tropical conditions and so these three crops are grown as protein-rich foods rather than for their soil restorative qualities. Temperate crop legumes fix between 100 and 225 kg nitrogen per ha per annum.

In evaluating the efficiency of agricultural systems in energy terms, Bayliss-Smith (1982, pp. 33–4) offered four different measures:

- The energy ratio: the edible energy produced by the system in a net form (i.e. excluding animal fodder), divided by the total human-derived energy input.
- The gross energy productivity (GEP): the total food energy produced by the system, in consumed and other forms, divided by the total population. This shows the gross energy production per person per annum, from which the daily energy productivity may be calculated.
- The surplus energy income: the energy not consumed directly by people, e.g. in the form of crops fed to animals.
- The energy yield in terms of net food output per ha.

Calculation of these four measures for a series of different farming systems enabled direct comparisons between them to be made, as shown in Table 1.6. This comparison shows how the application of industrial technology results in substantial increases in energy yield. By substituting machine power for manpower a huge increase in GEP is achieved. Surplus energy income also rises but as a proportion of GEP it is lower than in pre-industrial societies. The overall efficiency of energy use, the energy ratio, declines as the degree of dependence on fossil fuels rises, though, as shown in Figure 1.4, a semi-industrial system can

Table 1.6 Energy measures for seven agricultural systems

Agricultural system	Energy yield (MJ/ha yr)	Gross energy productivity (MJ/person day)	Surplus energy income (MJ/person day)	Energy ratio (output/input)
Pre-industrial				
New Guinea	1,460	10	2.3	14.2
Wiltshire (UK) (1826)	7,390	80	2.4[a]	40.3
				12.6[a]
Semi-industrial				
Ontong, Java	14,760	38	5.3	14.2
South India (1955)	42,280	44	8.6[b]	13.0
			4.0[c]	10.2[d]
South India (1975)	66,460	36	no data	9.7
Full-industrial				
Moscow collective	8,060	59	4.1	1.3
South England (1971)	44,860	2,420	18.8	2.1

a   Farm labourer's household
b   Peasant caste farmer's household
c   Untouchable caste household
d   Subsistence rice cultivation only
(*Source*: Bayliss-Smith, 1982, p. 108)

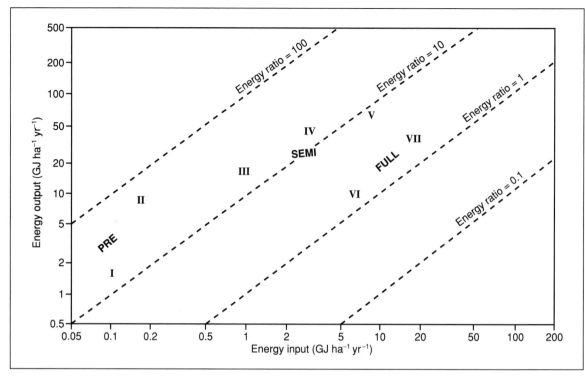

Figure 1.4 Comparisons of various farming systems based on energy use: I New Guinea; II Wiltshire, England, 1826; III Ontong Java Atoll, South-West Pacific; IV Wangala, South India, 1955, V Wangala, South India, 1975; VI Moscow Oblast collective farm; VII Southern England, 1971 (based on Bayliss-Smith, 1982, p. 109)

be almost as efficient as a pre-industrial one. Once agriculture relies heavily on mechanisation and purchased inputs, very little more additional energy is gained from farming than is expended in production. It must be acknowledged, though, that attempting to classify agricultural systems on the basis of energy inputs and outputs is just part of a process of differentiation that needs to consider a broad range of ecological, demographic, economic and social characteristics if it is to be more holistic.

1.6 Climatic change and agriculture

The preceding discussion has tended to refer to the distribution of climatic parameters in terms of their constancy or variation within known and understood bounds. Hence, it has been possible to produce maps of agri-climatic zones and soil types based upon data that reflect norms for recent

decades. However, during the last 30 years it has become clear that not only have there been world-wide climate changes occurring throughout the last 10,000 years, which have undoubtedly affected the distribution of crops and livestock, but also that recent short-term climatic changes may be affecting agricultural distributions (Mendelsohn, 1998). This has given rise to several studies assessing the potential agricultural impacts of global climatic change (Figure 1.5) (Parry and Livermore, 2002).

Global climate scenarios are usually derived from General Circulation Models (GCMs), which have been used to forecast the effects of an altered atmosphere on macro-scale climatic properties. Historical or spatial analogues and incremental changes to the observed weather record are also used to specify climatic change scenarios (Bootsma *et al.*, 1984; Easterling *et al.*, 1992). Various critical scenarios have been portrayed for agriculture as a consequence of predicted climate change during

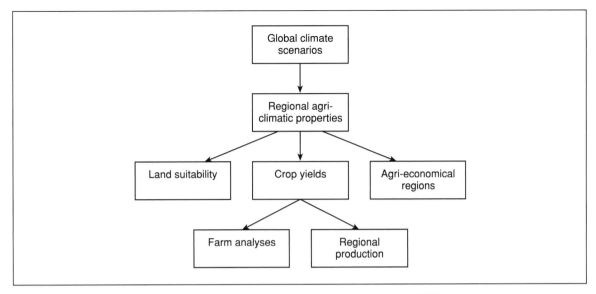

Figure 1.5 The agricultural impacts of global climatic change (based on Brklacich *et al.*, 1997)

the twenty-first century. For example, one prediction is for a general reduction in crop yields in many tropical and sub-tropical regions for most projected increases in temperature, and a general reduction, with some variation, in potential crop yields in most mid-latitude regions for increases in annual temperatures of more than a few degrees Centigrade (Fischer *et al.*, 1995). Perhaps the most serious current predictions relate to future failures of food supplies through diminished supplies of water. Such failures are projected for the Sahelian region of Africa, south Asia and large parts of Latin America as a consequence of shifting rainfall belts. Approximately one-third of the world's population (1.7 billion) already live in countries that periodically experience significant deficits in water supplies, and some reports predict that the population affected will rise to 5 billion by 2025. Moreover, in central Asia, north and southern Africa, because of a combination of higher temperatures and pollutant runoff, decreases in rainfall will be associated with declining quality of water that is available. Against this portrayal of impending disaster, it is possible that some regions may benefit from predicted warming, which may enable new crops to be grown, e.g. extending the cultivated area northwards in parts of Canada.

Impacts of long-term rises in sea level, possibly linked to global warming, are already apparent in some parts of the world. For example, along the east coast of China, mean sea level rose at a rate of 1.0 mm y^{-1} from 1920 to 1987 (Chen, X, 1991). It is estimated that in the Yangtze delta the sea level will rise between 50 and 70 cm betweeen 2000 and 2050. This is of particular significance around Shanghai where 95 per cent of agricultural land is below the high astronomical tide level (Chen and Zong, 1999). This area already has a long history of sea inundation and there is now a substantial programme to mitigate potential hazards induced or intensified by the rising sea level. The main measures are improvements in drainage quality and capacity, renewal and increase of pumping facilities, development of new crops tolerant to a higher groundwater table, and construction of a flood barrier.

Tropical regions are especially vulnerable to potential damage from environmental changes because the prevalence of poor soils covering large areas already provides significant problems for agricultural activity. Yet, relatively little research has been performed specifically on the agricultural impacts of climate change in the tropics, though results from the Developed World have been extrapolated worldwide. Indeed, agricultural impact studies have been performed at various spatial scales, from the regional (e.g. Cohen, 1994) to the national, and continental (e.g. Hulme et al., 1999; UKCCIRG, 1991) to the global (e.g. Rosenzweig and Parry, 1994). Impacts on specific major crop and livestock systems have also been performed (e.g. Easterling et al., 1993; Baker et al., 1993). For example, Blasing and Solomon (1983) concluded that a 1°C temperature increase would move the Corn Belt in the USA 175 km to the north and north-east of its present location. However, this prediction on its own offers little insight to the vulnerability of agricultural systems to changing conditions or to the capacity of agriculture to adapt to change (Chiotti and Johnson, 1995; Chiotti, 1998).

To obtain regional agri-climatic properties from broad global scenarios two processes can be employed (Hossell et al., 1996). The first spatially interpolates macro-scale data to a regional scale, and the second converts basic climatic parameters, such as maximum daily temperature, into agri-climatic properties such as growing-degree days. The actual methodologies employed have been numerous and it is difficult to evaluate their validity. However, results have then been applied to potential impacts of climatic change, notably upon land suitability and agri-ecological assessments. These have generally used resource-rating schemes to assign land parcels to broadly defined suitability or agri-ecological classes (e.g. Brklacich and Curran, 1994). Rating schemes have frequently used comparisons of basic climatic requirements, such as growing period or moisture supply, for broad categories of agricultural production in relation to specified shifts in selected climatic properties.

An alternative is crop yield analysis, which is conducted for specific crops and is usually more sophisticated than resource-rating schemes, being based on interactions among crop growth factors and generating estimates of output per land unit (Baethgen and Magrin, 1995). The outputs from crop yield analyses have produced studies at both farm and regional level. The former consist of whole-farm models to estimate impacts of changes in yield arising from global climatic change on cash flow and vulnerability of different farm types (Mooney and Arthur, 1990). Regional production and macro-economic models have also been used to estimate the effects of global climatic change on regional production potential and international trade in agricultural products (Fischer et al., 1995).

Slaymaker (2001) bemoans the lack of attention to the impacts of climate change upon land use, especially at regional and local scales, describing the relationship between climate and human activity as a subtly reflexive one with feedbacks between people and their changing environment that are difficult to predict. He argues that the impacts of social and economic forces upon land use are just as significant as those of climate change, but it is the potential impacts of the latter that are receiving the much greater share of research funding.

Similar sentiments are voiced by Chiotti et al. (1997) who argue that there is a strong need for a better understanding of the relationship between present climate and agriculture. Much of the conventional research on climate change impacts has been based on the neo-classical economic paradigm that assumes the market will encourage or discourage various adjustments. This tends to assume that the land will be devoted to the best economic use, with farmers having access to the best available technology and adjusting their farming practices to suit the changing and variable climate (Easterling et al., 1993). These assumptions ignore the constraints on farmers' choices and, until recently, have not engaged with the work of human geographers on agricultural restructuring and adaptation, even though there are examples from this work of farmers' responses to drought and famine in Developing Countries (e.g. Blaikie and Brookfield, 1987; Liverman, 1991). This shows how climate variation is just one of a series of factors impinging upon decisions made by the farm household.

Even in areas susceptible to extremes of weather and climate, farmers often tend to relegate the importance of climate as a major factor in their decision-making. This can be seen in work with farmers in the driest part of the Prairies in Canada, known as Palliser's Triangle. The farmers here recognise that their farming operations are sensitive

Table 1.7 Climate sensitivities and adaptive responses in southern Alberta, Canada

Farming system	Climate sensitivity	Adaptive response
Dryland	Drought Soil moisture Wind erosion Frost (early autumn) Hail	Minimum tillage Trash/stubble Chemical fallow Half rotation Crop share, crop insurance
Irrigation	Wind erosion Frost (early autumn) Hail Heat unit	Pivots Less tillage Some trash/stubble Chemical fallow
Feedlot	Summer heat Winter cold Blizzards Chinooks	Sprinklers Barns Rotational grazing Seeding grass/livestock breeds

(*Source*: Chiotti *et al.*, 1997, p. 212)

to particular climatic parameters, but have various adaptive responses that are generally regarded as a 'normal' aspect of farming operations. These strategies vary with different farming systems (Table 1.7). Overall, there has been a shift towards production of higher value crops and cattle in the region, but this may reflect a strategic adjustment to ensure economic viability rather than an adaptive response to climatic variability (Chiotti *et al.*, 1997; Kemp, 1991).

The current climatic regime has been treated as representative of baseline conditions and the scenarios for global climatic change as conceptually equivalent to treatments as in standard scientific approaches to field-level agricultural research (in which various plots are subjected to different treatments of chemicals or water supply) (Chiotti *et al.*, 1997). Brklacich *et al.* (1997) criticise this approach for its tendency to use sensitivity assessments for certain attributes of production systems to specified climatic perturbations in predicting responses to climatic change for selected systems. This type of approach makes certain, usually unstated, assumptions, namely: climate is the only condition that will vary; farmers will perceive the change in climate; agricultural systems are vulnerable to the changed climate; therefore farmers will choose to adapt to the altered climate. These

assumptions need to be challenged through use of new frameworks that place research on climatic change into a broader context of agricultural decision-making so that the latter becomes the key element in the research. There are also computational limits imposed by existing computing technology, scale problems when predicting from the global to the local, and the sheer uncertainty regarding the regional and local dimensions of climatic change.

There is relatively little evidence that farmers have responded to recent changes in climate by changing their farming practice, or that they have much knowledge of potential future climate change (Robinson, D. A., 1999). Initial evidence of farmers' responses in the Developed World to climatic changes over several decades suggests that factors other than environmental ones tend to be more influential in decision-making (e.g. Smithers and Smit, 1997). For example, Brklacich *et al.*'s (1997) pilot study in Renfrew County, Ontario, examined adaptive responses to climatic change over a 20-year period. During this time farmers believed that precipitation had decreased and climate was becoming less predictable. Specific adaptations to these perceived changes were modifications to crop varieties and types, adoption of alternative harvesting methods and modifications to infrastructure. However, many farmers made no explicit response to the changes, partly because their farms were already adapted to operate under a range of conditions and also because factors other than climate influenced their decisions. Nevertheless, when provided with predictions of future climate change based on increased CO_2 in the atmosphere, which could produce rising yields for grain and soybeans, farmers' responses were to predict a widespread adoption of crop varieties likely to benefit from longer growing periods.

In recent work on climatic change and agriculture, various studies have highlighted the role of technological innovation in the handling of climatic risks (Parry *et al.*, 1988; Smit *et al.*, 2000), and there has been research emphasis upon the attributes of agriculture that are most sensitive to climate, the types and combinations of climatic events that are most problematic for farming, the nature of farmer responses to climatic risk and uncertainty, and the role of the other forces as mediating factors in shaping these responses (Bryant *et al.*, 2000). However, more knowledge is required about the nature of agricultural innovations that have been induced by climate, and about the relationship between knowledge development and the forces that drive it. To date the key technologies mediating risk in the face of climatic variations are mechanical innovations (irrigation, conservation tillage, improved drainage) and biological science (hybrids). But the scale of deviation away from so-called normal conditions may define the experience of climatic change and it is this that may stretch the ability of technical innovation to provide 'solutions'. Moreover, high-tech agriculture has the capacity to influence climate adversely through increasing CO_2 emissions (Komen and Peerlings, 1997).

Work by Smithers and Blay-Palmer (2001) on the Ontario soybean industry suggests that some technical innovations permitted the crop to respond well to wide variations in heat, but that there was little evidence of progress toward more broadly based adaptability for inter-annual variations in weather conditions. In part this reflected the presence of factors limiting consideration of climate in development of the crop (Table 1.8).

Table 1.8 Factors limiting consideration of climate in the research and development process

Economic
- High cost of research
- Emphasis on profit versus curiosity-based research
- Increased domination of private breeders

Ownership of intellectual property
- Expensive to purchase rights to necessary genes/technology
- Limited accessibility
- Constrained innovations

Regulatory barriers
- Risk of developing broadly adapted varieties

Competing market needs
- Food and non-food niche market products
- Development of new technologies

(*Source:* Smithers and Blay-Palmer, 2001, p. 190)

It is only gradually being more widely appreciated that, even if farmers do perceive new opportunities arising from shifts in climate, there may be significant structural barriers that might restrict their adoption of crops well suited to the new climatic conditions. Holloway and Ilbery (1997) demonstrated this with respect to prospects for the introduction of navy beans in the UK. This crop, used for manufacturing the highly popular baked beans in tomato sauce, has been grown largely in North America, but could also be produced under warmer conditions in future in the UK (Holloway *et al.*, 1995; Holloway and Ilbery, 1996). Clearly it cannot be assumed that farmers in the UK would adopt navy beans or any other new crop as a simple response to global warming, although this is conventionally suggested by climate/agri-ecosystem modelling procedures (Hossell *et al.*, 1996; Smit, 1994). Any such potential adoption would be eventuated within a broader ongoing process of change on farms in which a key factor would be the role of food processing companies and retailers. For example, farmers may be keen to grow navy beans but could be prevented from doing so by processors unwilling to offer them contracts. This attitude by processors may relate to the views of supermarkets, which may be content with the nature of their current supplies of baked beans. Hence, for this particular crop, 'the effects of global warming, if it occurs, would be largely subsumed by a combination of structural resistances and a combination of processor and farmer decision-making behaviour' (Holloway and Ilbery, 1997, p. 354).

1.7 Classifying agricultural systems

Variations in the type of farm management have been summarised with reference to four main parameters (Smith and Hill, 1975): biological diversity, intensity of human management, net energy balance, and management responsibility. Differences in these have produced a continuum of farming systems, from maintenance of a semi-natural ecosystem, as in open-range grazing, to farming involving the creation of artificial environments such as glasshouses and hen batteries. In seeking to understand the spatial distribution of the various systems, geographers have utilised various types of classification. Indeed, classification has

been a significant element of agricultural geography for some time, and attempts to produce systems of world agricultural regions have a long history, generally based on the concept of a set of agricultural regions in which there is a recognised uniformity of agricultural production. In developing such classifications, three basic approaches can be recognised (Tarrant, 1974, pp. 112–45), described next.

1.7.1 Land classification

Land classification regions are based on the physical properties of land or its capabilities. The physical properties are usually ones relating to topography, soils and vegetation. In the UK such a classification was produced at the behest of the Scott Committee on Land Utilisation in Rural Areas in 1942, using the Land Utilisation Survey (LUS) as its basis (Stamp, 1940). This produced a simple three-fold classification of land, into good, medium and poor land, with some sub-categorisation (Table 1.9). This classification emphasised the current use of land, as revealed in the

Table 1.9 Land classification in Great Britain (1948)

Major category	Sub-category	% of total area
Good		37.9
	1. First class	4.1
	2. Good general purpose farmland	
	a. suitable for ploughing	15.2
	b. suitable for grass	5.0
	3. First class land, restricted use, unsuitable for ploughing	2.2
	4. Good but heavy land	11.4
Medium		24.6
	5. Medium light land	
	a. suitable for ploughing	4.4
	b. unsuitable for ploughing	0.4
	6. Medium general purpose farmland	19.8
Poor		35.2
	7. Poor heavy land	1.6
	8. Poor mountain and moorland	31.7
	9. Poor light land	1.5
	10. Poorest land	0.4
Built-up area		2.3
		100.0

(*Source:* Stamp, 1948)

LUS, as opposed to the land's inherent potential. Hence, subsequent classifications, especially in the land capability series prepared by the Ministry of Agriculture, Fisheries and Food (MAFF), have been based on a wide range of variables relating to soils (depth, structure, chemical composition and permeability) and other physical criteria (slope, precipitation, drainage, temperature, frost susceptibility and availability of groundwater). These variables provide an indication of the physical limitations in a particular area, and hence of land capability.

A similar basis has been adopted in land capability classifications in other countries, with classes graded from very suitable to highly unsuitable for agriculture, and mapped at varying levels of detail. In the case of the well-known classifications prepared by the Department of Lands and Forests in Ontario and the United States Soil Conservation Service (USSCS) it has been soil characteristics that have been especially prominent. In the case of the former, land is classified according to the costs of developing it for commercial agriculture. For the USSCS the classification focuses on the land's susceptibility to soil erosion, but tends to ignore general features of productivity. In Australia, the Commonwealth Scientific and Industrial Research Organisation (CSIRO) has produced land classifications since the 1940s, using a land-systems approach in which areas are defined 'within which certain predictable combinations of surface forms and their associated soils and vegetation are likely to be found' (Cooke and Doornkamp, 1990, pp. 20–1).

1.7.2 Land use classification

This focuses upon the use to which land is put rather than its physical characteristics. It was popularised by J. C. Weaver (1954a; 1954b; 1954c; Weaver *et al.*, 1956), who developed the idea of crop-combination regions in which it was recognised that regional production complexes usually include a range of crops rather than a monoculture. Thus the US corn, cotton and spring wheat belts are rarely absolute monocultures, and, even where one crop is predominant, there may be subsidiaries that can be recognised within a crop-combination region. In other cases the region can embrace the

crops grown in a crop rotation to include temporary leys or other areas of grassland. Weaver's classification used a simple statistical procedure (see Tarrant, 1974, pp. 122–5; Robinson, 1988a, pp. 296–8) to produce the type of maps illustrated in Map 1.4. In effect, the actual areal distribution of crops is compared with model arrangements (1-crop, 2-crop, 3-crop and so on) to determine the best fit. This best fit is then the crop-combination allocated to the spatial unit under consideration. The degree of best fit can be quite variable and the results are entirely dependent upon the crops considered in the model. So it can be crucial to decide whether permanent grassland should be included in the crop combination or only arable land or only those crops featuring in a crop rotation.

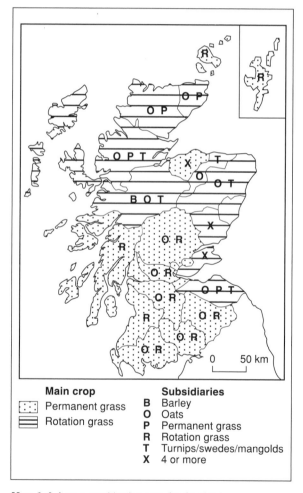

Map 1.4 A crop-combination map for Scotland

Such classifications may omit crops that are extremely important in financial terms but which only occupy a small area, as it is land use rather than other aspects of production that is usually being considered. However, conversions may be applied to convert crop areas into measures of labour input (e.g. standard man-days) so that low labour intensity crops covering large areas, such as permanent pasture or extensive production of cereals, do not automatically appear as the dominant element within a crop combination. It is possible to use gross margin and gross output data as conversions, but this usage is generally restricted by lack of readily available information.

The use of standard man-days is based on the reduction of all types of production on a farm to their standard labour requirements. This ignores variations in efficiency of different farmers as well as the effects of scale economies. It is possible to produce 'standard' figures on how many days of work per annum are required in cultivating a unit area of a given crop. In this way, labour-intensive crops can assume a much greater importance in any classification based on standard man-days. Using standard man-day conversions, farming activities involving livestock production can also be incorporated in classifications. However, Weaver *et al.* (1956) concluded that there was no suitable statistical method to enable them to combine both crops and livestock in a single index to create a map of farming regions. Hence other methods have been applied in establishing type-of-farming regions.

1.7.3 Type-of-farming regions

Land use and type-of-farming are closely related but sufficiently different to create problems when distinctions are drawn between the two. Chisholm (1962) argued that type-of-farming classifications should be based on individual farms, including a wide range of variables, notably the production and management of the farm as well as information on yields, crops and livestock. In practice, though, many of the widely used classifications have been based on a restricted set of variables (Aitchison, 1992). For example, the one most frequently used, devised by Derwent Whittlesey (1936), focuses on five criteria:

- crop and livestock associations;
- intensity of land use;
- processing and disposal of farm produce;
- methods and degree of farm mechanisation;
- types and associations of buildings and other structures associated with agriculture.

From these criteria 13 types of world agriculture were derived (Map 1.5; Table 1.10). These are essentially generalised descriptions, but they have been utilised for various purposes and usually with little modification (e.g. Symons, 1968). Nevertheless, suggestions have been made regarding the addition of more specific criteria that can be measured quantitatively (Helburn, 1957). It is the lack of available data and the complexity such criteria would create that has contributed to its lack of application (see Evans, 1996, for limitations of the UK's agricultural census).

At a regional level, various type-of-farming classifications have been applied (see Aitchison, 1992), following pioneering work by Baker (1926) on the agricultural regions of the United States. Many of these fail to employ systematic criteria on which to base their classification, though Geography's quantitative revolution in the 1950s and 1960s generated a range of approaches based on the Weaver method and various cartographic techniques (e.g. Adeemy, 1968; Birch, 1954; Edwards, 1992, p. 154; Scott, 1957). Of this work perhaps the most well-known is that of Coppock, who employed regionalisation extensively in his three major studies of agriculture in the UK (Coppock, 1971; 1976a; 1976b). He argued that it was the combinations of crops and livestock, termed enterprise combinations, which represented the primary distinguishing features of type-of-farming areas (Coppock, 1964a; Edwards, 1992, p. 136). He then used standard man-day conversions and the Weaver method to produce enterprise combinations for the UK's National Agricultural Advisory districts (covering around 40 parishes each), recognising seven types of enterprise: dairy cattle, beef cattle, sheep, cash crops, fruit, vegetables, and pigs and poultry.

Subsequently, more complex statistical analysis was used by other agricultural geographers to generate multi-attribute agricultural regions (e.g Ilbery, 1981; Robinson, 1981). The most favoured

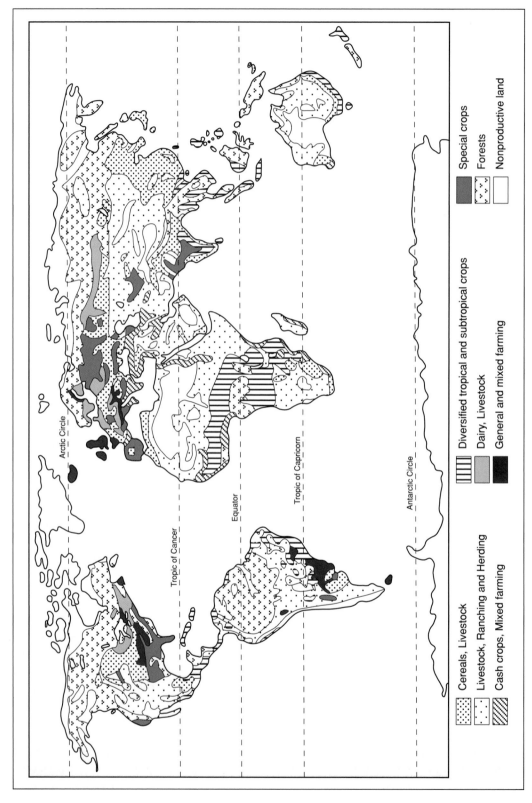

Map 1.5 Classification of world agricultural types (based on Whittlesey, 1936; and Mannion, 1995a)

Cereals, Livestock

Livestock, Ranching and Herding

Cash crops, Mixed farming

Diversified tropical and subtropical crops

Dairy, Livestock

General and mixed farming

Special crops

Forests

Nonproductive land

Table 1.10 Classification of world agriculture

Type[1]	Potential additional variables[2]
1 Nomadic herding	1 Degree of specialisation
2 Livestock ranching	2 Labour and capital ratios to land and to each other
3 Shifting cultivation	3 Sedentary as against migratory habits
4 Rudimentary sedentary tillage	4 Scale of operation
5 Intensive subsistence tillage with rice dominant	5 Land tenure systems
6 Intensive subsistence tillage without paddy rice	6 Level of living achieved
7 Commercial plantation crop tillage	7 Value of the land
8 Mediterranean agriculture	8 Value or volume of production
9 Commercial grain farming	
10 Commercial livestock and crop farming	
11 Subsistence crop and stock farming	
12 Commercial dairy farming	
13 Specialised horticulture	

(*Sources*: [1] Whittlesey, 1936; [2] Helburn, 1957)

technique for this purpose was principal components analysis (Robinson, 1998a, pp. 120–41), which replaced a set of agricultural variables (for example covering a range of information on crop types, livestock, the labour force, farm size, tenure and farmer characteristics) with a smaller set of components representing an amalgam of these variables. The outcome was a handful of key components comprising the basic differentiating features of farming. These components could be mapped to give an indication of the principal aspects of the geography of agricultural differentiation (Map 1.6). However, the subjectivity involved at various stages of the analysis raises questions as to the value of the results obtained, and there are problems of comparability between studies using different variables in the analysis.

1.8 Conclusion

This chapter has outlined the chief elements of the physical basis of farming. But it must be stressed that, although physical factors can exert controls upon agricultural activity, it is socio-economic and political factors that usually determine the detailed characteristics of a farm enterprise and hence the focus on these factors in agricultural classifications.

Important factors include tenure and land ownership, farm size, marketing, transport and labour supply, as well as a range of social and cultural variables intimately associated with the character of the farmer and the farm household. It is farmers' responses to the variety of ecological constraints presented in any given location, related to the complex interplay of socio-economic factors, that produce a range of different types of agricultural activity, so that a strictly ecological or environmental perspective does not provide a very coherent framework on which to base agricultural geography, though it forms the basis of approaches in other disciplines (e.g. Collinson, 2000; Dent and McGregor, 1994). In particular, the varied response by farmers to the nature of the land at their disposal has tended to be strongly influenced by a number of non-ecological factors, such as population pressure, technological innovation, the structures of social organisation and societal values. Hence agricultural geography embraces considerations of a broad spectrum of influences upon agriculture extending well beyond the physical and biological elements referred to in this chapter. Indeed, it has been the economic, political, social and cultural aspects of agriculture, as part of the broader agri-food chain, that have come to dominate agricultural geography.

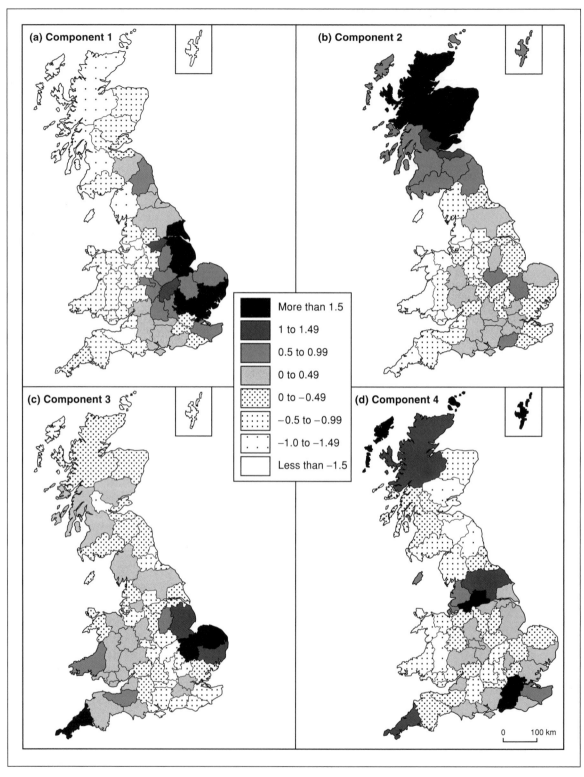

Map 1.6 Multivariate agricultural regions in the UK (the units represent standard deviations from the mean): (a) component 1 (+ arable versus cattle −); (b) component 2 (+ rotation grass/roots versus permanent pasture −); (c) component 3 (+ cash cropping versus beef cattle −); (d) component 4 (+ small farms versus large farms −)

Research on the physical underpinnings of agriculture has become the domain of ecologists, biologists and biogeographers, though human geographers have made contributions to studies of the impacts of selective breeding, biotechnology and genetic modification. In keeping with the current focus of agricultural geography the succeeding chapters deal with the key economic dimensions of agricultural change, emphasising the processes of globalisation and restructuring. However, the importance of the underlying physical and biological constraints is considered in terms of the ongoing concerns for the sustainability of agriculture in the light of increased knowledge regarding detrimental environmental impacts of farming and the increased ability of science to manipulate and modify plant and animal genes.

2 The changing focus of agricultural geography

2.1 'Traditional' agricultural geography

This chapter focuses on how the content of agricultural geography has evolved post-1945, thereby providing a context for the more extended consideration of key components of agricultural change in the rest of the book. Emphasis is placed upon how there has been a move from a 'traditional' form of agricultural geography to new approaches embracing different ideas from across the social sciences.

A standard definition of agricultural geography in the mid-1980s referred to 'the description and explanation of spatial variations in agricultural activity over the earth's surface' (Ilbery, 1985a, p. 1). This interpretation was based largely on consideration of two major avenues of enquiry that had dominated agricultural geography in the twentieth century:

- Location and context, in which emphasis was placed on the regional characteristics of agricultural activities, especially broad trends and tendencies (Coppock, 1968; 1971).
- Explanations of agriculture's great diversity, through consideration of relationships between the large number of relevant variables associated with social, economic, physical and historical factors affecting agriculture (e.g. Grigg, 1992a).

The regional focus in the first of these can be traced to the first time that a specialism specifically termed 'agricultural geography' played a leading role in the development of geography as an academic discipline. This was in the 1920s when agricultural geography was one of the principal specialisms that emerged as part of the growth of regional geography as the discipline's central paradigm (Johnston, 1997, pp. 44–52). An example of this was Baker's (1926) work on the recognition of 'agricultural regions' in different parts of the world. The region became the central focus of study for agricultural geographers, with both single-attribute and multi-attribute regions being recognised. Indeed, for the first half of the twentieth century agricultural geography involved regional delimitations following large-scale mapping of distributions of crops and livestock (e.g. Robertson, 1930) and the classification of agricultural systems (e.g. Whittlesey, 1936). Prevailing ideas on environmental determinisim emphasised the physical controls exerted upon the nature of agricultural activity. Description of agricultural variations was important, with land-use mapping of significance in some countries, a good example being the Land Utilisation Survey of Great Britain, begun in the 1930s by the geographer L. D. Stamp (1948).

Agricultural geography also played a leading role in disciplinary development in the early 1950s when the attempt to define multi-attribute agricultural regions was linked to statistical methods, initially by Weaver (as described in Chapter 1). This formed part of attempts to expand the use of statistical methods in geography. The focus of this work was upon regional changes in farm inputs, farm-size structures, farm incomes and agricultural marketing. Subsequently, work in agricultural geography, like many systematic specialisms in the discipline, became characterised by the use of statistical techniques. This was also part of a theoretical revolution through the use of structured models and economic theory (e.g. Henshall, 1967), with special emphasis placed upon the economics of agricultural production (Coppock, 1964b) and the use of sample surveys of farms (Emerson and MacFarlane, 1995; Errington, 1985).

Although the earliest of these models was devised by von Thunen in the early nineteenth century, it was not popularised within geography until the 1960s when various applications were proposed (Hall, 1968). The economic basis for much work in agricultural geography in the 1960s and 1970s can also be seen as a logical outcome from the formulation of general laws of agricultural location based on economic principles. Models based on von Thunen's ideas emphasised economic rent whilst more recent derivations, such as game theory and the application of linear programming techniques, also stressed the profit motive underlying many farming operations (Found, 1971; Gould, 1963; Thomas and Huggett, 1980).

Within the regional and statistical approaches, geographers devoted attention to both economic and physical environmental factors affecting agricultural development. They treated the diversity of production systems and complex patterns of spatial distribution as reflections of interaction between physical and economic variables. When behavioural approaches, popularised in the 1970s, added the personal characteristics of farmers to the equation, the resultant patterns of agricultural land use were viewed as the product of a complex inter-meshing of dynamic economic, physical and behavioural forces.

The nature of the role of economic forces in influencing farmers' decision-making is suggested by Tarrant (1974, p. 11): 'the economic facts of agricultural life never act in an entirely deterministic way but rather set limits within which farmers are able to operate; they define the freedom of choice.' Economic factors were cited in various studies as key underlying sources of spatial variation in agricultural practice (Morgan and Munton, 1971). That variation attracted the attention of geographers who attempted to explain its existence at various spatial scales. Generally following a positivist approach, this explanation included the formulation of general laws of agricultural location based on economic principles, including applications of von Thunen's model. However, the simplicity of this model meant that there were frequently large discrepancies between model-based predictions and reality. Hence geographers sought wider explanatory frameworks in which variables other than the strictly economic could be incorporated to explain spatial variation in agricultural systems and production. In some cases these explanations took an explicitly statistical form (e.g. Robinson et al., 1961), but more often relationships between causal factors were inferred in general terms on the basis of various forms of empirical evidence (e.g. Hart, 1956).

Only recently have there been more concerted attempts to express in more formal terms this interaction of causal factors, from across a broad spectrum. For example, Chaplin (2000) suggests that aspects of the co-evolutionary work of Nergaard (1993) can be applied to the role of economic and non-economic factors affecting farming. Co-evolution emphasises the mutual dependence between factors whereby change in one factor alters the context for the other, causing it to change and thereby signifying a continuous gradual evolution. The five main co-evolutionary components identified in this particular approach are: factors external to the farm business; farm resources; the farm household; the farm business decision-making process; and changes in farm business resource allocations (operation, initiation and evolution).

One recurrent problem for work on economic interpretations of regional differences in agriculture has remained the difficulty in obtaining suitable economic and social data. Although government departments often collect details about costs and profitability for individual farms or even for administrative areas, it is rarely made available in a form suitable for a geographer's needs. Coppock (1964b, p. 417), for example, cited this as one of the reasons for the relative neglect of economic aspects by geographers in the 1950s and early 1960s in favour of considerations of physical controls. Ironically, those working on historical change have an advantage, as historical farm and estate records can be of greater detail than those available for today's farms, for which farmers may be unwilling to release financial details of their operations.

Based largely on the work carried out in the positivist-based avenues of enquiry of the 1960s and 1970s, Bowler (1987) referred to the 'traditional themes' in agricultural geography as comprising work on data sources and regionalisation, farming types and the location of agricultural production, agricultural resources and behavioural factors. He also recognised four broad issues that had dominated international research in this field in the 1970s and 1980s, though it must be acknowledged that this largely reflected work on agriculture in the Developed World: the characteristics of industrialised farming systems, the loss of agricultural land, state intervention, and multiple job-holding or part-time farming.

Reference will be made to these themes and issues throughout this book, but only as part of their incorporation in the new agenda that has been pursued by agricultural geographers from the late 1970s onwards. This agenda has involved dramatic changes in the types of research undertaken, as part of wide-ranging paradigm shifts within human geography itself and the growth of multi-disciplinary enquiries, bringing expertise from throughout the social sciences to bear on agricultural problems. Various different ideas have been incorporated into agricultural geography in this period, initiated by the adoption of a behavioural perspective and followed by growth of political economy approaches in the 1980s, which reflected both the transformation of the discipline of human geography and also of agricultural production, the broader agri-food industry and patterns of food consumption, especially in the Developed World (Marsden, 2000a; Page, 2003).

2.2 Behavioural approaches

Most of the agricultural geography of the 1950s and 1960s operated implicitly within an empiricist and positivist framework that attracted much criticism from those opposed to this philosophy. For example, it was argued by one critic that the highly simplified economic approach popular in the 1960s produced a landscape 'occupied by little armies of faceless, classless, sexless beings dutifully laying out Christaller's central place networks, doing exactly the right number of hours farmwork in each of von Thunen's concentric rings, and basically obeying the great economic laws of minimising effort and cost in negotiating physical space' (Philo, 1992, p. 201). By introducing consideration of non-economic factors, such as farmers' motivations and decisions not based solely on profit maximisation, the focus of attention was then shifted from simplified models of farming activity. Nevertheless, the behavioural approach to agricultural geography was also highly empirical and positivist, focusing on farmers' decision-making (e.g. Wolpert, 1964), the diffusion of innovations (e.g. Hagerstrand, 1967) and the responses of individual farmers to changing economic stimuli (e.g. Hart, 1978). It was an approach tied closely to the emergence of behavioural geography in the 1960s (see Golledge and Stimson, 1997; Robinson, 1998a, pp. 374–8). This built upon work on human responses to physical hazards (e.g. Kates, 1962), and systematic analyses of the spatial outcomes of individual decisions, to develop a focus on the role of cognitive and decision-making variables (see Golledge and Timmermans, 1990). One of the central features of this approach was its ability to link environmental 'structure', decision-making and spatial outcomes, as shown in Figure 2.1.

In the UK the development of the behavioural approach to agricultural geography was closely associated with work by Ilbery (1982; 1983a; 1983b; 1983c; 1984) on the goals and values of hop-growers in the West Midlands. This research emphasised the characteristics and qualities of individual farmers, but relied greatly upon the researcher's ability to define, measure, model and analyse statistically the attitudes and revealed patterns of behaviour of farmers (Ilbery, 1978; 1985b), usually selecting a sample of farmers in a given area to study (Clark and Gordon, 1980). For instance, it was argued that it is farmers' reactions to, and perceptions of, changing economic circumstances that have to be considered if a realistic understanding of agricultural land use patterns is to be obtained (Ilbery, 1985a; Ward et al., 1990). As with much of the earlier post-war studies in agricultural geography, emphasis still tended to be

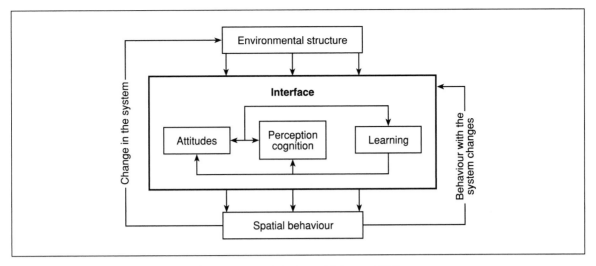

Figure 2.1 The people–environment interface (based on Golledge and Stimson, 1997)

placed on economic forces and upon quantitative measurements, thereby relegating the more interpretive humanistic concerns for individual identity and outlook to a minor role (Munton, 1986).

This behavioural approach also focused largely on the decisions of male farmers, and was often divorced from considerations extending beyond the farm-gate. The relationship of the male farmer to others in the farm household was generally ignored until political economy approaches in the 1980s investigated the strategies that individual farm households were adopting to deal with falling farm incomes and policy changes. However, a behavioural strand of research, or at least a re-emphasising of the importance of human agency in shaping the agricultural geography of a locality, has appeared in more recent work on decision-making relating to the implementation of agri-environmental policy, including the contributions by women farmers and farmers' wives (Evans and Ilbery, 1996; Gasson, 1994). In this work it is possible to recognise some influence of the so-called 'cultural turn' experienced within the social sciences from the late 1980s (Morris and Evans, 1999). This research has often tried to provide a balance between the impact of the state and structural controls on the one hand, and the role of the farmer as decision-maker on the other. In particular, it has added to knowledge of processes whereby

farmers assimilate environmental considerations. Also, because much of this work has been of an applied nature, it has had some feedback into policy modification and formulation (e.g. Whitby, 1994).

One of the important elements in the behavioural approach was consideration of farmer decision-making with respect to adopting innovations. This built upon the pioneering work of Torsten Hagerstrand, but was criticised for being too prescriptive, static and deterministic (Brown, 1981). Its underlying theory suggests an orderly, predictable and linear progression from awareness of an innovation to adoption, whereas in reality the process is unpredictable, uncertain and highly diverse (Ohlmer et al., 1998). The theory has also been criticised because of its tendency to emphasise the demand or adopter side of technological change rather than the supply or provider/promoter side. However, since the pioneering studies of the 1950s and 1960s the importance of the supply side has become apparent in the role of lead-user inventors (Von Hipple, 1998), change agents such as extension services (Van den Ban and Hawkins, 1988) and commercial marketing organisations (Unwin, 1988). Other factors, such as the influence of economic inducements, rural services and infrastructure may be inadequately accounted for by a focus upon individual decision-making behaviour (Ellis,

1988; Robinson, 1985a). Several recent studies have pointed out that external factors associated with political and institutional change are especially significant, particularly when there is significant policy shift or uncertainty (Allanson *et al.*, 1995; Hodge, 1996). Application of diffusion research to agricultural production innovations has also tended to be superseded by its applicaton to environmental innovation (Rogers, 1995).

2.2.1 Actor networks

Other models have been proposed recently that emphasise different elements in the decision process. These have included systems models, information models (MacFarlane, 1996), expert and decision support systems (Bowler, 1999a), and learning and knowledge transfer models (Garforth and Usher, 1997). The theory of reasoned action has been used to explain conservation attitudes and behaviour amongst farmers and others (Carr and Tait, 1991; Beedell and Rheman, 1999). Meanwhile, in the business sector, various studies have examined the role of organisational, human and business environment factors on the supply of and demand for innovation (see Golledge and Stimson, 1997).

In terms of approaches that operate within a system or network framework, perhaps the most popular has been actor-network theory, introduced to human geography, including agricultural geography, in the 1990s (Marsden *et al.*, 1993, pp. 140–7; Murdoch, 1995; 1998; Murdoch and Marsden, 1995; Thrift, 1996, pp. 23–6; Whatmore, 1997):

> Actor-network theory examines the complex composition of networks in the modern world and seeks to understand how the networks gain their strength and how they achieve their scope (Murdoch, 2000, p. 410).

This defines an 'actor network' as interaction between a set of human 'actors' and the circulation of non-human 'intermediaries'. Material objects may be active in configuring human social actors and their relationships, and so human and non-human actors may have equal status. Alternatively, a difference between the human and non-human may be retained, with non-human intermediaries allowing objects in the natural environment to be connected to social actors in ways different to the more familiar binary divide of society and nature.

Thrift (2000, p. 5) cites three key attractions of actor-network theory:

- It can be used as a means of producing a better understanding of both technology and nature, and their inter-relations.
- It problematises the act of representation (how meanings are constituted and communicated) by acknowledging that there are various forms of representation and that these are unstable and highly inter-related.
- It provides a means of understanding space as an order of partial connection or linkage within the network.

Use of the theory implies a belief that networks represent power relations in which power lies in the links that bind actors and entities together (Lockie and Kitto, 2000). Hence, power is exercised by complex associations of the social, the natural and the technological. Murdoch (2000, p. 411) argues that the theory might be a method that could be used to extend our understanding of agri-food chains by enhancing knowledge of all the multiple sets of contingent relations in operation in such chains/networks (Marsden, 2000a), though Lockie and Kitto (2000, p. 12) note that actor-network theory has tended to 'privilege the same state agencies and similarly large institutions that other approaches to food studies have focused on'.

Actor-network theory requires the researcher to identify intermediaries within a particular actor network, then account descriptively for the ways in which intermediaries are configured by, but also configure, human actants (Murdoch, 1998). The researcher must then explain how intermediaries provide (or fail to provide) stability in a network (Bowler, 1999b). The theory has its own terminology, as exemplified in Table 2.1, and focuses on narratives that deal with processes of construction, consolidation and stabilisation of dynamic actor networks. Dynamism is present because of the ways in which actors and intermediaries comprising a network change through processes such as enlisting and immobilising, and actors

Table 2.1 Actor-network terminology

Term	Explanation
Actor network	Comprises human and non-human entities which together may be referred to as actants. A network experiences processes of construction, consolidation and stabilisation. A network changes as actors contest and modify their power relationships
Non-human entity	An intermediary material object (which may be termed an 'actor') that can be active in configuring human social actors and their relationships
Intermediary actant	An object in the environment that can be connected to social actors in ways other than through the binary divide of society/nature
Hybrid actor network	Combines human and non-human actants
Acting at a distance	The ability of an 'external' actor to enrol or mobilise a 'local' actor into a network of control from a spatially distant centre of translation
Actor space	The spatial reach or concrete situation over which an actor has influence
Network lengthening	The increasing size or scale of actor network in which the spatial reach of the network is extended through the mobilisation of many interwoven actors and intermediaries

(*Sources*: Bowler, 1999b; Law, 1992)

within networks can contrast and modify their power relationships with each other over time.

Rural areas can be constructed as an actor-network, in which the 'local' can be connected to the national and the global through the connectivity of human actors who include individuals, farms, agencies, institutions, corporations and government departments. Connections are also developed via non-human intermediaries, such as scientific technologies and material forms, e.g. texts and skills. Hence, actors and intermediaries can comprise formal and informal networks, connecting through space to link the local with the global (Bryant, 2002). Through these networks, local events can be profoundly influenced by actors who operate elsewhere, 'at a distance'. For example, actors, in the form of politicians and civil servants, develop policy in the World Trade Organisation (WTO) that appears as regulations (the intermediary), which in turn affect the nature of farming in a particular locality (Murdoch, 1995).

An example of an actor network for recycling sewage sludge on farmland is given in Table 2.2, in which Bowler (1999b) has divided the actants between primary and secondary dimensions based on a subjective assessment of their power in the network. In this example, sewage sludge is a primary intermediary actant, which can configure but also be configured by other actants in the actor network. Various actors make decisions relating to the use of sludge for recycling on farmland, but scientific knowledge in its material forms comprises a second primary component of the network, influencing EU and national regulations and the development of technical equipment needed for spreading liquid sludge onto farmland. Control over scientific knowledge on the polluting effects of sewage sludge is a basis for power within the actor network.

Methods of regulation and pricing form two conflicting 'modes of ordering' of this actor network. Both modes enrol actants in stabilising and lengthening the actor network. Stabilising is associated with the way in which regulations exert direct control over the behaviour of producers and users of sludge, with guidelines and voluntary codes of practice informing that behaviour. Bowler (1999b, pp. 37–8) summarises pricing as a mode

Table 2.2 Actants in the actor network for recycling sewage sludge on farmland in the UK

Dimension	Actors	Intermediaries
Primary	European Union (DG XI) UK water companies Scientists Farmers OFWAT regulator	EU directives (91/271/EC) Scientific knowledge Books, papers, technical documents Sewage sludge
Secondary	Environmental groups Manufacturing industry Urban population Environment Agency Courts of law	Technical equipment UK national regulations UK national codes of practice

(*Source:* Bowler, 1999b, p. 32.)

of ordering as follows: 'water companies enrol manufacturing companies into the actor network through pricing the disposal of their effluents and enrol farmers through subsidising the cost of acquiring their sewage sludge, including the spreading costs'. It is, though, scientific knowledge that is the pivotal actant in this particular narrative of actor-network theory. This is because it is the 'expert knowledge' of scientists that forms the basis of regulation as one of the modes of ordering of the actor network, and control over such knowledge gives power to several actors including government policy-makers and the water companies. Similar findings have been made by Murdock and Clark (1994), who regard scientists as frequently being active collaborators in the power structure, and by Seymour and Cox (1992) with respect to problems of nitrification of watercourses (see also Hill *et al.*, 1989; Ward, 1996).

Bowler concluded that the sustainability of recycling of sewage sludge on farmland depended largely on the maintenance of regulation as the dominant mode of ordering, and the continuing success of water companies in representing the recycling of sewage sludge on farmland as a sustainable form of resource management. The durability of the existing actor network in this example is likely to depend on the impact of environmental groups who may seek to monitor and, if necessary, contest the practice. Their involvement may challenge existing control over scientific and bureaucratic knowledge.

2.3 Political economy approaches

From the mid-1980s, geographers have made a more systematic analysis of the political and structural frameworks within which agriculture nests. Initially this analysis was labelled as the political economy approach, to signal its Marxist-related content (Fine, 1994; Goodman *et al.*, 1987; Le Heron, 1988), though its direct links to Marxist thought were not always clear. The nature of this approach initially owed much to Blaikie's *The political economy of soil erosion in Developing Countries* (1985), in which emphasis was placed on the social relations of production and their determination of the nature of land use. The farm household was seen as being part of two kinds of social relations: the local and the global, with the impress of the latter increasingly assuming greater prominence.

In more general terms, 'political economy' refers to the management of the economy by the state, and therefore political economy approaches have emphasised the role of the state, at various levels, in setting the parameters within which both economic and social change occurs. The underlying ideas are based in general on the writings of Ricardo and Marx, and their concern with value, reproduction and distribution. It is the focus upon distribution that introduces the political dimension of political economy, thereby extending analysis beyond the purely economic to the sphere of political and social considerations. The focus on the political marks a distinct break with neo-classical economics in which rational choice and optimal allocations of resources hold sway (Barnes, 1990). However, there are various 'schools' of analytical political economy, which have led to a degree of conflict and uncertainty in its application. Moreover, within human geography, the term 'political economy' has tended to be applied rather loosely, treating it as an approach with a focus upon the role of the state, but with neither an explicit attempt to apply Marxist ideas to economic analysis nor adoption of Marxist dialectics (Bowler and Ilbery, 1987; Castree, 1996).

The political economy approach 'suggests that if the social relationships and processes of change within a given society are to be understood, then it is necessary to examine the nature of the economy and the power relationships that it sustains' (Mannion and Bowlby, 1992, p. 15). This entails considering divisions of property ownership, both land and capital; the structure and conduct of relationships between employers and workers; the structure and conduct of relationships between the sexes and between different ethnic groups; political groupings; and the organisation of state power. In translating this to a theorised approach to agriculture in the Developed World, Bowler and Ilbery (1987) emphasised a need to consider class and power relations, and the investment decisions of agrarian and industrial capital, with an 'extended' theoretical base applied to the broader empirical context. That context had to include those circumstances where agriculture impinges on the wider society.

They argued that traditional approaches to agricultural geography had disregarded the absorption of the production sector (i.e. the farm) into the larger food supply (or agri-food) system, which includes off-farm inputs such as suppliers of seeds, fertilisers and machinery (sometimes termed 'backward linkages'), and to food wholesalers, retailers, processors, distributors and consumers (sometimes

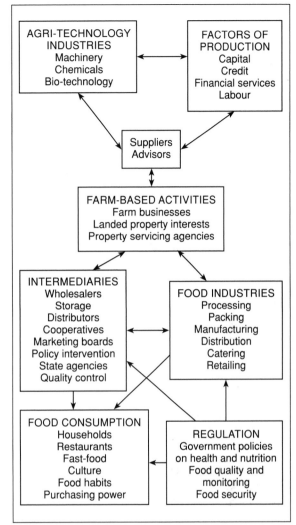

Figure 2.2 The agri-food system (based on Whatmore, 2002)

termed 'forward linkages' or 'downstream linkages') (Figure 2.2). Therefore, critical influences and components of the overall agri-food system were being neglected, especially the increasing involvement of large trans-national corporations (TNCs). Attention to these linkages has given rise to a broader research focus, acknowledging the farm as part of a wider agri-food chain, itself embedded in a multi-faceted web of economic, social, cultural and political dimensions all impacting upon farm-based production (Marsden, 1988; 1989; Page, 2003).

The role of the TNCs in the growing economic concentration in both the agri-inputs and food processing industries has been part of vertical integration across the whole agri-food 'chain', and hence the need to examine the totality of the chain rather than just farm-based production (Page, 2003, pp. 246–8). Advocates of the political economy approach argued that this totality needed to consider linkages between food retailers and processors, and their influence on the production sector. Furthermore, the closer relationships between changing consumption patterns of food and other sectors in the agri-food system needed more attention, for example the impacts on farming of major developments such as convenience frozen foods, contract production for supermarkets and media advertising. These have all contributed to the exposure of family-run farms worldwide to a broader range of external pressures, and have promoted new forms of production (e.g. McKenna *et al.*, 1999a). As a result, certain parts of agriculture have been absorbed within more general urban-industrial structures and processes, and new industrial-style complexes have emerged, sometimes referred to as agribusinesses (see Troughton, 2002).

As part of political economy approaches, geographers' attempts to gain a greater understanding of changes occurring on farms, especially in the Developed World from the late 1980s, have increasingly involved an engagement with structuralism as a means of generating greater theoretical coherence and rigour. Structuralism follows arguments presented by Karl Marx (1971), who argued that 'the mode of production of material life conditions the general process of social, political and intellectual life . . . Changes in the economic foundation lead sooner or later to the transformation of the whole immense superstructure' (p. 21). The 'superstructure' is the level of appearances at which spatial patterns of social exchange can be recognised. However, underlying this is an infrastructure of processes that produces an intermeshing of social, cultural, political and spatial organisation (Gregory, 1978, p. 100). Geographers applying Marxist ideas have focused upon both the infrastructure and the superstructure, but especially on economic processes operating within the capitalist mode of production. They have

generally argued that spatial patterns only offer clues as to the nature of processes rather than being suitable for their identification and understanding. Therefore they have turned to the need for theory to advance understanding of how the infrastructure functions and how it relates to the superstructure. Structures associated with social class initially attracted particular attention (Castells, 1977; Harvey, 1973), but subsequently there has also been work linking structures to local variations in pervasive infrastructural processes at both national and international level (Massey, 1996). There remains a substantial challenge, though, to link changes occurring at a local scale (e.g. on individual farms) to abstract theories concerning the movements of capital, the labour market and the role of the state.

The analysis of the economic, societal and political structures within which the agri-food chain operates has placed great emphasis on the need to understand agriculture as part of the global agri-food system (Goodman and Watts, 1994). This analysis has had direct links to Marxist theories of structuralism and the nature of relationships between state, capital and labour. Its adoption has led agricultural geographers to address theoretical considerations of class and power relations, and the investment decisions of agrarian and industrial capital. An initial framework for this approach was suggested by Marsden *et al.* (1986a), with a focus on four elements:

- Uneven development, whereby capital penetrates different regions at different times, as exemplified in the process of regional specialisation within the European Union (EU) where 'the intensive capitalisation of farm production in some regions at the expense of others, provides a contemporary example of the unevenness of capital penetration and the superimposition of new spatial divisions of capital and labour' (Marsden *et al.*, 1987, p. 299).
- Geographical and historical specificity, whereby locality studies are required in order to understand the process of uneven development. Historically, different areas developed different agricultural systems and

this initial regional uniqueness is important in understanding the spatial ramifications of the agricultural restructuring process (Marsden *et al.*, 1986a; 1986b).

- Conceptualisation of the family labour farm, whereby the different transformations it makes to capitalisation need to be explored. It is important to determine whether the basis of agricultural production remains sufficiently different from other sectors of the economy for small- to medium-sized family businesses to survive as the modal group of operators.
- Agriculture and state policy, whereby government policies have supported the family farm but inadvertently encouraged the penetration of agriculture by corporate capital. This and other contradictions need to be examined at particular points in time in different nation states.

These four elements are ones relatively neglected in traditional approaches to agricultural geography, but which have formed the foci for research since the mid-1980s. In recommending attention to these foci when political economy approaches were first popularised, Bowler and Ilbery (1987) argued for a new definition of agricultural geography based on the following three dimensions:

- the concept of the agri-food system, so that links between production and processing, distribution and marketing of agricultural produce become more significant;
- a broadened empirical context, including circumstances where agriculture impinges on the wider society via such issues as resource allocation and rural employment;
- an extension of the theoretical base to encompass perspectives developed in cognate disciplines.

In another interpretation, by Cloke *et al.* (1990), political economy approaches to agriculture were summarised as shown in Table 2.3. The four columns in the table represent four key steps within this particular perspective, as follows:

1 In its most general form the political economy approach focuses on different theorisable layers of production, primarily the process of capital accumulation, different production systems and sites of production.

2 When applying this approach to agriculture emphasis is placed upon production in agri-commodity chains (from suppliers of inputs to farming through production to processing, marketing and sales) and production on the farm. Traditionally geographers have stressed the economic aspects of the three different levels recognised in Table 2.3 whilst neglecting political dimensions that provide a structure or context within which the economy operates.

3 In redressing previous lack of attention to politics, policy and the roles of the state and supra-national organisations, the approach recognises six foci representing amalgams of political and economic forces:
   a) internationalisation (perhaps now superceded by globalisation) in which production, trade and finance are part of the accumulation process;
   b) different state policies determine opportunities for organisations to participate in this process;
   c) the operation of key organisations and the strategies they employ, e.g. with respect to investment or at the micro-scale in their use of labour and technology;
   d) the politics of regulating competition and dissecting complex relationships within production systems;
   e) the nature of management systems and the relationships between technology and labour;
   f) the local politics of production at farm level or in processing, which can include considerations of gender and ethnic conflicts and employer–employee relations; the macro-scale political dimensions dealing with resource allocation and international relationships.

4 There are various aspects to the incorporation of a spatial dimension within this approach. In particular the nation state is a key unit of analysis, enabling examination of the spatial impacts of state policy and internationalisation. Furthermore, each

Table 2.3 Development of the political economy perspective

Level	General political economy perspective	Present position for agriculture	Suggested extensions applicable to agriculture	Spatial dimension
Level 1	Capital accumulation	Capitalist accumulation process	Internationalisation process; role of the state	Spatial expression of capital accumulation process internationally, nationally and regionally
Level 2	Production systems or commodity chains	Agri-commodity chains	Key organisations & investment/labour strategy; politics of competition/ distribution of value	Peculiarities of agro-ecologies & agro-commodity chains shaping geographic arenas of accumulation
Level 3	Production sites	Farm-level production	Technological & management systems; politics of production	Production systems and bio-physical processes

(*Source*: Cloke *et al.*, 1990, p. 16)

agri-commodity chain rests on different production systems that will shape the geography of accumulation at different spatial scales.

The political economy approach therefore provided opportunity for a more substantive development of theory than was previously the case in geographical studies of agriculture, but with a need for empirical exploration too in order to understand change. Although Table 2.3 is ahistorical, the approach does emphasise the 'temporality' of accumulation processes and of particular commodity chains, set within the broad context of both the stability and 'crises' of capitalist development. A class-based view of this political economy approach can be found in the works of Cloke and Thrift (1987; 1990), linking rural change to more general economic transformations, but specifically by viewing class relations and class groupings as both the end-product of previous rounds of economic restructuring and an agent in shaping future development.

Examples of the concerns for backward and forward linkages and an appeal to structuralist arguments can be seen in the mid-1980s in research on the growth of agribusiness in North America and Western Europe (e.g. Marsden *et al.*, 1986a; 1986b; Wallace, 1985). However, it quickly emerged that it was the scale at which these enquiries were performed that was likely to be of considerable importance to geographers, for example with four scales being recognised at which the process of uneven development of agriculture occurs (Marsden *et al.*, 1987): regional; sectoral, e.g. between arable and livestock; internal organisation of individual production units; and differing divisions between the units of production and other agencies involved in the food chain, e.g. manufacturers of inputs, food manufacturing, retailers and financial institutions.

It is also worth emphasising that the application of the political economy approach by geographers has coincided with a period of crisis in agriculture, especially in the Developed World, leading to widespread agricultural restructuring (e.g. Gray *et al.*, 1993). Therefore the approach has been applied to studies of substantial change, providing opportunities for foci on various links in agri-commodity chains. However, even if attention is only directed at the level of the individual farm business, the approach emphasises the way in which this 'nests' within the wider circuits of the global economy and geopolitics. Hence it is argued that this macro-scale level of structural change tends to shape changes at the local level.

For example, a food processing company's responses to the increased need to compete with rivals at a global scale may have dramatic influences upon farm-gate prices and hence lead to related adjustments to farm production and land use.

The wider agri-food chain has been the new focus of attention, primarily as an integrating conceptual framework and a means of exploring linkages between agricultural production and urban and industrial food systems. Subsequently, this challenge has been taken up through the stimulation of the development of an increasingly globalised food system. Much of this work has been carried out under the umbrella of a political economy approach, though it should be recognised that this has been a very broad concept under which a number of different themes and ideas have been pursued, notably those paying attention to the 'diversity of social relations and cultural practices shaping accumulation and regulation' (Marsden et al., 1996, p. 312).

Moreover, there has been a range of work in agricultural geography from other perspectives, frequently ignored by the proponents of the political economy approach (Marsden, 1998a). For example, Morris and Evans (1999) highlight the substantial amount of work on the evolution of agricultural policy, the related development of work on the 'post-productivist transition' (in which farmers have been encouraged to take on roles additional to their traditional function as producers of food), and research on recent agricultural change, including issues relating to people working in agriculture, and new thinking on culture and animals, linked to ideas developed in other areas of human geography (Anderson, 1997; Yarwood and Evans, 1998; 1999; 2000).

2.3.1 Critiques of the political economy approach

Buttel (1996) criticised the initial work on agriculture in the political economy perspective for its excessive concentration upon the agricultural production sector, dominated by the farm enterprise, farm household, land ownership, farm labour and agricultural technology. Explanations based within agriculture itself were also predominant rather than

considerations involving factors exogenous to the agricultural sector. Buttel's arguments form part of three broad criticisms of the political economy approach:

- It relegates farmers to the role of insignificant decision-makers faced by constraints rather than having choices to make: 'in order to understand agricultural restructuring there is a need to examine the individual farm family members as active participants and not simply as passive subjects of inevitable structural processes' (Whatmore et al., 1987a, pp. 120–1; 1987b).

- It is not well suited to studies of a sector of production dominated by family-run businesses. Even where behavioural dimensions have been added to political economy studies there have been limitations in their ability to consider non-economic decisions on allocation of resources.

- There are difficulties in reconciling the theory with empirical information, especially as the former over-emphasises the significance of macro-economic factors.

In recognising these limitations during the 1990s, some agricultural geographers modified their approach by incorporating a behavioural component in their work to obtain a better understanding of internal family dynamics. Nevertheless, from a post-modernist perspective (increasingly popular in human geography in the 1990s), such a modification still fails to be sufficiently sensitive towards the differences between people and places (Dear, 2001). One outcome of this has been a call for more consideration of the cultural aspects of farming, to draw upon the theories underpinning postmodernist studies (Morris and Evans, 1999). This recognises that there has been little research on the nature of farmers and farming as cultural constructions, despite various strong portrayals of farming as fulfilling particular cultural functions: as 'guardians of the countryside' (see Ilbery, 1992), maintainers of the 'pastoral myth' (Short, 1996), those responsible for the 'theft of the countryside' (Shoard, 1980), or as the remnant 'others' in a countryside populated by ex-urbanites (Yarwood and Evans, 1998). This work is yet to consider whether research should begin with farming and farmers as cultural constructions or add cultural dimensions as an ingredient. In short, there is a need to see wider cultural views of farming in society. Morris and Evans (1999, pp. 354–5) suggest three foci for this:

- Cultural constructions of different groups within the farming community, to overcome the previous neglect of farm workers, tenant farmers and women in farming. Links between farmers and agricultural advisors also merit closer inspection.
- Different constructions of farming as an activity. This could include portrayals in literary texts and images, film, various types of writing, the national press (e.g. Wilson, 2001), policy documents and promotional materials. Related images and constructions of food could also be worth pursuing (e.g. Bell and Valentine, 1997).
- Animals in farming have generally been regarded simply as products or commodities in the food-production system with little concern for their association with local folklore and culture. There has been some work on the distinctiveness of different breeds of farm livestock in the cultural landscape (Evans and Yarwood, 1995; 2000; Yarwood and Evans, 1998), but there is scope for further research on the role of livestock in agriculture and the shifting societal view of animals that are reared for slaughter.

In the 1990s the focus of political economy approaches shifted towards global agri-food restructuring, using the argument that modern agri-industrial commodity chains are really just a version of the flexible globally sourced, globally integrated form of production of the large TNCs (Buttel, 2001). It is questionable how far this analogy may be taken, but undoubtedly the broader concern for the full breadth of the agri-commodity chain, globalisation processes and geopolitical systems has added a new dimension to agricultural geography.

This new dimension is a far cry from the micro-sociological studies of the diffusion of agricultural innovations popular in the 1970s. Indeed, political economy approaches have effectively transformed

the nature of agricultural geography, broadening its scope, fostering inter-disciplinary work and providing opportunities for engagement with processes operating well beyond the confines of the farm-gate. However, there have also been limitations, notably that too much of this work has been reliant on a structural determinism that overlooks farm- and farm-household dynamics, and neglects key components of farm-based decision-making, not to mention certain institutional and technical arrangements of farming. Fortunately, there has been a rediscovery of the importance of culture and locality by some agricultural geographers as part of the so-called 'cultural turn' taken by human geography from the late 1980s (Whatmore, 2002), and also by rural sociologists, notably those associated with Wageningen Agricultural University in the Netherlands (Long and Long, 1992; Van der Ploeg, 1992). This work has been sceptical of the power of global macro-structural forces and especially of their homogenising capacity. Instead, it has tended to emphasise the ways in which farmers develop locally based adaptive strategies reflective of local culture, agri-ecology and household resources, in order to survive and adapt to changing economic circumstances. This has been termed the 're-localist viewpoint' and has been more popular in Europe than North America, though work by some American anthropologists presents similar arguments (e.g. Barlett, 1993; Salamon, 1992).

The emergence of alternatives to political economy approaches raises the question of whether the political economy perspective has reached its empirical and conceptual limits, as new approaches to rural development problems develop that emphasise commodification (in which social relations appear as things) or 'relating farmers' variable actions to an understanding of simple commodity production' (Arce and Marsden, 1993, p. 296; Buttel, 2001). Related work, developing similar themes has focused on the relationship between the exchange process, social life and its politics (e.g. Appadurai, 1986) and, increasingly, on environmental considerations, which include sustainable development, valuation of environmental assets, the importance of environmental accounting and the use of market indicators (Harvey, 1996). To this can be added the growing volume of work on the effects of new forms of retailing and business organisation upon food production and marketing (e.g. Wrigley, 1992; Wrigley and Lowe, 1996; 2001). Hence, there is a tremendous diversity of approach within agricultural geography at the start of the twenty-first century, and a range of concerns far removed from the positivist-based spatial differentiation theme that was so dominant in the 1960s.

2.4 New theories to explain agricultural change

Within the broad sweep of work under the political economy umbrella there have been different emphases upon the role of related concepts and theories appropriate for aiding the understanding of contemporary agricultural change. Three concepts in particular have been utilised as means of enhancing understanding: food regimes, regulation theory, and a restatement of the so-called 'agrarian question'.

2.4.1 Food regimes

Food regimes as a concept can be viewed as part of an evolving geography of agriculture that, having embraced political economy and having extended beyond a concern for production on farms

to one dealing with the geography of food (e.g. Le Heron, 1993; 2002), now includes explicit consideration of a broad spectrum of agri-commodities within a historico-political context that includes geopolitics, business economics, consumption patterns, retailing, processing, farming and supply industries (Buttel and Goodman, 1989; Robinson, 1997a). The concept links international relations of food production and consumption to forms of accumulation and regulation under capitalist systems from the 1870s onwards (Friedmann and McMichael, 1989). It considers macro-forces of demand and supply acting as a 'growth ensemble in the international food sector', varying from country to country and 'differentiated from slower growing or declining food industries' (Le Heron and Roche, 1995, p. 24). So it acknowledges that researchers must consider the entire 'food system' from upstream supply to farms, through farm-based production to downstream manufacture, marketing and distribution. The concept has been championed as an organising scheme that can transcend simplistic international comparisons and country case studies. It can explicitly include aspects of the spatial reorganisation of production and consumption inherent within the notion of regimes (Le Heron and Roche, 1995). The principal proponents of the concept recognise three regimes, the third of which is in the early stages of development (Table 2.4).

Table 2.4 Food regimes			
Character	First regime	Second regime	Third regime
Products	Grain, Meat	Grain, Meat, Durable food	Fresh, organic, re-constituted
Period	1870s–WWI	1920s–1980s	1990s onwards
Capital	Extensive	Intensive	Flexible
Food systems	Exports from family farms in settler colonies	Transnational restructuring of agriculture to supply mass market	Global restructuring, with financial circuits linking production and consumption
Characteristics	Culmination of colonial organisation of pre-capitalist regimes; Rise of nation states	Decolonisation, Consumerism; Growth of forward and backward linkages from agriculture	Globalisation of production and consumption; Disintegration of national agro-food capital and state regulation; 'Green' consumers

(*Source*: Based on Roche, 1994, as derived from Friedmann and McMichael, 1989).

The first food regime represents production for the world's metropolitan core in North America and Western Europe, with the settler colonies supplying unprocessed and semi-processed foods and materials to the core. The introduction of refrigerated ships in the 1880s increased both the range of produce that could be supplied by distant colonies and the distance over which perishables such as butter and meat could be transported (Peet, 1969; Schedvin, 1990). This regime was undermined after World War One as agriculture in the core competed with imported produce, often under the protection of trade barriers. Its replacement reached its peak in the 1950s and 1960s, based on the development of agri-industrial complexes focusing on production of grain-fed livestock (e.g. barley-beef in the UK), fats and durable foods. This regime incorporated production in both the Developing and Developed Worlds and was associated with increased specialisation and geographical segmentation of production systems.

Friedmann (1987, p. 253) asserts that this regime reached a crisis point in 1973 as USA–USSR grain deals eliminated the American wheat surplus, as export competition between the USA and the European Community grew, and as traditional patterns of trade with Developing Countries (and former colonies) were disrupted (Friedmann, 1982). It has been argued that this crisis has produced a gradual loss in the growth potential of the established agri-industrial complexes, hastening the emergence of a third regime. The latter is based on the global sourcing of produce by footloose TNCs aided and abetted first by the General Agreement on Tariffs and Trade (GATT) and more recently by its successor, the WTO. For example, McMichael (1992a) argues that, in addition to restructuring performed by those TNCs involved in food processing and marketing, the changing role of the International Monetary Fund (IMF) and the GATT/WTO provided a new regulatory structure in which the new regime is emerging and operating. One aspect of this new regime recognised in some quarters is production of fresh fruit and vegetables for a global market (e.g. Roche, 1996). This is part of a complex mix of processes relating to globalisation and sustainability that could lead to the dominance of a clearly recognisable third food regime in the first decades of the twenty-first century.

The third regime has been described as agri-commodity production 'characterised by the reconstitution of food through industrial and bio-industrial processes via flexible global sourcing of generic crops and increasing importance of affluent foods (animal protein, processed foods, fruit and vegetables)' (McMichael, 1992a, p. 359). In the context of New Zealand, Le Heron and Roche (1995) suggest there is evidence that this regime involves a conflation of globalisation and sustainability tendencies (but see also Campbell and Coombes, 1999a). This may involve a reorientation of 'basic' foods and export crops to a more important role for the supply of inputs for 'elite consumption in the north' (McMichael and Myhre, 1991, p. 100). The third regime may also include extensions of technology associated with the second regime, such as the use of chemicals to simulate naturalness and control ripening of packaged fruit (Arce and Marsden, 1993) and the production of 'organic' fruit and vegetables to tap the rising wave of 'green' consumerism (Clunies-Ross, 1990). It may include the 'greening' of agriculture via the development of environmental dimensions within agricultural policy. Aspects of a new regime are apparent in the ways in which farmers throughout the Developed World are being asked to fulfil new and often contradictory roles. The result is a patchwork of productivist and environmentalist scenarios that are often present to different degrees on neighbouring farms (Robinson, 1997a).

Another label applied to this putative third regime has been 'post-Fordist agriculture' (e.g. Cloke and Le Heron, 1994). The use of this terminology emphasises the increased dependency of farmers on transnational finance capital, which propels them towards greater acceptance of new technology, including the latest biotechnical innovations, flexible labour arrangements and output of a specific product for a particular market. A close link is then established between the 'niche market' and changing patterns of food consumption, in which the increasingly wealthy middle-classes demand more high-quality foods of an organic or speciality nature (e.g. Cwiertka and Walraven, 2002). This change in patterns of consumption by one sector of society can be contrasted with the retention of more traditional demand for mass-produced and inexpensive food in other parts of society, and especially in Developing Countries. Therefore production characteristics from different regimes can be expected to exist side-by-side, perhaps even within the same farming enterprise. However, evidence of the flexibility, which is a pre-requisite of the label 'post-Fordist' when applied to industry, is lacking in many sectors of agriculture. This might lead to the conclusion that, so far, in much of the Developed World the so-called current 'international farm crisis' (discussed in detail in Chapter 3) does not represent a move to a clearly recognised new food regime, but rather a somewhat chaotic set of reactions to changing economic and regulatory conditions (Drummond *et al.*, 2000; Goodman and Redclift, 1989).

The food regimes concept implies that, despite the variation in national contexts, there is a common framework that can be utilised to help elucidate trajectories of local change. However, it is not clear from work on regimes just how or whether the contours of the new (third) regime can be clearly delimited from the previous one nor of how these general concepts advance understanding of the tensions between general tendencies and unique occurrences in particular locations. Moreover, Moran *et al.* (1996) contend that food regimes are of little value in explaining the experience of individual nations in the international food system. In part this is because important national and local processes are not incorporated into events at an international or global scale. Therefore there is a strong need to consider the continuation of regulatory influences from the past and to look at the organisation of agriculture at local, regional and national levels. The concept has been criticised strongly for offering only limited indication about regional and local details, especially regarding farm practices and regulatory mechanisms (McKenna and Murray, 1999). This criticism has been applied by Hollander (1995) with regard to food regimes' ineffective explanation of change in the Caribbean sugar economy.

Goodman and Watts (1994, p. 20) cite the concept's lack of reference to the role of human agency in effecting the changes described in the world's food systems. It has a 'preoccupation with economic factors and world economic systems and so fails to elucidate the differentiated integration of agriculture into global processes or to consider any role for such concepts as contestation and resistance' (Coombes, 1997, p. 36). Thus it tends to overlook the ways in which rural change is integrated into broader socio-cultural changes. Moreover, there is also a need for examples to readdress local experience and pay more attention to how and why modes of social regulation function in the ways they do.

In summary, there are four broad reasons for the relatively limited applicability of food regimes theory to changes in the agri-commodity sector of individual nations:

- its emphasis on food rather than agriculture and other rural production. This almost presupposes that power lies in the hands of TNCs;
- the centrality of the North American and European experiences in its conceptualisation and the neglect of the Developing World;
- the level of abstraction at which food regimes are conceptualised;
- the relative neglect of national regulatory processes.

As a result of such criticism, the concept has increasingly been used only as a departure point for new ideas on how the conflicts between global processes and local specificities might be better understood (Roche, 1999; Roche *et al.*, 1999). There is, though, potential for the general framework of the food regime to offer a basis for comparison between countries, regions and localities experiencing different aspects of the transition from the second to the third regime. This may be a valuable organising concept, as it is clear that a single and fixed spatio-temporal divide between different regimes is not a reality. Instead there are mixes of all three regimes present in different localities. There may be a transition from the second regime to the third, but it is a contested transition over which businesses and governments are vying for control.

Where the concept of food regimes does offer opportunities for further development of research is in terms of the linkages between land-based production and its suppliers and markets. Indeed, Marsden (1997) argues that there needs to be a move from a broad conceptualisation of food regimes to one which focuses on food networks, which are playing an increasing role in the social and political development of regions and nation states. Essentially, his contention is that there is a need for more focus upon the latter stages of the agri-food chain because of the significant impacts of these stages upon society and economy. This argument is advanced in Chapter 3.

Goodman (1999) pursues similar ideas in arguing that various forms of network analysis might be used to examine the food networks inherent within the agricultural production base. He identifies three areas of research that can illustrate the complexities of the networks: food 'scares' (such as Bovine Spongiform Encephalopathy), technology and organic production. These raise profound ethical issues associated with the new 'biopolitics' of food production and consumption.

To date, key research areas stimulated by the work on food regimes include:

- The relationship between international food regimes and agricultural structures, and the consequences of the new international division of labour in which TNCs have developed production facilities in Developing Countries, especially at the routine and low-skill end of the production process (Sayer, 1995; Sayer and Walker, 1992). The inequities in returns for labour between Developed and Developing Worlds are a key feature as is the low pay of unskilled workers in the Developed World's fast-food and food processing sectors (Jarosz, 1996).
- Analysis that extends beyond a concern for agricultural production to examine consumer behaviour and changes in marketing, distribution chains and food processing (Herod, 2001; Kelly, 1999).
- The 'macro-explanation' of food systems, linking accumulation strategies of

international and national capital to employment and consumption norms, types of state regulation and legislation, and household livelihood and consumption practices (Katz, 2001; Young, 1997).

It can be argued that the notion of food regimes includes certain aspects of regulation theory, which has been used recently as a framework within which agriculture and rural development have been related to broader social, economic and cultural changes.

2.4.2 Regulation theory

The growing strength of the agri-food industries, the development of biotechnologies and diminishing opportunities for profits on farms from increased labour productivity have shifted sites of food production away from farms and into factories (Guthman, 1998). Profits are being redistributed away from agricultural producers to those who control and add value in the processing, distribution and retailing links in various commodity chains (Buttel, 1994). It is argued that this industrialisation of food provision, forming the basis of the production of 'durable foods' and the increasingly prevalent convenience foods, transcends the boundaries of Fordist and post-Fordist production, where 'Fordism' is the term used to describe a regime of capital accumulation covering the period from 1945 to the 1970s, typified by assembly-line industrial production (Marsden, 1992).

This movement away from one era to another is also symbolised in the changing nature of the regulations governing production, e.g. policies enforced by different arms of government or other regulatory bodies operating at a variety of spatial scales. There are still regulations encouraging farmers to maximise production, and others dealing with agricultural overproduction vis-à-vis demand, but these now exist alongside ones linking overproduction and support for the environment, environmental quality, competing claims on land use, and issues of food quality, purity and taste (Flynn and Marsden, 1992; Pritchard, 1999). In recognising these changes some geographers have focused attention upon the importance of studying and theorising the changing nature of regulation (Jessop, 1990).

Several 'regulation theories' have been developed since the influential work of the French economist, Aglietta (1974), in the mid-1970s. In essence these theories aim 'at understanding the current transformations in advanced capitalist economies, using comparisons between nations, at the sector or aggregate level' (Boyer, 1991, pp. 3–4). To do this they consider a mix of social, political and economic factors in addressing three central questions (Boyer, 1990):

- Why do periods of fast and stable growth turn into relative stagnation and instability?
- Why, during the passage of time, do major crises take different directions?
- Why do growth and crisis assume significantly different national forms?

In attempting to answer these questions, regulation is conceived in broad terms, not just as governmental rules but also 'the conjunction of mechanisms working together for social reproduction, with attention to the prevalent economic structures and social forms' (Boyer, 1991, p. 20). Regulation theory has also endeavoured to understand the links between the political economy of rural change and concurrent cultural representations, political strategies and social conflicts (Goodwin, 1995). This emphasises how people and social agencies deal with and seek to direct economic change (Phillips, 1998, p. 38).

There is a strong Marxist element within the theories, using Marx's law of the tendency of profit to fall over time as the key to the periodic 'crises' in capitalism. However, this is neither to imply that 'regulation' is simply determined by the requirements of global capitalism nor that natural events can be predicted deterministically from a theory of crisis (Lipietz, 1992). Instead, change has to be understood by considering particular sets of relationships (Leborgne and Lipietz, 1988, p. 264):

- the governance and organisation of labour within the firm, e.g. on the farm;
- the macro-economic principle which relates production conditions to consumption;

- the set of socio-political institutions and norms through which the interests of capital and labour are mutually adjusted.

From these, and from consideration of key factors affecting economic development, various models of development have been constructed representing changes or transitions in the post-Fordist era, with various labels, e.g. neo-Taylorist, Californian, Saturnian. However, there is much argument over the extent to which the different experiences of industrialised countries constitute a co-ordinated world system, and what political and economic power relations might produce a coherent period of development (Tickell and Peck, 1992; Webber, 1991).

One interpretation makes a clear distinction between one period of economic stability and another by referring to the way in which states try to regulate their economies to resolve fundamental crises in capitalism by producing 'a multifaceted configuration of economic and socio-political institutions and norms' that generate stability (Esser and Hirsch, 1989, p. 419). This stability was typical of the Fordist regime of accumulation, characterised by relatively stable processes of capital concentration within national frontiers, the formation of new mass industries (notably cars, electronics and domestic consumer goods industries), bureaucratic and centralised trades unions, comprehensive wage settlements, a degree of incorporation of union representation in industrial decision-making, the expansion of the welfare state, and involvement of the state in the planned production of suburban physical infrastructure (Low, 1995, p. 209). The demise of this stable regime is attributed to falling rates of profit and a 'crisis' typified by destabilising situational relationships in the global economy, which affected the 'regulation' imposed or developed by national, regional and local institutions. It can be argued that a transition to a post-Fordist mode has occurred, typified by new methods of labour organisation and production technology, new roles for women, reduced involvement of the state in welfare provision and at the local level, and a significant increase in the importance of service industries, amongst others.

With respect to the agri-food sector different terminology has been used to illustrate this move from one form of regulation to another. One example, food regimes, has been discussed above. Another, the move from productivism to post-productivism, is discussed in more detail below (Section 3.3). However, this latter terminology does have distinct parallels with the move from Fordism to post-Fordism in that it implies a move to different work/production practices, different relations of production and changes in the regulations governing production. Hence, for example, post-productivist farmers may be rewarded for producing environmental goods (as well as traditional output of food and fibre), whilst post-productivist regulation may involve new aspects of regulation, e.g. food quality, and new sources of regulatory control, e.g. supermarkets and food safety agencies.

Criticisms of Aglietta's work, and of regulation theories in general, have addressed the causal element in the theories, contending that it is too reductionist to relate economic growth and decline to the development of institutions governing the relationship between wages and labour and associated balances of class power (Brenner and Glick, 1991). For example, 'despite efforts to move regulation theory away from the totalising discourse of structural Marxism, in the end the struggle to introduce contingency, indeterminacy, and process merely serves to highlight how far the ensuing account is tied to its initial assumptions' (Murdoch, 1995, p. 737). This critique has also questioned the extent to which the very different reactions of national institutions to developments in the world economy can be combined into a single generalised model. Nevertheless, geographers have been strongly attracted to the notion of recognisable periods of stability in the world economy and of a transition in world capitalism that implicates the quasi-autonomous behaviour of individual states. This is a common attraction of both food regimes theory and regulation theories: for example, Moulaert and Swyngedouw (1989, p. 330) claim 'each regime produces a specific mode of spatial organisation profoundly different from the previous one. Each regime creates new or renewed forms of spatial crisis'.

Initially, geographers' response was to focus upon a 'space of regulation' in which a changing economy had spatial consequences impressed upon regions and localities by the mode of regulation originating at global level and mediated by nation states (Low, 1995, p. 210). However, the debates on regulation theory and a regulationist programme of research are complicated by the way social scientists have blurred the focus of the theory and have greatly expanded its boundaries. This 'expansion' has taken various forms. As suggested above, one in particular, food regimes, has been utilised by agricultural geographers.

There is also some dispute over the relationship between regulation theory and 'real regulation' as developed by Clark (1992). Real regulation is based on theories of the state and legal interpretism, and emphasises the way in which social practice and legal administrative controls limit the extent to which nation states are able to readily alter their mode of regulation (Marden, 1992; Pritchard, 1996). This contrasts with some interpretations of regulation theory that stress economic imperatives and state-controlled regulation (Le Heron, 1993) as opposed to the broader 'real regulation' conception that recognises the importance of wider social processes and existing regulation, structures and bureaucracies. It is this real regulation that Moran *et al.* (1996) contend is germaine to understanding agricultural development in their case studies of France and New Zealand. Their work considered not only the enactment of legislation, but also the underlying formulation process (incorporating the bureaucracy, social action and lobbying) and resultant interpretation via planning processes, social contestation and the courts (Figure 2.3) (Lewis *et al.*, 2002).

In the case of farming, lobbying in its widest sense has often been vital to the protection of farmers' interests, as seen in the establishment of producer marketing boards or legal controls favourable to the survival of particular specialisms or groups of farmers, for example the French agricultural syndicates (Cleary, 1989). Therefore, the power of farmers to shape the conditions of agricultural production by participating in and influencing the agri-commodity chain should not be overlooked. This power represents one aspect

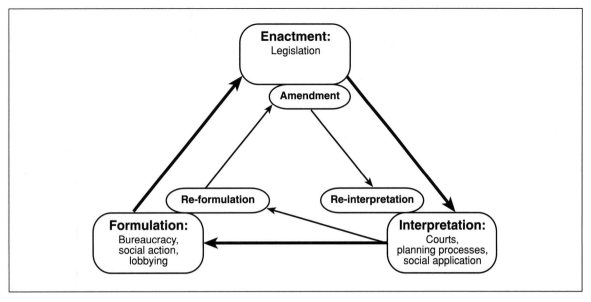

Figures 2.3 Real regulation (based on Moran et al., 1996, p. 249)

of 'regulation' that merits more study, as does the growth of regulation in the countryside by individuals not involved in agricultural production and with a non-rural focus, e.g. non-government organisations (NGOs), banks (Lowe et al., 1993).

Clark (1992) argues that in contrast to the American New Deal regulation of the 1930s, which established institutions to regulate markets and the national economy, newer regulatory agencies have specific agendas, with narrow objectives, and they exist to administer individual rights. In so doing they are less concerned with managing economic relations and more on developing performance standards. Hence, a distinction tends to be made between more traditional forms of regulation of the food system, often based on notions of public interest, and a developing private-interest regulation reflecting the increased power of the food retailers (Flynn and Marsden, 1995; Flynn et al., 1994). For example, one aspect of traditional regulation has been based on the actions of environmental health and trading standards officers in local government, whilst, in the 1990s, the regulation of food quality has become part of the controls exerted by large retailers (e.g. supermarkets) over consumer interests (Harrison et al., 1997). In part, the latter reflects the way in which the state has

partially abrogated responsibilties for overseeing certain aspects of control over food quality. For example, Guthman (1998) sees the new types of regulation as playing a significant part in the development of 'organic' farming through a new assertion of standards for food quality and safety, often through quasi-public (e.g. NGOs, marketing boards) and private institutions (e.g. retailers) (Marsden and Wrigley, 1995), or outside the state through trade organisations and private networks (e.g. Marsden and Arce, 1995). This new regulation of food is typified by eco-labelling, 'appellation' and other forms of branding. Within this changing instititional context of regulation, 'organic' regulation has emerged.

Clark (1992) also refers to the need for more attention to the workings of the regulatory state in terms of its administrative form, style and manner of regulation, by focusing on the nature of administration *per se* rather than theories of the states. This may involve more attention to local modes of regulation, and relationships between the central and local states. However, the power of national and supra-national governments to regulate the food industry and the food system in general must not be ignored. Many countries still contravene existing multilateral trade agreements and, whilst some regulations have diminished in importance, others have been strengthened, for example ones relating to hygiene, health and environmental factors.

TNCs have gradually increased their emphasis on food safety, 'freshness' and 'green' values as a means of enhancing their economic advantage. This can lead to economic goals taking over from more ecological values even in the context of 'organic' production. However, there is often scope for local interests to subvert global forces through exercise of various forms of control, e.g. in determining practice at the farm level or by implementation of regulatory mechanisms. That is, the global imperative does not always subsume the local interest, especially where the latter may still represent large numbers of independent small producers. So, whilst McDonalds may be a global brand recognised throughout the world, its existence has not eliminated the small, private trader selling local foods.

This contest between forces operating at different scales is addressed by Swyngedouw (1997), who asserts that 'the theoretical and political priority... never resides in a particular geographical scale, but rather in the process through which particular scales become reconstituted' (p. 141). He refers to this as 'glocalisation', which combines global trends in food economies ('McDonaldsisation') with local and regional aspects (e.g. local organic production) that contribute to the shaping of power and productive relations within these economies. In other words, there are a series of nested, interconnected and overlapping relationships between global, regional and local. He advocates the abolition of the 'global' and the 'local' as conceptual tools (p. 142), adopting an argument similar to that of Hoggart (1990) who proposed that 'rural' should be removed as a theoretical concept for analytical purposes. Others have accepted 'global' and 'local' as remaining important both theoretically and empirically, but as part of a set of glocalisation processes (Cox, 1997; McKenna *et al.*, 2001). An excellent example of this inter-meshing of the global and the local is provided by Jarosz and Qazi (2000) with respect to the apple industry in Washington State. They show how three aspects of this relationship have brought about wide-ranging changes in farm structure, production technology, labour processes and relations, and the composition and settlement patterns of farm labour. Hence 'the global' can be seen in Washington's Red Delicious apples being exported to over 60 countries, but there are local dimensions to how the global is manifest within the state. In this case the nature of the interaction has depended on three things: the social construction of value in fresh apples (how the apples are portrayed in advertising around the world), the changing structure of the apple industry (new technology, the role of the TNCs), and the changing social relations of production as they relate to transnational wageworkers (the cheap Mexican and Mexican-American labour who pick and pack the crop).

2.4.3 The agrarian question

Although not stated explicitly in most of the work by geographers under the political economy umbrella, some political economy approaches, principally by Marxists working in various other disciplines, have links to the 'rediscovery' of the agrarian debates of the German Social Democracy movement of the 1890s and the subsequent Russian Norodnik debates involving Lenin and his populist critics (Kautsky, 1920; Lenin, 1982). Much of this rediscovery returned to Karl Kautsky's *The Agrarian Question* (first published in 1899), drawing parallels between the 1890s and the 'crisis' facing agriculture in the last quarter of the twentieth century (Byers, 1982; Kautsky, 1988; 1994).

In the 1890s (towards the end of the so-called Great Agricultural Depression), Kautsky analysed falling prices for agricultural produce, declining farm rents and profits, and the rise of global market integration and international competition. As summarised by Watts and Goodman (1997), he concluded that:

- there was no tendency for the size distribution of farms to change over time, i.e. small 'peasant' farms were not being displaced by capitalist enterprises;
- technical efficiency was not a pre-condition for survival, though an ability for self-

exploitation (e.g. greater use of farm family labour) might be;

- changes driven by competition and market integration transformed agriculture, largely by shaping the production mix of different enterprises, deepening debt-burdens and through promoting labour shedding, but not by radically reconfiguring the size distribution of farms.

This late-nineteenth-century farm crisis facing both the European peasantry and their landlords was largely resolved by intensification of production and the appropriation of some farming functions by capital in processing and agri-industry (Goodman and Redclift, 1981). Kautsky contended that, despite the growing influence of agri-industrial capital as the motor of change, certain characteristics within farming hindered agrarian capitalism, e.g. agriculture's biological character and rhythms and the ability of family farms to resort to self-exploitation in the face of falling prices for agricultural produce.

There are certain parallels between the changes affecting agriculture in the 1890s and those occurring one hundred years later. There are significant developments in the application of science, technology and capital that are affecting the food processing, farm input and farm finance systems, offering new opportunities for agri-business and posing threats to the maintenance of family farming. Moreover, agricultural issues are prominent on the world political stage once more following the central role of agricultural tariffs and subsidies in the Uruguay Round of the GATT negotiations, the proposed reforms to the European Union's (EU's) Common Agricultural Policy (CAP), scares over food safety, and neoliberal reforms widespread in many agri-commodity producer nations in the Developing World. Against this background and the pervasive effects of economic globalisation, there are many local and regional specificities that are producing particular tensions between the global and the local, shaping the outcome of restructuring currently widespread amongst agri-food systems in many parts of the world.

Kautsky's analysis, and that of Lenin (1982), which focused on processes of social differentia-

tion and class formation within the agricultural sector, dealt largely with the notion that, because the peasantry in Germany and Russia were so numerous, then its future political allegiances and the nature of its economic development would be crucial to the future direction of the two states (Bernstein, 1996). However, their work, and that of the political economy approach in the 1980s, tended to neglect factors external to the agricultural system. The roles of the state, politics and culture were suborned in favour of a narrow economism, which also neglected upstream and downstream links to and from agriculture and broader economic influences. Hence, in the rediscovery of the agrarian question, there has been a need to address this limitation by injecting a framework that encompasses global markets, geopolitics and international regimes. This has taken various forms, including the closely related regulation and food regimes theories decribed above, analysis of global commodity systems, and a wide range of work looking at how economic processes now operate at a global scale (Dicken et al., 1997).

One key aspect of the agrarian question that has been revisited recently relates to the view that agriculture is a collision point between nature and capital in which capitalists 'cannot use what is within their reach in every way they might like to' (Henderson, 1998, p. 73). This has been used as an argument in explaining the perpetuation of the family farm and its related variants, with nature presenting an obstacle to large-scale capitalist expansion rather than an opportunity. This is considered in more detail later on in an examination of the reasons for the survival of the family farm (Chapter 6).

The work of the political economy approach since the early 1980s has shown that one of the key differences between the current agricultural crisis in the Developed World and that of the 1890s is the extent to which the current crisis is threatening the existence of family farming (Cox et al., 1989). Hence, there is a need to re-examine Kautsky's arguments and to assess their validity with respect to the new inroads to traditional farming structures that are being made in an era of globalised capital. This is presenting opportunities for developing the theories underlying food regimes, regulation and the agrarian question to gain greater understanding of contemporary restructuring of the agri-food system.

2.5 Conclusion

This chapter has charted the history of agricultural geography since its emergence as a specialism within the discipline of geography in the 1920s. Emphasis has been placed upon its transformation from the early 1970s through a series of shifts in thinking, including behavioural and political economy approaches. These have altered the focus of study by viewing farming in the context of wider economic, social and political processes. New ideas such as food regimes, regulation theory and a reconsideration of the agrarian question have contributed to changes in how agricultural geography is practised. A critical development has been the different types of questions asked by the various different approaches, with a strong tendency towards growing concern for processes that operate on a global scale, but which have distinct regional and local impacts that vary from one locality to another. It is this theme that is now taken up in more detail by examining geographers' contributions to the study of this 'globalisation' (or 'glocalisation'?) and 'restructuring', and especially their work on the contrasting problems of over-production in North America and Western Europe versus scarcity and famine in parts of Africa.

3 Globalisation of agricultural production

3.1 The nature of globalisation

In the last three decades both food production and distribution have been radically restructured in favour of a more global scope and character, with TNCs playing an increasingly important role, especially in activities 'upstream' and 'downstream' from farms. This 'globalisation' is shaping our lives in profound cultural, ideological and economic ways (Goodman and Watts, 1997). Indeed, the concept of globalisation has become part of the standard vocabulary within the social sciences, with a general acceptance of the notion that we are experiencing a new and qualitatively different phase of capitalist development. As with many new terms, though, globalisation has been utilised in different ways by various writers so that it has acquired multiple meanings. Perhaps the four most common interpretations of the term are as follows:

- The worldwide spread of modern technologies of industrial production and communication (including capital, information, money, production and trade) regardless of frontiers. This has been termed by Ohmae (1996) as the 'borderless world'.
- The networking of virtually all the world's economies, representing a widespread move in the direction of functional integration (Castells, 1996).
- The linking of and inter-relationships between cultural forms and practices that occurs when societies become integrated into and dependent on world markets (Wallerstein, 1991).

- The convergence and homogenisation of capitalist economic forms, markets and relations across nations (Boyer and Drache, 1996).

None of these definitions has been uncontested, with criticisms that they exaggerate the true nature of the economic changes occurring. For example, some economists have taken the 'global' part of the process to signal the 'end of geography' (O'Brien, 1992) or the 'death of distance' (Cairncross, 1997), by virtue of instantaneous information flows rendering space and location irrelevant. Indeed, geographers have highlighted this in references to 'time–space compression' (Harvey, 1989) and 'time–space shrinkage' (Allen and Hamnett, 1995). However, 'globalisation does not mean the homogenisation of social and economic relations across space' (Bryson *et al.*, 1999, p. 24). Geography has not ceased to be important simply because we live in a world in which events happening at a particular location are shaped by developments occurring thousands of kilometres away. Indeed, movements of capital, money, information and cultural exchanges occur because there remain substantial differences between places, regions and countries. So there is a complex two-way flow between the global and the local, the understanding of which has been a major focus of geographical research.

At the heart of globalisation is the way in which the geographical outcomes of economic processes have increasingly become the function of links and dependencies that extend well beyond local, regional or even national environs. Instead they incorporate diverse, multi-faceted interactions between people and locations in many different parts of the world. This can perhaps be seen most readily in the sphere of consumption, where Coca-Cola, McDonalds, Levi jeans or Microsoft computer operating systems, for example, offer an essentially uniform experience wherever they are consumed in the world (Ritzer, 1996a). This does not mean that everyone is affected in the same way by the availability of these 'global' products (Allen and Thompson, 1997), but there is far greater 'interconnectedness' in terms of product availability and consumption. Hence there is a trend towards greater social and cultural homogenisation, in part mitigated by gross inequalities in wealth distribution both globally and nationally (Ritzer, 1996b). This has been expressed in the (contested) argument that we are moving to a situation in which, ultimately, there will be no national products or technologies, no national corporations or national industries as every factor of production can move effortlessly across political borders.

Globalisation also embraces a range of changes within the evolving transformation of capitalist development. One label applied by David Harvey, and widely taken up, is 'flexible accumulation' (see Knudsen, 1996). Flexibility applies to production, labour utilisation, consumption and relations between the state and the economy. This is generally interpreted as capitalism's search for a new technological, social and spatial solution to the problems of over-accumulation experienced in the 1970s and early 1980s under the previous Fordist regime. The distinctive characteristic of flexible accumulation is its global dimensions. These dimensions have provided a dramatically increased circulation of goods, objects and cultural artefacts (Harvey, 1989), resulting from increased local instabilities in the process of capital accumulation, which in turn sustain the drive for greater flexibility and circulation of capital. The pace of change has also been accelerated because the goods being produced and circulated are not just material goods but also information and items with particular cultural or aesthetic characteristics. Through their inherent qualities these can be rapidly consumed or disposed of, contributing to a new consumerism associated with cultural images, symbols

and signs, of which agri-commodities are also a component part. Hence, labels such as 'green', 'fresh', 'organic' or associations with particular localities or traditional images have become part of what Lash and Urry (1994) refer to as the 'aestheticisation of consumption'.

The emergence of a global economy has had several vital consequences, as summarised by Amin and Thrift (1994, pp. 2–5) into the following seven main points (see also Daniels, 2001):

- A pronounced increase in the power of finance over production. Finance capital can move seamlessly and very rapidly around the world, having major impacts on national and regional economies and the fortunes of individual companies.
- The pivotal role of knowledge as a vital factor of production. Hence there are significant economic benefits in producing an educated and skilled workforce.
- Technology has become trans-nationalised, especially in economic activities that are knowledge intensive, e.g. financial services or telecommunications. This advantages producers and institutions with the resources to manage technology.
- Globalisation of technological change, the mobility of finance capital, and transportation and communication has been accompanied by the emergence of global oligopolies, i.e. production in particular industries/sectors is dominated by a small number of firms.
- The ability of individual nations to regulate their own economic development has been limited by the emergence of transnational institutions that co-ordinate and regulate aspects of the world economy. The most well-known examples are the IMF and the WTO.
- The electronic flow of information has stimulated more extensive cultural flows, reconstituting the meaning of cultural symbols and identities.
- The outcomes of globalisation appear as new (global) economic geographies. These new

geographies have been characterised in several ways, including 'the borderless global economy' (Ohmae, 1990) and 'the new global division of labour' (Massey, 1996).

This seven-point summary implies that the development of globalisation during the last two decades is qualitatively different from the internationalisation that characterised the first food regime, which involved the extension of a firm's activities across national boundaries. Globalisation implies a degree of purposive functional integration among geographically dispersed activities. Thus TNCs with a globalising strategy centralise strategic assets, resources, responsibilities and decisions, but operations, which are based in several countries, are aimed at tapping a unified global market (Bartlett and Ghoshal, 1989). However, the home base or country of origin of a TNC may remain important to the nature of its operations and is likely to affect the character of its globalisation. In recognising this, the process of globalisation is summarised by Bonnano (1993, p. 342) as involving 'the recomposition of production processes across national boundaries in such a way that they transcend the locus determined by the physical limits of the nation state'. Therefore this also transcends the multi-national capitalism of the second food regime because it affects the international organisation of productive sectors at both spatial and sectoral levels. Its effects have been to intensify agricultural specialisation at regional and production-unit levels, whilst, at sectoral level, it has brought increased transformation of agricultural products from items destined for immediate consumption into inputs for the greater food manufacturing system (Constance and Heffernan, 1991; Friedland, 1991). This chapter focuses on work by geographers and other social scientists investigating this role of globalisation in changing the character of agri-food production.

As globalisation is an all-embracing term for various different forms of commercial operation, it therefore covers a number of different manifestations and can vary in form between industries. There are some commodities for which the producing corporations are globalised, whilst in others the retailers are globalised but the production has a local, regional or national focus. Friedland (1994a; 1994b) highlights the case of fresh fruit and vegetable production, where production is increasingly disparate and occurs in many localities worldwide, but there are only a small number of genuinely global corporate distributors and many national ones.

In general, globalisation processes imply developments beyond national-level agri-food oligopoly or the extension of transnational food firms into overseas markets. Instead, 'competition is now being expressed through processes such as decreasing the costs of, and increasing the corporate control over, sources of raw materials and components through global sourcing' (Buttel, 1996, p. 24). Inter-firm sourcing, via subsidiaries spread across the world, is also occurring, as are changes in corporate organisation and behaviour involving intra- and inter-firm coordination through strategic alliances, e.g. the alliance between Monsanto, DeKalb and Calgeze in the development of transgenic (genetically modified) crop varieties.

There has been considerable debate about the impact of globalisation upon the nation-state and the declining influence of state action. However, nation states still maintain strong controls over monetary regulation, social control and foreign policy, the latter usually being closely linked to attitudes towards free trade and the debates associated with the Uruguay Round of the GATT, the WTO and the establishment of regional trading blocs. In considering aspects of the maintenance of nation-state controls and globalisation, Goodman (1994) developed a typology of world-scale processes in which he distinguished between the concepts of globalisation, trans-nationalisation, multi-nationalisation and internationalism. He argued that, whilst each process subsumes some elements of 'the global', they have different dominant logics and have variable implications for economic development and social welfare. His scheme is summarised in Table 3.1.

Goodman's portrayal of these processes can be considered alongside the more generalised food regimes concept outlined in the previous chapter (see Table 2.2). For example, internationalism

Table 3.1 Typology of world-scale processes

Process	Dominant logic	Economic governance and corporate organisational form	Spatial dynamic	Diagnostic
Internationalisation	Exchange	Market over hierarchy. National corporations and 'champions'	National production systems in an expanding international trading system. Rising significance of intra-industry trade. Triadisation of trade, FDI and financial flows	'Golden Age' growth of international trade. Securitisation. Globally integrated financial markets. Time/ space compression. Increasing separation of financial markets and real production sphere
Multi-nationalisation	Production	Hierarchy. Central coordination and ownership of relatively autonomous overseas affiliates. 'Multi-domestic' strategy. Centralised investment, production and marketing	Access to raw materials; plantations, mining, energy, commodity traders. Export substitution production, for innovation rents. Intra-firm sourcing. Acceleration of FDI.	International oligopolistic competition. Product-cycle diffusion of in-house R&D via standardised mass production. Low-cross-border integration of production. Intensified global competition: merger wave. FDI concentrated in Triad and E. Asian NICs. Truncated globalisation
Trans-nationalisation	Production	Hierarchy. Centrally controlled, vertically integrated transnational production systems	Global intra-firm division of labour. 'Footloose' locational decisions driven by internal firm-specific criteria. Homogenisation of space.	Standardised mass production and mass customisation (and variants). 'Global composites'. Dominance of FDI-related trade
Globalisation	Innovation	Heterarchy. Non-market forms of inter-firm collaboration	Valorisation of localised techno-economic capabilities and socio-institutional frameworks. Mutual reciprocity between regional innovation systems and global networks	Collaborative R&D networks (G–7 space). Technological convergence. Strategic alliances in advanced tech-sectors. Inter-firm processes of permanent innovation and co-operation

Notes: italics indicate the reconfiguration and mutation of processes. FDI: foreign directs investments.
(Sources: based on Goodman, 1994, and Buttel, 1996)

might be regarded as representing the apogee of the second food regime, for it is based on commodity exchange in world markets as best illustrated by the post-1945 expansion of powerful national corporations. Multi-nationalisation, which can be seen in the contemporary world agri-food industry, involves more central coordination and ownership of relatively autonomous overseas affiliates. In contrast, trans-nationalisation is more readily associated with footloose, flexibly specialised firms in which production systems are centrally controlled and vertically integrated.

Buttel (1996) argues that Goodman's typology offers a possible means of disaggregating processes of world-scale integration in order to develop a better understanding of globalisation. He suggests that it offers a better 'yardstick' for analysing globalisation than merely typifying it as the development of 'a footloose, stateless form of "flexible" production involving inter- and intra-firm multiple sourcing' (p. 30).

The preceding discussion, and indeed much of the literature on globalisation, tends to portray the outcome of current processes as a global world-economic integration. Yet, this may be a gross over-simplification of what is a complex, uneven and fragmented set of processes in which spatial economic differentiation is clearly manifest. In particular, reinforcement of the 'global triad' of North America, Western Europe and south-east Asia is occurring through their concentration of the growth of trade and capital investment. The contrasts with most parts of the Developing World are stark, but there is pronounced differentiation of 'the South' too, as sub-Saharan Africa and parts of Central America and the Caribbean become progressively more marginal to the world economy. This exclusion of some countries from the 'globalised' world economy suggests that globalisation is at best partial and contingent rather than overarching and inexorable (Mohan, 2000; Ritzer, 1996b). However, the presence of regions marginal to the global economy is regarded as highly significant in the food regimes literature, which emphasises their role as providers of cheap labour or sinks for capital investment in alternative ventures, e.g. tax write-offs or low-yield but 'safe' primary production.

3.2 Globalisation and agri-food production

In considering how globalisation processes affect agricultural production, processing and consumption, Marsden (1997) lists seven key changes in the dynamics of agri-food networks:

- It is not only agribusiness firms and their relationships to the farm-based sector that are the dominant component in agri-food networks (Tansey and Worsley, 1995), it is the downstream element in the network where value is added, and hence the growing power of 'near-consumer' agencies, e.g. consumer 'watchdogs' and food safety organisations.
- It is the non-farm sector that increasingly has control over the nature of the food that we consume.
- Globalisation of food is paralleled by local and regional effects such as local diets, food scarcity and abundance which have an impact upon globalisation.
- Technological developments and their social effects must be placed in a broader context, for example with respect to considerations of food quality which act as a regulator of the food supply chain.
- Whilst global processes operate through corporate manufacturing and retail capital, national processes can still act independently as a regulator of the food supply chain.
- The demands of sub-sections of the market, especially in the Developed World, act as 're-regulators' of the food market.
- The aforementioned processes provide new spatial patterns of production and consumption in a dynamic that drives uneven development.

In simple terms, the progress of globalisation in linking production and consumption in different parts of the world by new means can be seen in the way in which trade in certain agricultural products has become global. Perhaps the best example of this is trade in fresh fruit, fresh vegetables and cut flowers which, in the early 1990s, reached the equivalent of 5 per cent of global commodity trade, or roughly comparable to that of crude petroleum (Jaffee, 1994). In asserting that this trade in fresh horticultural products was 'truly transnational', Watts (1996) has argued that the key prerequisite to globalisation in this context is vertical integration through contracts rather than through control and ownership of the means of production (see the example of H. J. Heinz in New Zealand in work by McKenna et al., 1999a).

Argentina, Brazil, China, Kenya and Mexico account for over 40 per cent of world trade in perishable horticultural items from Developing Countries. However, the fastest growing contribution to these exports is coming from Africa, assisted by low costs of production, complementarity to European seasons, relatively short flight times and increasing ability to supply produce of the quality and quantity required by international markets (Barrett et al., 1999, p. 160). The expansion in this trade is also a reflection of changes in both consumer food demands and food retailing in Developed Countries, placing a growing emphasis on fresh, 'healthy' produce. Moreover, in the UK, between two-thirds and four-fifths of fresh horticultural imports from Africa are made through supermarkets that control 80 per cent of all food sales. Therefore the policies and operations of the supermarkets are central to the future of this trade. Particularly critical has been their policy to offer year-round supplies of fresh produce lines, thereby requiring sourcing from different areas around the world. This significant role of the changing nature of retailing and consumption patterns within the context of globalisation is considered further later in the chapter.

At this point it is important to re-emphasise that globalisation has not been an inexorable linear trend, but rather a partial and contingent process with considerable variation in the extent and nature of its impacts. In terms of trade in agri-products the long-term dominance of North America and Western Europe has been reinforced, as has the significance of south-east Asia with respect to trade in tropical foods (Thompson and Cowan, 2000). Significant marginalisation and differentiation has occurred within the Developing World whilst, throughout the world, increasing differentiation of agriculture at regional and local levels is occurring. Moreover, globalisation has not been the only large-scale set of processes affecting both production and consumption of food. For example, considerations relating to sustainability have grown considerably since the World Commission on Environment and Development (the Brundtland Report) in 1987 produced a definition of sustainable development that has acted as a rallying call for many opposed to what they see as pernicious aspects of globalisation.

It has been argued, though, that aspects of both globalisation and sustainability have combined to deepen and intensify the potential exposure of agriculture to large-scale capitalist forces, with the adoption of 'international norms, standards and practices as well as production to meet consumer preferences' (Le Heron and Roche, 1995, p. 25). Combinations of globalising tendencies and concerns for sustainable development have been central to international dialogue in the GATT and the WTO and within the reforming agenda pursued with respect to the EU's CAP. This dialogue has frequently placed those championing globalisation on the opposite side to those pursuing sustainable development. However, many large retailers and food processors have acquired some of the elements normally associated with sustainability and 'alternative'/'local' food economies in their attempts to enhance sales of 'green' foods, e.g. organic fruit and vegetables, and 'healthy' foods. Ironically, globalisation processes have also offered new opportunities for agri-commodity production at a local level, whereby locally produced quality products and services, strongly associated with a particular geographical locality, are transferred to regional and national markets (Gilg and Battershill, 1998; Kneafsey and Ilbery, 2001; Marsden, 1996).

Another limitation of the globalisation thesis is that most TNCs still source their food domestically (not globally) and then process and distribute it internationally through TNC firms and retail chains. Overseas subsidiaries of large agri-food TNCs generally operate fairly independently of the parent company in the form of what is termed a 'multi-domestic', again applying 'domestic' or local sourcing. Indeed, rather than production and supply within the agri-food chain being the quintessential set of globalising processes, it is finance and cross-border flows of capital that are more truly global. In this respect it is finance that has had the greatest influence on the agri-food sector in the form of TNCs acquiring existing firms operating in established markets and through mergers amongst TNC food firms. This has enabled TNCs to move into lucrative foreign food markets, though often at the cost of increasing their indebtedness and reliance on the behaviour of the financial markets (O'Brien, 1994). Indeed, on examination of the largest food manufacturers and processors in the world (e.g. Nestle, Philip Morris, Unilever), they are revealed primarily as TNCs operating subsidiaries in a number of countries through multi-domestic strategies. They do not resemble the flexibly specialised firms found in many other industrial sectors. Moreover, they have generally expanded through acquisition rather than increasing capital investment in their own plant. In the USA, for example, the agri-food industry is still largely associated with domestic vertical integration and contracting rather than being part of globalised commodity chains (Wells, 1996). Meanwhile, there has been continuing exploitation of the global division of labour, through use of cheap labour in primary stages of production (in Developing Countries) and value-adding near the point of consumption (processing plants in Developed Countries).

Three types of external capital have become especially important in helping to absorb agriculture into the broader food supply system (McMichael, 2000):

- Agri-food companies. These produce manufactured foods, which may be canned, frozen, part-cooked and/or pre-prepared. Indeed, most foodstuffs are part of some form of value-added processing before they are purchased by the consumer. There has been some purchase of farms by these processing companies in the USA, notably the operation of gigantic feedlots by ConAgra and Excel (Schlosser, 2002, p. 138), but generally their link to farms is maintained via forward production contracts. These contracts tend to favour larger farms located

close to the processor. Production of fruit, vegetables, pigs and poultry has been most affected by such contracts.
- Food retailers. The emergence of large supermarket chains throughout the Developed World, and increasingly in the larger cities in the Developing World, has dramatically altered the relationship between farmers and retailers. The supermarkets are generally able to determine the market price for both food processors and farmers. This often links places on a global basis so that processors and farmers are experiencing new levels of competition.
- Financial services. These have been particularly important in areas where the value of farmland has become very high, as the operation of credit facilities has been closely linked to farm indebtedness.

The new patterns associated with these different types of investment continue to link places of production, e.g. the Caribbean, and consumption, e.g. the urban markets of the Developed World, in an unequal division of power and capital accumulation. For example, McMichael (1992b, p. 113) refers to the 'internationalisation' of the Pacific Rim's food systems leading to 'the reconstruction of supply zones such as the United States, Australia and Thailand, especially in the development of integrated animal protein-complexes sponsored by Japanese capital'. In Marxist language this fosters a series of dialectical relationships: production–consumption, globalisation–localisation, deregulation–reregulation, devaluation–valuation, powerful spaces–dependent spaces, cultural distinctiveness–social exclusivity.

These effects, or globalisation processes (McMichael, 1994), are not solely a phenomenon of the last two decades. The global divisions of labour that are inherent in globalisation have been a feature of food production since the sixteenth century (Bonanno et al., 1994), though a crucial impetus came with the demise of the Bretton Woods system in the early 1970s and the decoupling of the US dollar from gold in 1971, ending national currency regulation via the gold and dollar standard. This helped foster new institiutional mechanisms of control and co-operation in the global movement of capital: and hence the emergence of 'flexible accumulation' (McMichael, 1992c). For example, the IMF has assumed increased importance as some aspects of policy-making have been displaced from state to supra-national institutions.

In recognising the importance of globalisation in creating the 'new world food order' of the proposed third food regime, Friedmann and McMichael (1989) refer to this declining significance of national regulation in shaping the character of agricultural production (see also Bonnano, 1993). This has even extended to previously heavily protected products such as sugar cane in Australia (Robinson, 1995). This is a view that stresses the global implications of what could now be termed the 'post-GATT era', referring to the conclusion of the Uruguay Round of the negotiations of the GATT, which focused on trade in agricultural produce (McMichael, 1992a).

There was an intimate link between globalisation and the Uruguay Round because underlying the Round, and the various multinational conflicts in the associated negotiations, were the demands made by transnational food companies and their global food production and processing operations (Grant, 1993; McMichael, 1993). The latter provided a stimulus towards trade liberalisation and represented a challenge to the nation state or multistate organisation as the major force reshaping regional agricultures. So there were challenges to existing agricultural and food supply regimes, and threats to the alliance between the state and farmers that had been in place since World War Two (Ufkes, 1993). This is discussed further in Chapter 4.

The free-trade regime, supported by several key negotiators in the Uruguay Round and subsequently through the WTO, is likely to further assist institutionalisation of the mechanisms and norms of a system of global regulation. In unifying the market, efficient producers will be helped and especially the global accumulation strategies of the TNCs, enhancing capital mobility and eliminating control of market and production sectors by national companies. Indeed, the activities of the WTO to reduce national subsidies to farm production and trade are helping to establish an environment that greatly enhances the geographical scope of agri-food capital. It can be argued that TNCs were restricted in the previously unliberalised agri-commodity markets (Friedmann and McMichael, 1989), so it is not surprising that they have been strong supporters of trade reform in agri-commodities whereby economic trade relations would be regulated by global institutions favouring global accumulation.

3.3 From productivist to post-productivist agriculture?

3.3.1 Agricultural restructuring

Having outlined the globalising processes affecting the broad agri-food system, it is now possible to focus more closely upon the agricultural production component of that system, and to consider that the nature of agricultural development during its so-called post-1945 'productivist' phase has been explained in a variety of ways. Essentially, economic explanations have emphasised the commercialisation of farming as part of the introduction of supply and demand relations within the market economy (Vandergeest, 1988). Technological development was an integral driving force within this process, encouraged by the state through support of research and advisory services. This state support was geared to encouraging farmers to raise production. However, there are more socially orientated explanations, such as commoditisation theory, which focus primarily on social formation and on farm inputs rather than outputs. These emphasise the way in which farm households became involved in the market economy by purchasing manufactured goods and so became part of commercial exchange transactions to fund such purchases. They tend to assume an essentially Marxist view of the capitalist system, postulating that family farms will become increasingly subsumed by external capital, e.g. agri-industrial complexes. Subsumption may be real in the sense of a direct change in ownership or it may be formal in that effective control passes to external capital, as in the growing power of supermarkets and food processors over farmers through the increased use of contract production (Hart, 1992; Marsden *et al.*, 1987).

It can be argued that the uneven penetration of agriculture by external capital observed in both the economic and social explanations is part of the process of uneven development, with capital moving through time and space into sectors where profits will be greatest (Harvey, 2000, pp. 75–83). However, there would appear to be significant flaws in some aspects of these ideas, as the persistence of family farming (albeit in a variety of forms) seems to negate parts of the subsumption argument (Blunden *et al.*, 1997). Moreover, there is the added complexity that small family farms, termed simple or petty commodity producers in some Marxist-based literature (e.g. Friedmann, 1986; Moran *et al.*, 1996), frequently hire waged labour, whilst large 'agri-businesses' can be family-owned and family-operated. In tackling these paradoxes raised regarding the nature and maintenance of family farming, social scientists have revisited the writings of Kautsky and his work on the 'agrarian question', as outlined in Chapter 2.

In examining the evolving nature of productivism, both from economic and social perspectives, geographical literature dealing with the related impacts of globalisation upon agriculture has referred to various characteristics of the associated restructuring of farm-based production. These characteristics have been a response not only to globalisation but also to broader technological, economic and social forces that have affected the nature of farming activities. Restructuring itself has many meanings and has been used in different contexts, but generally implying 'qualitative changes in the relations between constituent parts' (Lovering, 1989, p. 198), for example changing relationships between consumption and production. Thus, in rural areas it can be argued that spaces of production (e.g. farming) are being increasingly supplanted by spaces of consumption (e.g. recreational pursuits, retailing of 'rural' products) (Marsden, 1998b). When applied specifically to farming, another view is that restructuring is 'concerned essentially with the repositioning of local agricultural production within the wider food system, which is now dominated by international corporations both upstream and downstream from the farm gate, and within a global economy where protectionist trade barriers are gradually being dismantled' (Symes and Jansen, 1994, p. 7).

In work by Marsden *et al.* (1986b; 1987), focusing upon capital accumulation in the British farming industry, restructuring was recognised as having four principal characteristics:

- Overproduction of agricultural goods in the face of stable demand. Technological improvements and state support have accelerated tendencies towards concentration and centralisation of capital in different parts of the food production industry (Busch *et al.*, 1989). For farmers in many parts of the Developed World this has meant increased indebtedness (e.g. Wilson, O., 1995; 1996).
- 'Downstream' environmental consequences have led to major landscape modifications (e.g. Ghaffar and Robinson, 1997).
- Increased differentiation has occurred between farms and also between regions. A 'dualism' has been created between marginalised family farms on the one hand and large heavily capitalised businesses on the other, albeit with significant local variation (Munton and Marsden, 1991). This in turn affects the direction and nature of capital penetration.
- The transformation of the family farm has created a wide range of different types of farm under this umbrella category. In particular, the farm household has had to seek alternative sources of income and capital. This has contributed to the growth of part-time farming (Gasson, 1988a), especially near large urban centres, and the more recent increase in farm-based tourism (Walford, 2001).

Thus there are several different facets to the repositioning of local agricultural production within the wider food system, including changes in the application of technology on farms, changing labour relations, and new external relations upstream and downstream of the farm. However, one significant conceptualisation of the changes that has been implicit in much geographical literature, especially that from British researchers, is that since the mid-1980s there has been a policy-related development that has carried agriculture from its dominant post-war phase, termed 'productivism', into a new era, termed 'post-productivism' (Wilson, 2001).

3.3.2 Productivism

Productivism, sometimes termed 'development-alism' (e.g. McMichael, 1996), generally refers to the state-managed policy framework present throughout the Developed World that supported an acceleration in two key processes in the industrialisation of agriculture (Bowler, 1992a):

- Appropriation, whereby certain parts of the agricultural production process were transformed into a more industrial activity and utilised as purchased farm inputs, e.g. mechanisation in which the horse was replaced by the tractor, and the application of chemical fertilisers whereby manure was superceded by synthetic fertilisers. Not all farm-based processes have been affected and appropriation has proceeded in irregular phases of innovation and diffusion.
- Substitution, whereby agriculture functions as the supplier of raw materials to industry, but increasingly, agricultural products have

been substituted by industrial products, e.g. the use of synthetic fibres rather than cotton. These relations between agriculture and industry are greatly affected by the nature of state intervention and support but operate across national boundaries.

Recent developments in biotechnology are examples of appropriation and substitution. They are leading to 'textured vegetable protein becoming a competitor for meat products, isoglucose becoming a competitor for sugar, and ethanol becoming a petrochemical substitute' (Ilbery and Bowler, 1998, p. 64). In turn this is producing a demand from agri-food firms for crops with improved processing characteristics.

Whilst recognising the significance of these two key processes, it is possible to summarise the changes occurring within productivist agriculture in a more direct manner, in terms of three structural dimensions: intensification, concentration and specialisation, each of which has contributed to raising output (i.e. maximising production) (Bowler, 1985a; 1994). The primary and secondary consequences of their operation are shown in Tables 3.2 and 3.3. The three together may be regarded as representing the industrialisation of modern farming (Troughton, 1986), a central component of productivism, at the heart of which are appropriation and substitution.

Table 3.2 The industrialisation of agriculture: primary process responses

Structural dimension	Outcome
Intensification	Purchased inputs (capital) replace labour and substitute for land
	Increasing dependence on agri-inputs industries
	Mechanisation and automation of production processes
	Application of developments in biotechnology
Concentration	Fewer but larger farming units
	Production of most crops and livestock concentrated on fewer farms, regions and countries
	Sale of farm produce to food processing industries – increasing dependence on contract farming
Specialisation	Labour specialisation, including the management function
	Fewer farm products from each farm, region and country

(*Source*: Bowler, 1985a)

Table 3.3 The industrialisation of agriculture: secondary consequences

Structural dimension	Outcome
Intensification	Development of supply (requisites) cooperatives Rising agricultural indebtedness Increasing energy intensity and dependence on fossil fuels Overproduction for the domestic market Destruction of environment and agri-ecosystems
Concentration	Development of marketing cooperatives New social relations in rural communities Inability of young to enter farming Polarisation of the farm-size structure Corporate ownership of land Increasing inequalities in farm incomes between farm sizes, types and locations State agricultural policies favouring large farms and certain regions
Specialisation	Food consumed outside region where it was produced Increased risk of system failure Changing composition of the workforce Structural rigidity in farm production

(*Source*: Bowler, 1985a)

Within 'intensification' some researchers have made a distinction between the key symbiotic and commensalistic technologies closely associated with major increases in output:

- Symbiotic. This includes biological, chemical and bio-technological developments, the introduction of hybrid seeds, and irrigation, which have allowed farmers to use new or previously unavailable resources. In recent years the best examples have been within biotechnology, which promises to revolutionise crop and livestock production and so make symbiotic forms of technology an increasingly important source of change in the organisation of agriculture. An excellent earlier example in the USA was the widespread development of irrigation on the Great Plains, making use of previously unavailable water resources feasible and greatly increasing per-unit area productivity (Bowden, 1965). Yields from irrigated land there are typically between two to six times greater than on unirrigated land. Around half of all irrigated land in the USA is now in the Great Plains. Symbiotic technologies may encourage increases in the numbers of farms or the farm population by virtue of making additional resources available.

- Commensalistic. Technologies that produce increased efficiency in the use of resources, reducing a need for farm labour and eventuating in a decline in the number of farms through farm consolidation. Such technology includes the tractor and other machinery that increases the work capacity of human labour, thereby permitting farmers to accomplish more efficiently processes that they were already performing. It can be measured quantitatively as the value of machinery, equipment and energy consumed per value of farm sales (Murdock and Albrecht, 1998, pp. 306–7).

Three broad responses to the operation of this changing economic context of farming can be recognised:

- farming becomes only a minor contributor to the overall income of the farm household;
- increased diversification is pursued in which both new on- and off-farm enterprises may

Table 3.4 Elements of farm business adjustment

Element	Description
Farm enterprise	Changing the emphasis of the farm enterprises (e.g. expanding sheep production while contracting dairy enterprises)
Labour	Usually by substituting family labour for hired labour in order to reduce costs, but could be an increase in hired labour
Business structure	Usually by changing from sole operator to a partnership to reduce tax
Tenure	Either by buying land that was previously rented or by selling owner-occupied land and leasing it back
Size	Buy or sell land either to expand the farm business or to finance restructuring
Economic centrality	Increase (or decrease) income from off-farm sources, thus changing the economic centrality of the farm business to the family household
Diversification	Increasing income from non-farming enterprises based on the farm (e.g. bed and breakfast or farm shop)

(*Source:* based on Munton, 1990)

be developed. This is termed a survival strategy;

- an accumulation strategy is pursued through corporatisation or agribusiness development involving complex relationships with finance and industial capital (Marsden *et al.*, 1989; McMichael, 1999).

These responses have been further refined in Munton's (1990) 'elements of farm business adjustment' shown in Table 3.4. However, analysis of this adjustment or restructuring has been refined by some researchers in terms of a recognition of a significant step-change in the character of agricultural production, and indeed of the role of agriculture within the agri-food system. This is termed the post-productivist transition.

3.3.3 Post-productivism and its critique

The term post-productivism was popularised in the 1990s, perhaps its earliest definition, referring to agricultural adjustment and restructuring, especially by family farming households, being made by Munton (1990, p. 10; quoted by Evans *et al.*, 2002, p. 315):

The post-productionist (*sic*) period that agricultural policy, farmers and the food industry are now

entering will mean that the margins of profitability will become tighter and the overall logic of the agricultural treadmill (involving increasing stocking levels, scale and level of subsumption) will be increasingly questioned.

Hence, initial usage of the term 'post-productivist' referred to certain key changes in farming, especially relating to adjustment strategies pursued by farmers, e.g. in response to calls for greater environmental regulation, and through developments such as reshaping of gender roles in family farming businesses (Symes, 1991) and the possible demise of family farming (Marsden *et al.*, 1992a). Both Shucksmith (1993) and Symes (1992) identified 'post-productivist' changes in farm household behaviour throughout Europe, and especially (for the latter author) with respect to changes in Eastern Europe associated with the demise of collectivism and the growth of decentralisation. This particular concern for developments in Eastern Europe has not been pursued and, instead, it has been changes in the agri-food sector in north-west Europe and North America that have been at the heart of most literature applying the term post-productivist. Some of this work has ascribed a definite temporal dimension to post-productivism. For example, Halfacree (1997) refers to a transition traceable to the 1970s whilst G. Clark *et al.* (1997a; 1997b) relate it directly to the reforms to the CAP of 1992 and the conclusion of the Uruguay Round of the GATT talks. Other writers have introduced a definite spatial dimension, linking particular 'spaces' to post-productivism (e.g. Marsden, 1998a).

From the miscellany of definitions applied to post-productivism, Evans *et al.* (2002, p. 316) highlight that of Ilbery and Kneafsey (1997): 'a shift in emphasis away from quantity to quality in food production; the growth of alternative farm enterprises, conceptualised as "pluriactivity"; state efforts to encourage the development of more traditional, sustainable farming systems through agri-environmental policy; the growing environmental regulation of agriculture; and the progressive restructuring of government support for agriculture.'

One approach is to view post-productivism as the direct antithesis of productivism, that is a reversal of trends towards intensification, concentration and specialisation, into trends of extensification (e.g. set-aside), dispersal (e.g. de-concentration of farm production) and diversification (e.g. pluriactivity) respectively (Ilbery and Bowler, 1998). This is strongly contested by Wilson (2001) who argues that there is little empirical evidence for farm dispersion; the extent of diversification may be limited and is not necessarily linked to environmental benefits or to decreased agricultural production; and extensification often exists alongside intensification and, if it takes the form of land abandonment, as in the Mediterranean, it may lead to environmental degradation, the opposite of one of the key post-productivist indicators. Indeed, more research is required on the extent to which such trends can be observed before it is legitimate to claim the existence of anything remotely worthy of the term post-productivism.

An alternative conceptualisation is provided by Wilson (2001):

- The loss of the central position of agriculture in society, and a reduced importance of the farm and landowning lobby in parliament and government.
- Changing public attitudes to farmers – from regarding them as stewards of the land to destroyers of the countryside. In turn, this relates to changing notions of the role of the countryside, in which agriculture may be perceived as a threat to maintaining or developing the countryside in a desirable manner.
- The agricultural policy community has changed to embrace environmental and other non-agricultural concerns. This has been typified by the advance of agri-environmental policy, reduced subsidies, tighter pollution regulations and changes in property rights – all part of new forms of governance of rural areas, though some see 'greening' as an incrementalism aimed at farm income support rather than environmental conservation, or certainly not radical reforms.
- There is a challenge to the second food regime, what Wilson terms the 'Atlanticist Food Order', in the form of a rapid dismantling of protectionist state and supra-state policies. This has created both uncertainties and new opportunities, e.g. as part of non-standardised demand for high-quality goods and services.
- New forms of agricultural production, with less emphasis upon securing national self-sufficiency for agricultural commodities.
- Commodification of former agricultural resources, including land, wildlife habitats, barns and cottages, by urban migrants to rural areas.

Morris and Evans (1999) argue that work on the post-productivist transition has comprised three chronological research components:

1 Post-productivism has been used as a means of summarising aspects of agricultural adjustment previously conceptualised as survival and accumulation strategies (Marsden *et al.*, 1989), elements of farm adjustment or farm business development paths (Bowler, 1992b; 1999c). This has been a convenient means of acknowledging differential responses made by farm households to new conditions in the farm sector (Ilbery and Evans, 1996).

2 The establishment of the characteristics of the transition. These have included:

a) a shift in emphasis away from quantity towards quality in food production (Marsden, 1998b; Morris and Young, 1997a);
b) the growth of alternative farm enterprises as part of pluriactivity (Evans and Ilbery, 1993; Ilbery, 1991);
c) state efforts to encourage a return to more traditional, sustainable farming systems through agri-environmental policy (Ilbery *et al.*, 1997; Wilson, 1996);
d) the growing environmental regulation of agriculture (Robinson, 1991a; Ward *et al.*, 1995);
e) the progressive withdrawal of support for agriculture (Winter, 1996).

3 Relatively little theorisation of the transition has occurred, though Ilbery and Bowler (1998, pp. 70–1) have provided a process-oriented view in which there has been a simple reversal of previous productivist components of change, as described above, through moves from intensification to extensification, concentration to dispersion, and specialisation to diversification:

a) From intensification to extensification. This has been fostered by policies promoting reduced purchase of non-farm inputs, in part to reduce levels of environmental pollution and to restore natural habitats.
b) From concentration to dispersion. The polarisation of farms into a few large agribusinesses and a dwindling number of smaller, more marginal 'family' farms may be reversed by policies that provide a less differentiated reward system. However, there is little current evidence of this occurring.
c) From specialisation to diversification. The impacts of the cost-price squeeze and reductions in price supports are promoting searches for different types of agricultural and non-agricultural diversification. This

may reduce specialisation whilst also increasing the amount of income in the farm household drawn directly from traditional farm-based production.

These changes may then provide links to thories of regulation, as they are associated with new regulatory conditions (Clark, G. *et al.*, 1997a; 1997b). However, post-productivism is as yet only an incipient development in need of a more careful evaluation, especially as most farmers remain firmly productivist, especially those who have a strong commercial orientation (Walford, 2002). Ideas from other 'transitions' may be usefully applied here, e.g. work dealing with Fordism and post-Fordism (Goodwin and Painter, 1996; Jessop, 1994). It needs to be established more clearly whether the extent of non-food production activities on farms is sufficient to constitute a 'transition'.

Partially as a test of whether the trends towards post-productivism postulated by Bowler, Ilbery and Wilson are observable, Evans *et al.* (2002) focused upon five key categories emergent from the literature on post-productivism, arguing that the complexity of the underlying processes helps to render as invalid the application of such an all-embracing term as 'post-productivist'. Their five categories were closely related to the quote from Ilbery and Kneafsey cited above:

1 The shift from quantity to quality in food production. The notion of quality varies tremendously between products, individuals, regions and countries. It is socially constructed by consumers as well as having regulatory connotations, e.g. the application of hygiene requirements, which themselves may be linked to indicators which are socially constructed. Notions of quality can be applied to both production and food for the mass market and niche markets. Hence there is not necessarily a special link between quality issues and post-productivist niche-market foods. Quality assurance procedures are playing an increasing role in the mass food market, notably through large food retailers establishing new supply chains based on particular quality-assurance

schemes (Morris, 2000a; Morris and Young, 2000).

2 The growth of on-farm diversification and off-farm employment (pluriactivity). There is only limited evidence that farming systems are becoming less specialised, and indeed many upland farmers in Europe are becoming ever more reliant on sheep (Winter *et al.*, 1998). However, there is certainly some strong evidence that many farmers are having to give more attention to non-agricultural and novel agricultural enterprises as part of an overall diversification of their enterprise. Hence there may be a tendency towards increased specialisation on some (often large) enterprises whilst others are diversifying. However, at least some of this diversification is perhaps more appropriately termed 'productivist' in that it is highly commercially orientated farm-based production. Some parts of the productivist farming system may also be largely immune to the types of diversification held to be 'post-productivist' (Boulay and Robinson, 2001). This is discussed further in Chapter 6. Clearly, more research is needed on what aspects of farm diversification can be regarded as post-productivist.

3 Extensification and the promotion of sustainable farming through agri-environment policy. Extensification has been promoted in the EU since the 1980s and can also be seen in North American farm policy for at least as long. Measures have included attempts to decrease stocking densities of farm animals; limits to cereal production, e.g. set-aside of arable land; and agri-environmental incentives to curb the rate of intensification. Although these collectively would seem to offer real changes in favour of extensification, Evans *et al.* (2002, p. 320) offer three critiques:

a) Stocking levels have been little affected by measures encouraging reductions (Winter and Gaskell, 1998). Farmers have resolutely tended to pursue productivist thinking in this area, and policies have been less than rigorous.

b) Problems of slippage and selectivity are well-known with respect to set-aside (Robinson and Lind, 1999), and it is quite likely that this aspect of policy may be deleted in the EU in the near future.

c) Agri-environmental policy includes a diverse collection of measures that can promote extensification and greater sustainability. However, throughout the Developed World the expenditure devoted to such policy has been minimal and has not challenged the predominance of conventional productivist activity. Food production has tended to continue in productivist mode whilst some environmental benefits are yielded from the application of agri-environmental policy on a selection of farms or parts of farms. Variants such as organic farming and integrated farming systems offer an essentially production-based alternative that might be termed 'neo-productivist' rather than post-productivist (Evans *et al.*, 2002, p. 321). These alternatives are much more of a radical break with conventional systems when compared with the 'tweaking' of conventional farming that agri-environmental policy tends to represent (Clark, J. *et al.*, 1997).

4 Dispersion of production patterns. This is a difficult process to identify, especially when there appears to be much evidence that specialisation accompanied by concentration of production is readily apparent for many types of farming (Bowler, 1986; Walford, 2002; Walford and Burton, 2000). In some areas there has been growth of small-scale, part-time or hobby farm activity, but the relationship between this and the concept of 'dispersion' needs to be examined more closely if this aspect of post-productivism is to be readily understood.

5 Environmental regulation and restructuring of government support for agriculture. The best evidence for the development of a post-productive agriculture rests with the growth of policy that has encouraged farmers

to act in ways that are not productivist. Yet, despite the emergence in the 1990s of agri-environmental policy, many of the general agricultural policy adjustments in Western Europe, Australia and North America in the past two decades have been concerned primarily 'with making agriculture more competitive, able to respond to the challenges and opportunities of the world market' (Evans et al., 2002, p. 324). There have certainly been more environmental regulations in agriculture, especially to control farm pollution (Lowe et al., 1994) and to protect wildlife sites (Winter, 1996), but these represent only a piecemeal re-regulation of farming that is, at best, a highly spatially variable and limited tendency towards a post-productivist set of regulations.

With respect to the first of these cateories, the growing concern amongst the general populace with the quality of food can be related to a series of inter-related factors (Evans et al., 2002, p. 317):

- There has been a rise in consumer concerns about agriculture's negative impacts on the environment, food safety, the welfare of farm animals and rural economies. This has had a substantial impact on the production of foods that can be shown to have some measure of being 'environmentally friendly'.
- In purchasing 'quality' foods certain groups of consumers have been able to differentiate themselves. Thus quality foods are associated with the enhancement of 'cultural capital', as distinguished by Bell and Valentine (1997).
- The development of quality foods has represented a marketing opportunity for producers in terms of adding value to their product and in developing supply-chain differentiation from competitors (e.g. Nygard and Storstad, 1998).
- Trends towards a growing concern for quality within the major food retailers has led to the growth of 'quality assurance' across various economic sectors. This has been associated with new approaches to the management of supply chains by retailers in order to ensure

their market share and maintain a competitive advantage.
- As a result of various 'food scares' the focus upon quality within supply-chain management has become part of a crucial 'insurance policy' for the major food retailers.

In Chapter 4 more consideration is given to this growing impact of consumer concerns for quality upon the agri-food system, especially in the Developed World.

Wilson's (2001) extensive review of productivism and post-productivism argues that there are a number of limitations within the work on the perceived transition towards a post-productive agriculture. The principal problems he identifies are:

- although the work on this topic comes from a range of disciplinary backgrounds, there is a need to include a broader set of environmental, economic, social and cultural dimensions;
- the work has been dominated by political economy and structuralist approaches that may have provided only partial answers/ interpretations;
- an almost exclusively UK context has been used so that the applicability of the concept beyond the UK remains to be proven, although see work by Holmes (2002) which applies it convincingly to the Australian rangelands.

Within these arguments Wilson contends that notions of both productivism and post-productivism extend beyond agriculture, with links to other ideas on the presence of distinct 'regimes' and regulationary arrangements (e.g. Halfacree, 1997; Le Heron, 2002; Marsden, 1999a). He suggests that productivism and post-productivism have been conceptualised on the basis of seven inter-related dimensions (Table 3.5): ideology, actors, food regimes, agricultural production, agricultural policies, farming techniques and environmental impacts. The table illustrates the multi-dimensionality of the transition, though within it is hidden the lack of consensus as to whether or not productivism has actually been superseded by

Table 3.5 Dimensions of productivism and post-productivism

Dimension	Characteristics
Productivism	
Ideology	Central hegemonic position of agriculture in society Ideological security Agricultural fundamentalism rooted in memories of wartime hardships Agricultural exceptionalism Belief in farmers as best protectors of countryside Countryside idyll ethos/rural idyll Main threats to countryside perceived to be urban and industrial development 'Rural' defined in terms of agriculture
Actors	Agricultural policy community small but powerful, tight-knit and with great internal strength 'Corporate' relationship between agriculture ministries and farming lobby Relative marginalisation of conservation lobby at fringes of policy-making core
Food regimes	Atlanticist Food Order dominated by USA Fordist regime
Agricultural production	Industrialisation (agribusinesses) Commercialisation Securing national self-sufficiency for agricultural commodities Intensification Surplus production Specialisation Concentration Increase in corporate involvement Farmers caught in agricultural 'treadmill'
Agricultural policies	Strong financial state support Conservative faith placed in ability of state to plan and orchestrate agricultural regeneration Encouragement to farmers to expand food production Government intervention Protectionism Price guarantees/financial security for farmers Agriculture largely exempt from planning controls Security of property rights/land use rights
Farming techniques	Increased mechanisation Decline in labour inputs Increased use of biochemical inputs
Environmental impacts	Increasing incompatibility with environmental conservation
Post-productivism	
Ideology	Loss of central position of agriculture in society Move away from agricultural fundamentalism and agricultural exceptionalism Loss of ideological and economic sense of security: farmers branded as destroyers of countryside Changing attitude of public toward agriculture; agriculture as villain Changing social/media representations of the rural

(cont'd)

Table 3.5 (cont'd)

Dimension	Characteristics
	Changing notion of countryside idyll; contested countrysides Main threats to countryside perceived to be agriculture itself Loss of security of property rights 'Rural' increasingly separated from agriculture; new social representations of the rural
Actors	Agricultural policy community widened; inclusion of formerly marginal actors at the core of the policy-making process Weakening of corporate relationship between agriculture ministries and farming lobby Changing power structures in agricultural lobby Counterurbanisation; social and economic restructuring in countryside Increasing demands placed on rural spaces by reconstituted 'urban' capitals in terms of new manufacturing and service industries
Food regimes/ market-related forces	Challenge to the Atlanticist Food Order from the early 1970s Post-Fordist agricultural regime; non-standardised demand for goods and services; vertically disaggregated production Critique of protectionism; free market liberalisation; free trade Increased market uncertainty New consumption-oriented roles of agriculture Changing consumer behaviour
Agricultural production	Critique of industrialisation, commercialisation and commoditisation of agriculture; critique of corporate involvement Less emphasis on securing national self-sufficiency for agricultural commodities Extensification Dispersion Diversification; pluriactivity Farmers wishing to leave agricultural 'treadmill' Move from agricultural production to consumption of countryside
Agricultural policies	Reduced financial state support; move away from state-sustained production model Demise of state-supported model of agricultural development which placed overriding priority on production of food New forms of rural governance Enhancement of local planning controls Encouragement for environmentally friendly farming; greening of agricultural policy Increased regulation of agricultural practices through voluntary agri-environmental policies Move away from price guarantees; decoupling Increasing planning regulations for agriculture Loss of security of property rights
Farming techniques	Reduced intensity of farming Reduced use or total abandonment of biochemical inputs Shift toward sustainable agriculture Replacing physical inputs on farms with knowledge inputs
Environmental impacts	Move toward environmental conservation on farms; critique of notion of production maximisation Re-establishment of lost or damaged habitats

(*Source*: Wilson, 2001, pp. 80–1)

post-productivism. This echoes Goodwin and Painter's (1996; 1997) observations regarding the unevenness of transitions in regulatory regimes. However, a major problem is a lack of agreement in the various studies over what constitutes post-productivism.

Wilson (2001, p. 85) contends that the seven dimensions of the transition are largely defined through exogenous forces of agricultural change, and that therefore more attention should be given in future to the endogenous perceptions and attitudes of actors involved in decision-making processes. This will involve a move from 'unidimensional political economy conceptualizations' to injection of an actor-oriented and behaviourally grounded approach that moves away from structuralist arguments. This will have to tackle the basic question of 'whether values of actors directly involved in processes of agricultural/rural change reflect the postulated shift toward a post-productivist agricultural regime' (p. 87). Initial answers to this question suggest that farmers and other key actors actually have a marked plurality of views and represent diverse communities which cannot easily be labelled as either 'productivist' or 'post-productivist'. Nevertheless, their overall 'position' is broadly productivist (Burton, 1998; Ward *et al.*, 1998; Wilson, 1997a), as will be discussed further in Chapter 10 with respect to the adoption of agri-environmental schemes.

Even so, there is also evidence to suggest that within Europe there has been increased adoption of post-productivist ways of thinking by farmers (IFLS, 1999). Such thinking is more typical of younger farmers with better education and better access to information. Yet, the apparently limited extent of such thinking by farmers and agricultural officials/policy-makers has led Morris and Evans (1999) to refer to the 'post-productivist myth'. At the very least, the idea that the productivist era is over should not be accepted and any transition to post-productivism should be recognised as more complex than has hitherto been acknowledged.

Furthermore, the term 'post-productivist' disguises the fact that most farming activity throughout the world is still dominated by the need to produce food for mass consumption. This is true for virtually all agricultural commodities with the exception of certain sub-systems of production geared to generating products for a smaller economically privileged section of society, e.g. specialist breads, expensively packaged vinegars and oils, unpasteurised bottled milk. Friedland (1997) recognises this ongoing production of food for mass consumption as comprising a mixture of Fordism, Sloanism (mass production but with the production process reoriented to deal with differentiated markets) and post-Fordism. This mixture has evolved unevenly for different products, with different combinations of the three processes occurring. For example, Friedland (1996) shows the changes with respect to production and marketing of salad crops like lettuce and tomatoes. In the early nineteenth century these were simply produced on a seasonal basis for limited geographical distribution. Production became Fordist in the late nineteenth century when they were conveyed by rail to remote markets (e.g. Robinson, 1983). In the USA this change became apparent for iceberg lettuce with the introduction of refrigerated rail cars, but has since been extended to other forms of lettuce, such as romaine, redleaf and oak (Friedland *et al.*, 1981), which are produced by large firms such as Tanimura and Antler who have helped to develop what is termed 'annualisation' of production or year-round production so that many fruit and vegetables grown in temperate climates can be supplied to markets at any time of the year. For some items imports have to be used to sustain this, for example strawberries from California, Spain and Israel supplying the UK during the UK's winter. These annualisations promote greater mass production and often incorporate industrial practices (both Fordist, Sloanist and post-Fordist) such as just-in-time delivery systems dependent on large globalised TNCs such as Dole and Del Monte.

Wilson argues that post-productivism is essentially a phenomenon of northern Europe and North America. In contrast, parts of the Mediterranean may not have even fully entered the productivist phase, though post-productivist policies can be imposed there through the CAP. However, the predominance of productivist thinking amongst Mediterranean farmers seems a powerful reason behind the low uptake of agri-environment schemes in these countries (Buller *et al.*, 2000).

For Developing Countries, Wilson (2001) identifies the presence of pre-productivist agricultural regimes. These are 'characterized by high environmental sustainability, low intensity and productivity, weak integration into markets and horizontally integrated rural communities' (p. 92). But this raises questions such as whether these countries have to pass through a productivist phase before re-attaining current levels of sustainability via post-productivism. The implication is that the set of criteria used to recognise post-productivism may not be applicable in most Developing Countries unless some element of the relative changes in action and thought of rural societies is included. This would avoid use of absolute indicators of post-productive change based largely on the UK experience. It could possibly be linked to future policies aimed at correcting environmental disbenefits from those Developing Countries where productivist agriculture has been introduced by external agencies through the Green Revolution or in the form of plantation agriculture.

As a means of building on current conceptions of post-productivism, Wilson (2001, pp. 92–5) suggests two main additions:

1 Territorialisation of actor spaces. This would give clearer recognition to the spatial inconsistencies involved in the co-existence of productivist thought and action and post-productivism. It appears that the first stages of the transition have involved the growth of a 'two-track' countryside (Ward, 1993) or an 'agrarianism – environmentalism spectrum' or 'super-productivism – rural idyll spectrum' (Halfacree, 1999). In Western Europe this is typified by, on the one hand, further evidence of intensification in East Anglia, the Paris Basin, Emilia Romagna, the Netherlands and the newly intensifying arable regions of former East Germany (Hoggart *et al.*, 1995; Wilson and Wilson, 2001); and Halfacree (1999) suggests that some farmers in these regions are becoming 'super-productivist' through increased intensification. On the other hand, agriculturally marginal upland

areas may champion the adoption of post-productivist action and thought. However, whilst a general polarisation may occur, there is also likely to be a broader spectrum of responses that encompasses a wide variety of adjustments by farmers (Robinson, 1994a). A simplified spectrum proposed by Marsden *et al.* (1993) recognised 'pressured', 'contested', 'paternalistic' and 'clientist' countrysides, though this is conceptual rather than spatial territorialisation. This further serves to question the utility of the dualism inherent in most writing on productivism and post-productivism. In developing this spectrum further, Marsden (2000b) proposes three categories that may also be applicable outside the UK:

a) The agri-industrial dynamic. Productivist, industrialised agriculture, typified by standardised products, capital intensity, sophisticated food-supply chains, high concentration of farms and large farm units.
b) The post-productivist dynamic. Rural space as consumption space, rural land subject to development, counter-urbanisation in pursuit of the rural idyll, commodified nature, and not necessarily supportive of sustainable environmental management.
c) The rural development dynamic. The potential of rural resources is realised through new roles for agriculture, locally and regionally embedded food supply chains, recapturing of the 'lost values' of rural space.

All three dynamics may occur at the same time and in close spatial proximity, so that the three individual dynamics are not spatially defined. This is similar to the idea that Fordist and post-Fordist modes of regulation may occur simultaneously (Goodwin and Painter, 1996; 1997). Hence, even in the postulated transition to post-productivism, productivist institutional forms, networks, ideologies and norms are not necessarily superseded (Wilson, 2001, p. 94).

2 Multifunctional agricultural regimes (MAR). Wilson (2001, p. 95) argues that this term is much more appropriate than post-productivism in conceptualising changes in contemporary agriculture and rural society. A MAR may follow the post-productivist transition and will be characterised by territorialisation of both productivist and post-productivist action and thought, incorporating a multiplicity of responses to the challenges of post-productivism, but with all responses co-existing.

This discussion demonstrates that there is now much agreement that there has been a strong tendency for the term 'post-productivist' to be adopted too readily and uncritically. The simplicity of the dialectic that it represents has detracted attention from more nuanced insights to change offered by regulation theory, actor-network theory and more culturally informed approaches. However, in their critique of the concept of post-productivism, Evans *et al.* (2002, pp. 327–8) offer another future avenue of enquiry that they feel offers a fruitful alternative. They champion consideration of ecological modernisation, which recognises environmental disbenefits arising from modern farming methods, but presents solutions that include pursuit of sustainable development, the precautionary principle (anticipation rather than cure), equating pollution with inefficiency, regarding environmental regulation and economic growth as mutually beneficial, and privileging the rights of future generations over market forces (see Giddens, 1998). These ideas have been developed in diverse ways from the early 1990s as an alternative social theory or as a means of characterising a new type of politics (Buttel, 2000; Mol, 1997).

There are a few studies that have specifically applied ecological modernisation to analysis of agricultural change (e.g. Frouws and Mol, 1997; Jokinen, 2000; Tovey, 2000). Tovey's work on Ireland's Rural Environment Protection Scheme (REAP) argues that ecological modernisation allows for an understanding of agricultural development that is non-productivist but not anti-production. This recognises the continued primacy of productivist imperatives but set in the context of a rapidly changing policy context in which production is being renegotiated to meet new demands and constraints imposed by concerns about public health, the environment and farm welfare. It remains to be seen, though, whether there will be a substantial body of work emerging from this ecological modernisation perspective, acting as a coherent alternative to the dichotomy postulated within the much vaunted productivist–post-productivist transition.

4 Agri-food networks

4.1 Food retailing and consumption

During the 1990s, human geographers showed much greater interest than hitherto in aspects of food consumption as opposed to their long-term foci upon the initial stages of its production. One aspect of this work on consumption has been a concern for the new culture of food consumption that centres on certain consumers' desires for a healthier diet and the rediscovery of traditional cuisines (Marsden, 1999b). This cultural dimension to food consumption patterns represents a counter-current to what has been termed the 'delocalisation' of the agri-food system associated with globalisation and the increasing similarity of lifestyles and habits in different parts of the world (Kuznesof *et al.*, 1997). The presence of a local dimension to consumption patterns within the globalised trend that is represented in the presence of the same chains of restaurants in all major centres (such as McDonalds, KFC, Pizza Hut and Burger King) has been recognised as a process of 'relocalisation' (Murdoch and Miele, 1999) in which mass consumption patterns are mediated by local specificities. Amongst the characteristics of relocalisation are concerns for the place or region of origin of food as part of a desire for authenticity, greater variety and concerns over the standards of mass production and processing practices in the wake of various food 'scares', such as those relating to BSE, *E. coli* and genetically modified (GM) organisms. It has also increased the amount of research that looks at the nature of the agri-food chain, generally treating it as a network of linkages from farmer to consumer (see Figure 4.1)

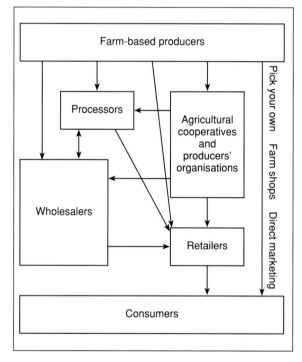

Figure 4.1 Simplified network of the linkages in an agri-food system (based on Kaldis, 2002, p. 86)

and acknowledging the importance of retailers and consumers in shaping the nature of farm-based production. The growth of interest in consumption offers the prospect of a more well-balanced investigation of agri-food chains in which the emphasis is shifted away from the previous dominant concern with farm-based production.

Specifically, the concept of 'food networks' has been developed as a means of considering both global-local diochotomies and relationships between human and non-human intermediaries in the social construction of food (Marsden *et al.*, 2000; Whatmore, 1994). Food networks comprise 'inter-relationships between the human actors in a commodity chain but extended to include the non-human intermediaries that bind the actors together in power relationships. Examples of non-human intermediaries include the contracts that link farmers to processors, the regulations that link processors and farmers to national politicians, and the international agreements that link TNCs to the WTO' (Atkins and Bowler, 2001, p. 46). A good example is the linkages between retailers in the Developed World and producers of cut flowers in Kenya (Hughes, 2000). A detailed example referring to the international trade in bananas is discussed in Chapter 7.

An example of the significance of changes to these various linkages is provided by the way in which the traditional internal market in the EU, based on competition within and across national borders, is increasingly being challenged by competition between firms and between different supply chains. Links between producers and wholesalers and/or retailers have become much more important as an efficient means of making goods available to consumers. Networks of producers have emerged to develop these efficiencies, including purchasing groups, franchises, and retailer co-operatives.

Initially, geographers tended to view food networks in terms of relationships between global forces and the variable local responses of those engaged in the production and distribution of food (Marsden and Arce, 1995). So locales were treated as particular niches within the food system that offered opportunities for powerful global concerns. The links between the global and the local, with respect to food networks, were often regarded as operating in two ways (Le Heron, 1993, pp. 32–3). First, global forces may find various forms of expression in local changes; second, local initiatives and resistance to global forces may alter global processes at the local level, thereby giving rise to the reshaping of these national or global processes.

This view of a dual impact tended to treat farmers and local food processors as inordinately subservient to global interests, with consumers as 'relatively passive recipients of the food produced by globalized food processors and retailers' (Atkins and Bowler, 2001, p. 48). This approach was criticised by Van der Ploeg and Long (1994) who protested that farmers were not merely passive recipients of global economic forces, but that they developed locally based adaptive strategies grounded in local culture, agri-ecology and farm household resources. They emphasised diversity of responses amongst local farming communities, but with those farm businesses that had developed in similar ways, and who were linked to local food processing industries and distributors, having the capacity to create 'agri-food districts', such as those recognised by Italian researchers amongst others (e.g. Fanfani, 1994; Iacoponi *et al.*, 1995). Subsequently, more focus has been placed upon consumer resistance to globalisation through various social movements that have formed the economic base for 'alternative' food networks (discussed further below). For example, Buttel (1996) recognised a wide range of social movements related to concerns associated with the environment, sustainable agriculture, community-supported agriculture, health issues, conservation of genetic resources, animal rights, rural social justice, consumer preference, organic foods, farmers' markets, non-traditional medicine, and ethnic cuisines. Other interpretations of the development of alternative networks, cited by Atkins and Bowler (2001, p. 49), are Watts' (1996) 'the politics of collective consumption', and Friedland's (1994a; 1994b) 'post-modern diet'.

In effect, these researchers are all identifying certain aspects of growing consumer resistance to globalised, industrial food products, as new political alliances and marketing networks develop between consumers and farmers in different parts of the world. From Buttel's list, two key trends can be distinguished. First, increasing numbers of consumers in the Developed World are willing to pay premium prices for food that has either been produced from animals kept under guaranteed welfare conditions or from fruit and vegetables grown without recourse to chemical treatments. Second, there has been an increased awareness of health risks associated with food. This concern has broadened from worries over direct links between diet and obesity, high cholesterol and heart disease to specific diseases such as salmonella poisoning, *E. coli* infection and new variant Creutzfeld-Jacob disease (nvCJD) linked to meat infected with bovine spongiform encephalopathy (BSE).

This mixture of de- and re-localisation processes has been translated into four key changes in consumer behaviour in the 1990s (Dawson and Burt, 1998, p. 163):

- a search for individualism in lifestyle through products and services purchased. This represents a counter to the tendency towards a uniform pattern of mass demand, forming an accentuation of local differences;
- higher expectations of the quality of products, services and shopping environments;
- non-price responsiveness to advertising and promotional methods;
- less direct price comparison, with an increase in product range and differentiation.

In part these changes have contributed to growing consumer resistance and reaction to processed, manufactured and fast food. Atkins and Bowler (2001, pp. 99–100) regard this as support for the contention that we are entering a third food regime. However, despite the attention given to the growth of this new sector of food retailing and consumption, it must not be forgotten that the share of food retailing by large supermarkets has grown throughout the Developed World and that fast-food consumption is also rising: 'In 1970 Americans spent about $6 billion on fast food; in 2001, they spent more than $110 billion. Americans now spend more money on fast food than on higher education, personal computers, computer software, or new cars. They spend more on fast food than movies, books, magazines, newspapers, videos, and recorded music – combined' (Schlosser, 2002, p. 3).

These two opposing trends, healthy (local) foods versus fast (global) foods, can be seen to have had effects on the changing content of diets around the world. Changing patterns of consumption in the Developed World post-1945 have

included declines in sugar, animal fats and milk products, partly through fears about the effects of their over-consumption upon health. However, there has been notable variation in trends: in Europe and the former Soviet Union consumption of livestock foods has risen, especially meat; in North America a major increase has been in consumption of cereals and various forms of sweetener, especially high fructose corn syrup; consumption of vegetable oils and oil crops has risen strongly throughout the Developed World. These changes are largely unrelated to changes in income and mark a fundamental break with previous patterns of consumption. In contrast, changes in the Developing World from the early 1960s have reflected growing incomes in certain regions leading to increased consumption of all foods, as happened in nineteenth-century Western Europe. In some countries this has brought about the start of a decline in the reliance on starchy staples, e.g. Thailand, Malaysia and parts of Latin America (Grigg, 1999). Increased consumption per capita of roots and cereals has been greatest in south and south-east Asia, often through increased reliance on imported grains (Kop and Wallace, 1990). Consumption of animal foods, and especially meat, has risen sharply from the early 1960s in Asia and Latin America. However, changes in diet have been least in sub-Saharan Africa, where there have been the least positive changes in real income per capita.

Directing attention towards the consumption end of the agri-food chain has necessitated a more concerted engagement with the notion of the commodity chain, circuit or network with its links between production, distribution and consumption. These links are mediated through a variety of social, cultural, political and environmental conditions that influence commodity movements (Fine *et al.*, 1996; Hartwick, 1998). However, to date, much of the attention has remained upon the particular dimensions of commodity chains – production, distribution and retailing – across particular commodities, rather than the horizontal dimensions such as the places that are the connecting nodes for commodity movements. Moreover, there is room for closer consideration of the complexity of consumption patterns. Previously, there has been a strong tendency to merely view consumption as a simple outcome of activities associated with product provision or a response to ways in which retailers, the state and consumer organisations 'determine' consumption choices (Marsden *et al.*, 2000). Indeed, the 'power' of consumers has often been overlooked as has the complexity of consumption within the agri-food chain (though note the recognition of this by Lockie and Kitto, 2000). With respect to this complexity, Cook and Crang (1995) have referred to the importance of two concepts:

- Commodity fetishism. This combines the consumers' lack of knowledge about production conditions with their construction of the place of origin of production. Thus, consumers often know little about, or are divorced from, the social relations of the production and provision of food commodities produced outside their own local area. Yet, consumers are being encouraged to gaze upon and collect signs and images about widely sourced food commodities (Ilbery *et al.*, 2000).
- Displacement. The majority of food products consumed in the Developed World originate elsewhere and arrive at the place of consumption via various production, processing and marketing networks that can encompass different parts of the world. Hence there is 'displacement' of a particular commodity from a place of production to a place of consumption. A fine example of this is the 'English' cup of tea, which depends upon networks of imperial connections before being consumed as a specific aspect of English/British social life and consumption. The notion of displacement is related to issues of authenticity regarding the relationship between food and a place of origin. Hence Jackson (1999) suggests that, instead of pursuing authenticity a more useful concern is to consider 'authentification' in which attempts are made to identify those making claims for authenticity and the interests that such claims serve.

Recent examinations of the geographies of consumption are treating consumption as a complex spatial form in which new landscapes of consumption are merging in response to post-Fordist forms of capitalism (Jackson, 1999; Jackson and Thrift, 1995). However, little work has been done specifically on geographies of food consumption, other than recognising impacts of changing taste and sensibilities towards particular types of food. These impacts are related to what has been termed 'symbolic consumption', in which individuals develop their own identities through what they consume (Jackson and Holbrook, 1995).

Cook (1994) illustrates this type of consumption, and its role in agri-food networks, with respect to the growing volume of 'exotic' fruit and vegetables sold by British supermarkets. He argued that there has developed a symbiotic relationship between the actual production of the fruit and vegetables and the 'symbolic' production of its meanings. For the producers the key factor has been the development of contract farming in which there is day-to-day negotiation of specifications by growers. Contracting has produced a partnership between the rural poor and private capital, highly advantageous to the contractor because of its delivery of produce at a predictable price, quality and schedule of supply. The less familar side to Cook's work is that of symbiotic production in which the supermarkets take produce from their suppliers and sell it to their customers. This is not a straightforward process as it involves implementing various aspects of successful marketing in order to promote sufficient sales of the new 'product' to meet the company's target. For exotic fruits this has often involved provision of instruction manuals for customers informing them of the nature of the unfamilar fruits and how to eat them. Incorporation of the fruits into company-sponsored magazines has also been important. The selling of exotic fruits has been used by stores as a means of differentiating themselves from competitors and showing their international credentials by having produce from exotic locations. Thus the produce has a meaning beyond that of just an item of food sold in a supermarket, infused with connections to producers, contractors and marketing, but unconsidered by the vast majority of the consumers.

The increased sale of 'exotic' foods in the supermakets of the Developed World has also had significant implications for producers in the Developing World, with new linkages between the rich and poor countries of the world established. Some of the new trading arrangements that have been created have prompted concerns for rural social justice in Developing Countries, and have given rise to 'fair trade' networks intended to safeguard the interests of the small growers. These networks link the growers to consumers in the Developed World who represent a relatively small social elite willing to pay generally higher prices for a fair-trade product compared with produce marketed by conventional means (Whatmore and Thorne, 1997). The role of these networks on agricultural production in the Developing World is considered further in Chapter 7.

The construction of identity through patterns of consumption has taken a particular form within some parts of the Developed World as consumers express a preference for 'quality' products and services, reflecting a growing interest in the origins of food, its freshness, the farming and processing practices through which food is produced, and the health and safety aspects of particular foods (Ilbery and Kneafsey, 1999). This preference has meant that some foods associated with traditional farming and processing practices have become highly desirable. In turn, this is linked to a perceived greater authenticity for certain foods associated with a particular place or region. This is related to a range of personal, product-related and structural factors.

In general, there appears to have been an increasingly positive perception of regional imagery associated with food, and also for products made by small businesses in rural areas, as small-scale producers are associated with traditional, 'handmade' production processes that stress quality. There is also evidence, though, that a minority of consumers link small-scale production to variability of quality. Hence different consumers place different constructions upon the quality of foods offered for sale. This variation in attitudes is revealed by Ilbery *et al.* (2000), who recognise differences on the basis of age, education, occupation and area of residence.

Food has become a central component in the construction of 'lifestyles' as large class-centred groupings are steadily replaced by a myriad of sub-groups of consumers (Bell and Valentine, 1997; Glennie and Thrift, 1996). The new patterns of consumption being created reflect a rising emphasis upon aspects of quality and convenience, with an increasing demand for 'healthy' foods and foods from market niches that may reflect ethnic variety and traditions (May, 1996). This has been referred to as a new model of consumption, in which there is a strong emphasis on taste and aesthetics as part of a new 'post-modern sensibility' (Jackson and Thrift, 1995, p. 207).

The growing public concern for the types of food that they eat, its origins and the nature of its transfer from farm to consumer has been termed 'reconnection' (e.g. Marsden, 1998a). This term symbolises the renewal of links between producer and consumer that some contend was lost during the period of post-war mass production and government regulation. Reconnection is also part of an increasingly diverse range of consumer demands nesting within highly competitive local, regional and global conditions and under different forms of state intervention. In particular, there is competition between the expanding global food sector dominated by TNCs/supermarket-based retailing and demands for sustainable production, 'natural' foods and local supply (Goodman and DuPuis, 2002). These two contrasting parts of the agri-food system are now examined further.

4.2 The corporate retailers

The previous section helps to support the contention that, in terms of forward or downstream linkages from farming, one of the key changes during the past three decades has been the growing importance of corporate retailers. They have become key players not only in selling food to the public but also in determining the nature and quality of that food via contract links to farmers, processors and packagers. Their increased power at a number of different stages in the food supply chain stems from their high market share of grocery sales: 69 per cent of European consumers purchase their groceries primarily from supermarkets (Tordjman, 1995); in Australia 61 per cent of all food and grocery purchases are from supermarkets

(Pritchard, 2000). Also, their ability to maintain regular supplies of food has helped them to restrict the returns to primary producers despite rising consumer expenditure. Two statistics illustrate this growing significance of downstream activity. In the USA the value of food sales nearly doubled between 1978 and 1988 whilst returns to farmers were virtually unchanged; and the 'added-value' to primary produce from industry, trade, retail and other food services is now three times larger than the farm (producer) value (Marsden, 1997, p. 172). As the large retailers have played an expanding role in selling food so they have increasingly defined the character of food, exerting a global influence by encouraging certain forms of exotic and globalised consumption (Grigg, 1993). Their command of the market is continuing to expand as the demand for their food grows in those parts of the Developing World where income levels are rising. This has been especially apparent in south-east Asia where processed products and pre-packaged food of high quality have begun to replace some staple foods (Arce and Marsden, 1993). Here the food market has been growing at between 4 and 7 per cent per annum and so there are obvious implications for expansion of this activity.

The growing power of the largest food retailers is seen in the USA and the EU where the ten largest chains account for approximately 30 per cent of total grocery sales. The dominance of these retailers is even greater within individual American states and certain EU countries (notably Denmark, France, the Netherlands and the UK) where the three largest retail chains account for close to 60 per cent of the grocery market.

As shown in Figure 4.2, in the UK grocery market in 1997 the leading five retailers (Tesco, Sainsbury, Asda, Safeway and the Co-op) accounted for 52.2 per cent of sales. This gives these firms tremendous influence over the food processing and wholesale sectors, and also over farm-based production. The latter can be dominated directly via the expansion of retailer-controlled contracts or by exerting controls on the quality of production. By developing efficiencies in this sourcing of the large stores, and by rapid growth of new, large outlets, the retailers have maintained and expanded high profit margins, especially in the UK where they were close to 10 per cent in the 1990s (Marsden and Wrigley, 1995; 1996; Wrigley, 1994). Figure 4.2 illustrates the high level of return to the supermarkets for lamb meat when compared with the price paid to farmers. Greater competition from the discount sector has restricted profit margins in Germany and North America. However, a common factor has been the retailers' use of product innovation and new value-added quality. The introduction of new varieties of processed and packaged foods has given large retailers advantages over smaller competitors, accentuated by economies of scale that have proved attractive to consumers. One result of this has been to reduce the numbers of grocery outlets, with a decline in the specialist independent sector (e.g. small multiple outlets) and specialist retailers (butchers, bakers, dairies and grocers). For example, between 1985 and 1993 in Italy, traditionally dominated by large numbers of small retailers, the numbers of supermarkets doubled, the numbers of hypermarkets increased six-fold and the numbers of retail outlets fell by 15 per cent (Marsden, 1997, p. 186). This has reflected not only economic changes within retailing but also the changing nature of consumption within society.

In Western Europe, supermarkets account for over 70 per cent of purchases of packaged foods, soft drinks and cheese, and over 40 per cent of fresh fruit, vegetables and wine. Only in the sales of fresh bread and meat do specialist stores retain a significant share. However, an important trend has been the growth of discount grocery retailers focusing on low-cost product lines. Indeed, the discount retailer, Wal-Mart, based in the USA, is the world's largest grocery corporation in terms of turnover and has bought a leading UK grocery retailer, Asda (Burt and Sparks, 2001). This purchase reflects the way in which the structure of food retailing has been subjected to the same types of international acquisition and merger that were prevalent within the food industry in the 1970s and 1980s. For example, the British grocer, Sainsbury, has expanded into the North American market through investments in Shaw's Supermarkets and Giant Foods (Shackleton, 1998). These developments often reflect the strategies of the financial institutions that own large retail

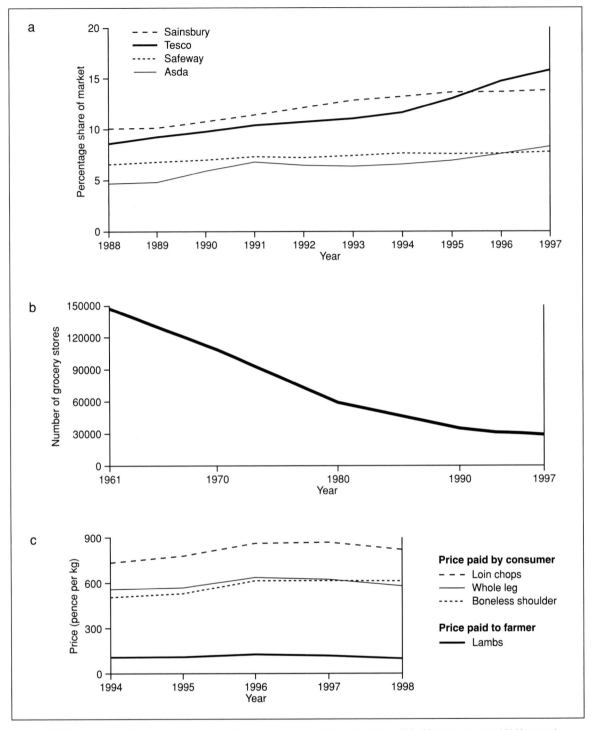

Figure 4.2 The growing share of the food market taken by supermarkets in the UK, 1988–97 (data source: UK Meat and Livestock Commission and Verdict Analysis)

corporations. These institutions are concerned with their share prices and dividends that can benefit greatly from the rising profits associated with increased sales and growth in the numbers of retail outlets (Wrigley, 2002). Hence the larger retailers have looked to new locations in different countries as the focus for future investment and for opportunities to spread into new markets. Cross-border alliances in the form of buying groups have assisted this process, e.g. Deurobuying brings together Asda (UK), Asko (Denmark), Carrefour (France) and Metro (Germany) (Atkins and Bowler, 2001, p. 93).

Supermarkets act as supply chain intermediaries, controlling the chain from producers to retail outlets but without owning the means of production. They have increasingly determined who produces what and have set the quality standards for production. In so doing they minimise their own risk by passing costs back to the supplier in respect of advertising, packaging, crop failure, unreliability and substandard produce. In countries where supermarkets have oligopsonic power in supply chains they have become 'masters of the food system' (Wrigley, 1991, p. 3; Wrigley, 2002).

In terms of regulation applying to food, the role of the supermarkets has become critical in many Developed Countries. These retailers have taken initiatives on issues such as food quality, authenticity and convenience in order to maintain their market share in the face of stiff competition. They have increasingly become key regulators of quality standards as the state has become more dependent upon the supermarkets to provide high-quality food. This has given supermarkets additional power over the suppliers of food products by means of developing their own regulatory systems (Marsden, 1997; 1998a).

In endeavouring to secure produce of the right quality the supermarkets have tremendous power over who grows their supplies and how they are grown, strongly influencing investment in the sourcing countries. They are, though, at the end of a long chain that includes growers, middlemen, agents and exporters. Often this chain has only been in existence from the 1970s if 'new' exports such as fresh horticultural produce are concerned, replacing the well-integrated marketing chains developed in colonial times to deal with plantation crops such as rubber, tea, coffee and cotton (Barrett *et al.*, 1999). The involvement of the supermarkets can be seen, for example, in Kenya where flexible contractual food supply networks have developed in which risk has been shifted from the retailer to that part of the chain located overseas.

It has long been recognised that the export sector of agriculture in Developing Countries was tied to the market needs of the Developed World. Therefore the whims of consumers in these markets can have profound impacts upon the growers and processors in the Developing Countries. However, something resembling this relationship has now become apparent for agriculture in the Developed World through globalisation processes, which have accentuated the power of the manufacturing and retail sectors (Ward, 1990). In effect, it has become the so-called downstream sectors that are increasingly shaping the nature of land-based production. Hence, it is decision-making by supermarkets, processors and other non-local agents that is determining the character of production and production practices. Frequently, such decision-makers are organised on a supra-national or global basis and hence the pertinence of the comparison with the colonially based relationship between growers in Developing Countries and markets in the 'mother country'. One consequence of this increasing importance of the downstream relationships is that an understanding of agricultural change has to place greater emphasis upon consumption rather than production *per se* and upon the supra-national processes affecting consumption. In particular, this has led to work on how market relations are constructed socially, culturally and politically within the context of global power relations (Goodman, 2001).

Marsden (1998a) refers to this in terms of considering the social economy of change, linking deregulation and reregulation of the countryside to arenas of commoditisation (see below) and to the networks and 'spaces' of the relevant agents or actors involved. He refers to three components of this social economy:

- A focus not only on patterns of uneven development, but also the ways in which combinations of market, public and community interests, and networks carry forward the process of rural development, i.e. a social dynamic.

- The different rural and agricultural trajectories or experiences redefine combinations of local resources in new ways. This is closely associated with the process of commodification whereby rural resources have attained a market value for exchange, i.e. the attachment of commercial value to items previously largely ignored by consumers or not offered for sale by entrepreneurs (Hopkins, 1998a; 1998b; Robinson, G. M., 1999). There has been increased commodification of the countryside, giving rise to a series of new markets for countryside products (commodities), including the crafting, packaging and marketing of 'pay-as-you-enter' national parks and theme parks, craft and food outlets and 'leisure experience' activities.
- Traditional economic relations may now involve different sets of social, political and regulatory 'actors' and agencies. For example, environmental payments may be handled by agricultural ministries rather than by environmental agencies. Policies may be area-based thereby affecting the spatial variability of economic development.

Harrison *et al.* (1997) relate the growing power of large food retailers to what they term as four regulatory domains acting in favour of the large retailers: competition and pricing policies; planning and environment; food law; and food quality. For example, in North America changes to antitrust legislation, including the relaxation of corporate regulation in the 1980s, and restructuring of retail capital have helped concentrate retail activity (Hughes, 1996; Wrigley, 1999; Wrigley and Lowe, 1996). In particular, it is possible to identify the significance of pricing in furthering the interests of this corporate sector (Atkins and Bowler, 2001, p. 94), as follows:

- As price determines a large part of purchasing behaviour with respect to food, large retailers can obtain price discounts from their suppliers when negotiating supply contracts to a supermarket chain on a long-term basis.
- Large retailers may be able to levy charges on their suppliers for giving their products access to the supermarket's large volume of customers. The retailers tightly control product ranges in order to balance maximising consumer choice versus limiting products to those with the greatest sales potential.
- As large retail corporations have a high volume of sales, they can discount the sales price of their goods, thereby undercutting small independent retailers and, for some products, their larger competitors. In the USA this discounting tended to be the factor underpinning the growth of Wal-Mart and Target.
- By introducing their cheaper, 'own-brand' labels, the larger retailers have taken a growing share of retail turnover from 'producer-brand' labels. In France, Germany and the UK the proportion of own-label food sales through supermarkets exceeds 20 per cent, and this is increasing as they extend it into sales of bakery items and meat.
- Supermarkets have adopted the 'just-in-time' supply system typical of many industrial sectors (Boyd and Watts, 1997), using computer technology to change the nature of ordering and stock management. This system, which also utilises centrally located warehouses, reduces operating costs. Moreover, cost saving has been further enhanced by supermarkets subsuming wholesaling and distribution functions previously provided by independent companies. As well as reducing costs at various points in the agri-food chain, efficiencies have been generated such as minimising the delay in the delivery of fresh products and ensuring maximum turnover from each unit of retail shelf space through continual restocking. The growth of online food retailing has also

benefited the largest supermarkets (Murphy, 2002).

The development of contractural arrangements between supermarkets and their suppliers has not only enabled the larger retailers to source materials globally but also with great reliability. They have applied the same principles in their contracts with farmers in Developing Countries as have been pursued in the Developed World, with tightly defined specifications for crop variety, method of husbandry, quantity, quality and timing of supply. This has had important, and not always beneficial, consequences for producers in Developing Countries (Rigg, 2001). To help ensure reliability, deals with large producers are generally preferred or consortia of smaller producers. This growth of global retail interests means that traditional supply chains have often been replaced by a highly complex pattern of sourcing in which the product sold to the supermarket customer comprises elements assembled from multiple locations prior to processing and packaging ready for sale. This has prompted Paxton (1994) to comment on the marked increase of 'food miles' or the growth in the accumulated transport of the components of food products prior to sale.

The huge growth of 'fast' food outlets has dramatically increased the demand for standardised and specified raw agricultural inputs to the processing stage. This has encouraged the extension of contract relationships between processors and farmers to supply commodities to specifications outlined by retail outlets. These can refer to qualities such as size, grade, colour, maturity and packaging specifications. In this way farmers are placed 'at the beginning of the fast-food assembly line, hence they may experience pressure from both processing companies and fast-food retail outlets' (Lyons, 1996, p. 244). This can marginalise smaller farms whilst favouring larger, more heavily mechanised operators.

Nevertheless, there are many different types of relationships existing between farmers and retailers, and between farmers and agribusiness processors. A traditional view is that both retailers and processors prefer to work with larger, more efficient farmers, but efficiency cannot necessarily be equated simply with size. For example, Miller's (1996) analysis of agribusiness processing in Tasmania distinguishes between the locally emergent and the globally oriented agribusiness. The former are firms that have grown mainly from local capital and which are owned and operated by either national or transnational companies. The globally oriented firms are essentially branch plants directed from transnational headquarters or with national administration based elsewhere in Australia, e.g. Glaxo, Johnson and Johnson, Cadbury's, McCain Foods. These global firms appear to have developed a more efficient contracting system with farmers, prompting these farmers to intensify production, though with some evidence of a recognition of long-term concerns for sustainability.

Farmers' responses to the growing stranglehold exerted by the supermarkets have included a greater recourse to direct marketing. This has included pick-your-own (PYO) schemes, farm shops, farmers' markets and purchase-on-trust roadside sales of produce, all of which have been labelled as part of an emerging 'alternative' food economy (Winter, 2002). Both PYO and farm shops have been popular in the USA since the 1960s, being reliant on car-borne consumers. However, their popularity has also grown in Western Europe as part of the increase in rural tourism (Bowler, 1992c, p. 192). More recent developments include vegebox schemes in which consumers contract to receive a box of seasonal vegetables (or fruit) from a farmer, delivered to their home on a regular basis. This desire for quality is also reflected in the growth of small, specialist food shops in the inner city. These often focus on sales of organic and speciality foods, thereby establishing an association 'between the differentiation of food product and the means of purchase by the consumer' (Atkins and Bowler, 2001, p. 100). Dawson and Burt (1998) assert that this may lead to a more fragmented retail structure in which there is more emphasis upon retail distribution networks serviced by regional food factories. Alternatively, regional and niche markets may be subsumed within the supermarket sector as part of globalising tendencies.

4.3 The alternative food economy

4.3.1 The nature of the alternative food economy

Recently there has been development of the idea that there are recognisable 'alternative food economy(ies)' (AFEs) that represent a reversal of long-established trends in the agri-food sector through a focus on quality, health, environment and fair trade. A central component of AFEs is, as described above, the emergence of a new category of consumer, motivated by ethical and health concerns rather than by price, packaging, appearance and ease of food preparation. In part, this new consumer is related to new forms of production and consumption, in which these two activities may be linked through new types of marketing, e.g. farmers' markets (Holloway and Kneafsey, 2000) and 'buy local' campaigns. There are particular aspects of the agri-food chain that represent part of the AFE, notably new forms of association between product and place, food and environment linkages (as illustrated in food labelling initiatives) and new forms of marketing that emphasise the 'local' character of the food and direct connections between food production and consumption. Changes in agricultural policy have been significant facilitators of AFEs, though policies can also act as a constraining factor, and hence contribute to a considerable amount of uneven development of AFE initiatives.

Recent literature on the changing nature of the agri-food sector has emphasised the growing significance of certain types of production and consumption, in which it is possible to define both 'conventional' and 'alternative' modes. The typical labels appended to the conventional mode are: modern, standardised, rationalised, mass production, economic, manufactured, disembedded and externalised. The 'alternative' labels are postmodern, specialised, traditional, craft/artisanal, non-economic, natural, organic, embedded and internalised. 'Alternative' also refers to a capturing of the economic value of the sales of novel, 'alternative' or better quality products at the rural, local and primary end of the food supply chain.

Such 'local' connections include direct sales at the farm gate, farmers' markets, box schemes, pick-your-own (PYO), farm shops, mail order, home delivery (including via the internet) and local outlets' direct delivery.

At one level this may suggest the presence of a simple binary opposition between the conventional and the alternative, in which the conventional is clearly associated with the post-1945 industrialisation of agriculture, the production of foods for mass consumption and the emergence of food consumerism associated with large-scale supermarkets, fast foods, ready-to-cook meals and an international 'packaged' cuisine available throughout the Developed World and also in many parts of the Developing World too (Fine and Leopold, 1993). The 'alternative' may be viewed as a reaction to this globalising 'industrial' tendency, but it can take various forms and hence is not readily definable, especially as some of its constituent elements are closely related to aspects of production and consumption within the 'industrial' sector. For example, there has been a substantial recent growth in sales of organic produce in the UK's major supermarkets, retail outlets usually identified with the conventional food economy. Moreover, the production agreements between the supermarkets and the organic producers tie the latter to producer-retailer relationships characteristic of the 'conventional'. Hence, the binary opposition between conventional and alternative may not be clear-cut, raising questions of how the different modes should be conceptualised and their differences analysed. In part, the commercial obstacles reflect the extent to which 'alternative' food supply chains represent a challenge to established interests. These interests in turn are diverting 'alternative' production towards industrial-style 'conventional' retailing modes.

4.3.2 Local connections and identities

If a key element of AFEs is the emergence of a concern for and a valuing of where food is produced then AFEs can be seen to have grown in the EU over the last decade, partly through the development of EU regulations for protecting products with distinct geographical origins. Hence there has been the promotion of regional branding as part of the development of 'alternative' niche markets for high-quality, specialist food and drink products with a distinct region of origin. This is part of several attempts to authenticate the links between product and place, giving rise to some distinctive regional associations with particular types of food, referred to by Kneafsey *et al.* (2001) as 'culture economies'.

In the UK a renewed emphasis upon a local link between production and consumption can be seen in the Countryside Agency's 'Eat the View' campaign launched in 2000. This is aimed at encouraging consumers to buy rural products (primarily food and craft products) that are derived from 'sustainable' production systems. Approved labelling systems are proposed in order to support the scheme. Another example is the recent call from the Council for the Protection of Rural England which has urged supermarkets to have at least 5 per cent of their food stocks from local suppliers by 2005. This is part of the concern to reduce 'food miles' or the distance travelled from source of supply to retailer (and the high energy consumption that long distances can generate).

Interest in local connections has also been fuelled by successive food 'scares', e.g. salmonella and BSE, and concerns over health and environmental issues related to GM foods. Whilst concerns over food safety remain high, more people are interested in knowing about where and how their food is produced, hence a rising interest in purchasing 'local' food, as illustrated in the UK recently in the growth of farmers' markets and vegebox schemes (Holloway and Kneafsey, 2000).

Proponents of farmers' markets contend that they have the potential to deliver a wide range of economic, social and environmental benefits through their ability to make direct connections between food production and consumption and their emphasis upon the 'local'. There are currently around 400 farmers' markets in the UK, 250 of which are now members of the National Association of Farmers' Markets. The rules for retailers at the markets usually require production to be within 20 to 50 miles of the market or, for London markets, within 100 miles of the M25. Farmers' markets in the UK account for between £65 million and £100 million in turnover. The benefits to farmers are the potential for improved returns through use of these additional outlets. More general benefits are the local economic multiplier – possibly as much as £1 for £1 spent, reduced 'food miles', and the social value associated with direct contact between producer and consumer. However, there are also limitations related to the limited capacity of the markets, the need for the producer's presence at the market, and the limited range of products offered. Three-quarters of the markets in the UK are only held on a monthly basis and so represent a highly limited outlet for both producers and consumers.

There has also been a small but potentially significant growth in direct sales of farm produce to consumers in the USA and some parts of Europe by small, specialist producers. For example, farmers' markets have grown from just 300 in the USA in 1974 to over 2500 in 1998, though in France it could be argued that every one of the country's 6000 weekly markets is in some sense a farmers' market in which the vendors are usually also the farm-based producers. In the UK in 1998 there were just 15 equivalent outlets serving over one million shoppers each week. New York has 25 such markets, the longest established located in Union Square. These generally have local systems of regulation ensuring a certain standard of quality. This contrasts with many urban street markets in the UK dominated by specialist wholesalers and retailers rather than farmer-/grower-operated outlets.

One attraction of farmers' markets for the producers is the fact that they can often sell produce not amenable to supermarket stipulations, such as traditionally hung or cured meats. In the UK farmers' markets are generally self-regulated or organised by local authorities to promote locally produced organic and conventional produce. The incentive for local authorities is the potential represented by farmers' markets in contributing to a revitalisation of town centres deserted by supermarkets which have moved to out-of-town or suburban locations.

It is possible to identify a number of benefits associated with this increased focus upon locally produced food: greater access to and consumption of fresh food, 'fairer' prices for farmers and consumers, greater local diversity of production, reduced food miles, increased local employment opportunities, less packaging, and more direct interaction between consumers and producers. However, there are several limitations on the growth of reliance upon local production. These include maintenance of the dominant 'supermarket culture' with its offer of 'one stop' shopping, conservatism involving resistance to innovation/change of habit, insufficient variety and/or quantity in local foods, lack of daily availability, insufficient local processing (most marked for abbatoirs) and lack of availability of capital to improve local facilities. There may also be problems linked to planning constraints, with a more flexible approach to food-related developments being needed. In addition, local foods tend to include a high element of speciality foods and value-added foods rather than staple foods, which limits their marketability.

A potent example of how farming is adjusting to the changing nature of the market is with respect to the impact of socially constructed quality criteria. These criteria are manifest in numerous farm, food and rural tourism quality assurance schemes (QAS). These have been developed by various regional groups, retailers and representative organisations (Morris and Young, 1999). Increasingly QAS are being promoted in conjunction with growing public awareness of ethical, environmental and safety implications of industrial-style farming systems. Added to this, certain sections of the population regard consumption of certain types of QAS-backed products as an indicator of good taste and sophistication (Bell and Valentine, 1997).

Some farmers have been able to take advantage of the way in which QAS has often been associated with particular places or regions (Ilbery and Kneafsey, 1998). This reflects the way in which some sections of society have become increasingly concerned with how food products have been produced and where. This may reflect both a desire to satisfy worries over health and safety, but also a need to consume 'real' food that may be linked to a perception of a past rural way of life associated with use of traditional raw materials, a traditional method of production or a recognisable geographic origin (Gilg and Battershill, 1998; Roest and Menghi, 2000).

The existence of such trends has encouraged the EU to implement regulations 2081/92 and 2082/92 giving a protected designation of origin and a protected geographical indication (Ilbery et al., 1999a). This may result in a particular tying of local quality products (QPs) to specific territories and cultural markets, such as regional landscapes, cultural traditions and historical monuments (Bessiere, 1998), and may be of particular benefit for some regions, especially in poorer, economically marginal rural areas, where there may be new opportunities to control the types of economic activity occurring and retaining more of the related economic benefits (Ray, 1998; Robinson, 1994b). As shown in Table 4.1, indicators of quality can be related to four inter-linked criteria.

Table 4.1 Indicators of quality products

	Indicator	Characteristic
1.	Certification	A QPS gains recognition via a quality mark or symbol from some external body, e.g. a producer group, so that self-regulation of quality is possible
2.	Specification	Refers to raw materials from the local area, ownership and production method, e.g. small-scale workshop, authentic recipes.
3.	Association	Either geographically with a region or local environment or historically with a tradition or culture.
4.	Attraction	Refers to consumers' desires for particular aspects of design, texture, taste and premium prices.

(*Source:* based on Ilbery et al., 1999a)

In a related development, one of the 'adjustment strategies' pursued by farmers in the last decade has been involvement in the development of speciality food products (SFPs), such as wines, farmhouse cheeses, smoked and cured meats, salad dressings, speciality cakes and biscuits and preserves, which are frequently associated with authenticity of geographical origin and traceability (Marsden, 1996). Ilbery (1999) refers to this as part of the 'relocalisation' of the agri-food system and therefore representative of agricultural dispersion (as opposed to concentration associated with intensive production systems). He argues that the so-called 'lagging regions' in Europe could benefit from the increasing demands for SFPs 'especially if they are tied to a regional image and notions of sustainability and environmental friendliness' (p. 282) (see also Bryden, 1994). SFPs have been defined as products 'differentiated in a positive manner by reason of one or more factors from the standard product, and recognised as such by the consumer, and can therefore command a market benefit if it is effectively marketed' (SFSG, 1993, p. 3). This is also asserted by Battershill and Gilg (1998) with respect to the quality of such foods, when sold directly from farms (*vente directe*) in north-west France. They claim that this valorises traditional low-intensity farm production and so has a role in conservation/sustainability. The economic potential of this activity is also great, with an estimated one-quarter of French farmers processing and selling their own produce, accounting for between 6 and 8 per cent of the country's food sales, including up to one-fifth of wine sales. The farms participating in *vente directe* were generally small, with 'unusual' enterprises, not profit-maximising small profit producers, run by well-educated individuals who had non-farming experience, adopted low-intensity methods and possessed high conservation values.

One frequently effective strategy has been the linking of the product to a given place or region so that customers purchase not only an item of food or drink but also an image linked to a location that is deemed desirable and possibly highly evocative. This can be seen in Bell and Valentine's (1997) suggestion that 'we are where we eat', in the sense that the link between product and place

can be so strong that 'almost any product which has some tie to place – no matter how invented this may be – can be sold as embodying that place' (p. 155).

The link between product and place is seen clearly in the French wine appellation system where the wine is a direct expression of the geographical individuality of its production locale (Moran, 1993). Hence, there is a need for a strong regional identity to be attached to the wine if it is to sell at the best prices in international markets. Indeed, internationally renowned wines such as Chateauneuf-du-Pape help create the image of their region of production. This specific link between product and region has been further strengthened by the European Commission through Regulations 2081/92 and 2082/92. The former introduced protected designations of origin (PDOs) and patented geographical indications (PGIs). These directly relate quality to a particular geographical environment, or attribute a specific quality to a product from a given region but not necessarily through its natural environment. Regulation 2082/92 introduced certificates of special character for quality products produced with local raw materials and/or a traditional mode of production. PDOs and PGIs are dominated by cheese and drink products and have been strongly motivated by protectionism. Some of the producer groups have been formed purely to achieve registration.

Ilbery (1999, p. 283) notes that the SFP sector in the UK employs over 20,000 people and has an annual turnover of nearly £3 billion. It is dominated by small- and medium-sized enterprises (SMEs), which either utilise traditional recipes and/or innovative ideas to make high-quality products. Over 60 per cent of the producers employ five or less employees, selling the majority of their products either through their own retail outlets or local catering firms. Mail orders and multiple retailers are also important outlets. Many of the SMEs focus on niche markets, but increasingly aim to sell the SFPs outside their regions of origin. Nevertheless their size restricts their ability to compete in national and international markets. To assist in overcoming this limitation a national network of fourteen regionally based speciality food groups has been created, launched by the Ministry of

Agriculture, Fisheries and Food (MAFF) in 1991 in a six-year grant scheme. Their development is coordinated and managed by a national body, Food from Britain.

4.4 Conclusion

This brief overview of AFEs, focusing on strategies adopted by farmers, illustrates that the notion of 'alternative' systems of production and consumption are not always clearly distinguishable from conventional systems. A simple binary classification is difficult to apply because of the various linkages between producers and consumers within the agri-food chain. Nevertheless, it is possible to identify elements within the chain where there have been substantive changes in favour of new objectives by both producers and consumers. These objectives relate primarily to concerns with health, quality and environment, but with the use of place and a

rediscovery of 'the local' as key ingredients in the forging of new types of linkage between producer and consumer. Further consideration to a key feature of AFEs, aspects of the production of organic foods, is given in Chapter 10.

Whatever trajectory is taken in future by AFEs, or for that matter by any other aspect of food production and consumption, it is clear that the constraints and opportunities provided by policy considerations will be highly significant. Thus, having referred to the impress of social and economic trends upon changing production and consumption patterns throughout the first part of this book, it is appropriate now to formally recognise that these trends themselves are shaped in a variety of ways by different policy contexts. Traditionally, agricultural geographers have addressed policies affecting farm-based production, but as the next chapter illustrates, such policies are increasingly being subsumed within the broader context of rural development.

5 Government and agriculture in the Developed World

5.1 The goals of agricultural policy

For the majority of the post-war period, agricultural policies in the Developed World have had as their primary objective the production of a high proportion of food requirements from domestic resources in order to reduce both external dependence and the risk of food shortages. Therefore policies have contained precautionary measures that have been applied directly to agriculture's fluctuating production by regulating prices and marketing in order to achieve stability (Tarrant, 1992). Within these policies both utilitarian goals and equity goals may be recognised (see Figure 5.1). The former refer to the contribution made by agriculture to the national economy, and have been developed much further than the equity goals (Self and Storing, 1962, p. 218). Utilitarian goals have been embraced by policies stressing farming's contribution to economic growth and stability, as in agriculture's role in improving the balance of payments and stabilising domestic food prices at a satisfactory level. In some countries additional utilitarian goals have been to provide a strategic reserve of land to meet unforeseen future demands, and maintenance of agriculture as a land use to be retained against the tide of advancing urban development. Concerns for economic stability have been linked to equity goals for social stability in rural areas and suitable remuneration for farmers and farm workers. Yet such aims have generally failed to prevent substantial reductions in the amount of farm labour or continuation of disparities between the wages of agricultural labourers and industrial workers.

State intervention has tended to concentrate on three broad aspects of agricultural activity that can influence farmers' behaviour:

- A focus on the general economic environment for agriculture, seeking to influence farmers' economic behaviour by effectively increasing or decreasing the risk associated with a particular decision (for example, raising the government's guaranteed purchase price reduces a farmer's risk when producing a given crop and is therefore likely to generate more production).
- Offering specific financial inducements, though not all farmers will respond in the same way to these.
- Introducing specific regulations, such as production quotas, that directly dictate farmer behaviour.

Tilzey (2000) refers to these policies as part of a process of structural causality whereby structures, comprising determinate economic, political and social relations, exist and are reproduced through the medium of contested and negotiated human agency. He regards the coherent body of policy structures that have been developed post-1945 in many parts of the Developed World as the outcome of negotiation and contestation, that is the 'social mode of regulation' (Drummond and Marsden, 1995a; 1999). However, these structures are not automatically taken up by the social agents (primarily the farmers) to whom they apply. Instead they are mediated, negotiated and contested by these individual agents. Hence agricultural policy is adapted according to the particular social

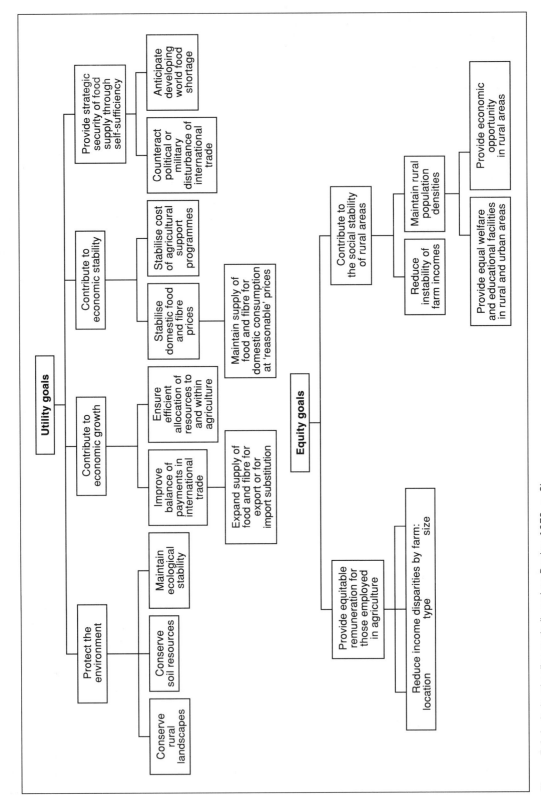

Figure 5.1 Agricultural policy goals (based on Bowler, 1979, p. 9)

and economic circumstances of individual farmers. Other characteristics of farms also affect this adaptation, for example the environmental impacts of policy will vary according to the particular ecological characteristics of the individual farms. However, 'the policy structures provide the basic and usually unavaoidable parameters within and through which the great majority of agents must act if they are to secure the conditions necessary for economic and social reproduction' (Tilzey, 2000, p. 281).

The combination of goals shown in Figure 5.1 has been translated into four basic types of measure designed to influence various elements of the farming system and the economic environment in which it operates (Bowler, 1979, p. 21):

- the growth and development of agriculture, e.g. subsidies on the costs of inputs, preferential taxes and interest rates, financial assistance to farm-based co-operatives, financing agricultural research;
- the terms and levels of compensation of production, e.g. management of demand and supply, direct payments;
- the economic structure of agriculture, e.g. financial assistance for land consolidation schemes, farm amalgamation grants, land reform schemes, retirement grants, inheritance laws, farm wage determination;
- the environmental quality of rural areas, e.g. pollution control measures, agri-environment schemes, land zoning ordinances.

As described below, there are numerous possible combinations of measures, frequently containing conflicting aims or without sufficient power to overcome constraining factors. Moreover, agricultural policies have tended to be separate from rural policies affecting the broader economic, social and environmental context of which farming is also a part. This has tended to be the case both in Western Europe and North America, where the focus on stimulating production has generated disbenefits to other elements within the rural milieu.

Research on agricultural policy has dealt with farm inputs, outputs and farm organisation. Work on inputs has included studies on both land and labour, for example work on farmer retirement schemes in France (Naylor, 1982). Research on output and production is discussed throughout this book, but it is worth emphasising that productivist policies in Europe and North America have tended to promote increased regional specialisation, by enabling farmers to take greater advantage of producing those items for which they hold an economic edge. This can be seen by the way in which EU milk quotas have intensified the regional localisation of dairy farming. Policies dealing directly with farm organisation have been less common, though the French government has intervened in the land market, with some effect, in attempts to improve the size structure of farms in western and southern regions (Winchester, 1993, pp. 97–8).

The operation of agricultural policy has been examined by geographers at a number of scales: from impacts upon decision-making on an individual farm through to the multinational contexts within which agreements on agricultural trade are made and broad levels of farm subsidies are negotiated. This chapter largely deals with examples of work at two different scales, starting with the supra-national arrangement of the CAP and moving to changes in the broad international context both as a response to and a shaping of globalisation processes.

5.2 The Common Agricultural Policy of the European Union

5.2.1 Overview

The EU is the world's leading importer and second ranking exporter of food. Its share of world food imports is around 20 per cent, compared with that of the North American Free Trade Association (NAFTA)'s 13 per cent. Traditionally the majority of the EU's member states were net importers of food, but since the enlargement of the Union, beginning with the accession of Denmark, Ireland and the UK in 1973, exports have expanded more rapidly than imports to comprise around 15 per cent of the world total. Exports have risen steadily under the aegis of the support system put in place

Table 5.1 The changing objectives of the Common Agricultural Policy

Original objectives	Objectives stated in Agenda 2002
• Increase agricultural production through the rational development of agriculture towards the optimum utilisations of resources • Ensure a fair standard of living for farmers • Stabilise agricultural markets • Guarantee continuity of food supplies to consumers • Ensure reasonable food prices for consumers	• Improve competitiveness of EU agriculture • Guarantee of food safety and food quality • Provide a fair standard of living for the agricultural community and stability of incomes • Integrate environmental goals into the CAP and develop the 'countryside stewardship' role of European farmers • Create complementarity or alternative income and employment alternatives for farmers and their families, on- and off-farm • Contribute to economic cohesion within the EU

in pursuit of the five original objectives of its Common Agricultural Policy (CAP) present in Article 39 of the Treaty of Rome, signed in 1957 (Table 5.1)

Special clauses were inserted into the Treaty to allow the authorities to establish marketing organisations, monetary restrictions and control of trade both internally and externally. Pride of place went to market regulations for cereals because of their importance in the financial structure of European farming and through their increasing significance as a source of animal feedstuffs (e.g. the 'barley beef' revolution). Initially, policies concentrated upon external protection against cheap imported food, but, gradually, internal common pricing policies were applied so that by the early 1980s, only potatoes, wool and agricultural alcohol remained outside the scope of this structure (Fennell, 1979).

The CAP has operated by means of a dual control system. One aspect is the application of levies and customs duties at frontiers so that imports from non-member states cannot be sold within the EU for less than the desired internal market price. The other involves intervention or the purchase by an Intervention Board of supplies surplus to the market when prices fall below an agreed level. One of the difficulties experienced by this system has been the setting of the intervention price so as to guarantee the farmer a satisfactory return on production, but without it being so high as to

promote overproduction. Critics can point to the well-publicised olive oil, milk and wine 'lakes', and grain, beef and butter 'mountains' as evidence of the lack of an adequate discouragement to producers once the internal market has been satisfied. Hence the intervention system became something of a licence for overproduction rather than its intended role of 'safety net'. Ultimately, the problem of surplus production became a political issue, with the cost of storing surpluses one factor in contributing to the growth of a reform agenda in the 1980s.

The structure of the EU as a protectionist organisation has also led to strong criticism from other countries, as seen in the early 1990s in the GATT negotiations. The system of levies has prevented external producers from being able to undercut domestic producers. This has had the effect of artificially supporting inefficient EU producers, a charge frequently aimed at some of the smaller Mediterranean producers and French 'peasant' farmers. The CAP support system has led to higher food prices in the EU than on the world market. To encourage EU producers to export there have been subsidies available to enable them to compete more effectively in foreign markets. Again, this can tend to protect inefficient and costly production within the EU from foreign competition.

The cost of supporting this system is met by the European Guidance and Guarantee Fund (usually known by its French acronym, FEOGA), which receives its finance from member governments. The Guarantee Section of this Fund is responsible for the pricing arrangements of over 90 per cent of farm output in the EU, and administers price support for over 70 per cent. In contrast the Guidance Section represents less than 10 per cent of total CAP expenditure, with funds for improvements to the structure of production, processing and marketing, and environmental measures on farms. In the late 1980s FEOGA accounted for 80 per cent of the overall EU budget, but social and development funds have since increased in importance so that funding to agriculture is now around half of the total. The CAP dispenses over €40 billion per annum. When indirect subsidies such as price supports and tax-breaks for farmers are added to direct payments, EU farmers received €104 billion in aid in 2001 compared with around €50 billion in subsidies in the USA. Farmers in the EU get about 35 per cent of their income from subsidies, compared with 21 per cent in the USA and just 1 per cent in New Zealand (*The Economist*, 13.7.02, p. 13) (see Table 5.2). Although France is the highest gross contributor to FEOGA, in net terms, following a complex system of rebates, it is Germany, Britain and the Netherlands who are the major funders (Table 5.3).

The system of common agricultural prices operated by the CAP depends on the rates of exchange between the national currencies remaining stable. The European Monetary System (EMS), introduced in 1979, cushioned the impact of

Table 5.3 Member state contributions to the EU's Common Agricultural Policy, 2000 (€ billion)

Country	Gross contribution	Net contribution
France	8.5	−2.32
Germany	4.9	4.37
Spain	4.8	−2.53
Italy	4.5	0.02
UK	4.0	2.34
Greece	2.3	−1.19
Ireland	0.8	−1.19
Netherlands	0.7	1.07
Denmark	0.6	−0.54
Austria	0.5	−0.04
Belgium	0.5	0.62
Sweden	0.4	0.42
Finland	0.4	−0.23
Portugal	0.4	−0.08
Luxembourg	0	0.06

(Source: *The Economist*, 13.7.02, p. 36)

parity adjustments, but a complex correcting mechanism had to be operated giving rise to differences between official and 'green' parities, e.g. the 'green pound' (Burtin, 1987). This enabled the unity of the market to be obtained, but it proved expensive, distorted competition, limited structural adjustment of agriculture and jeopardised optimal allocation of resources (Tarrant, 1980a). With the creation of the EMS, existing member states (except for the UK) agreed to align their currencies and the European Currency Unit (ECU) was introduced, with the ECU being valued with respect to the pre-existing units of account. However, the logic of currency alignment has now been incorporated within the process of Economic and Monetary Union (EMU), which aims at furthering the idea of a single EU market. This has produced a new currency, the Euro, with 11 member states (the 15 EU members as of 2002, excepting Denmark, Greece, Sweden and the UK) beginning to replace their national currencies with the Euro from 1 January 1999, and Euro notes and coins legal tender from early 2002.

One of the inherent weaknesses in the CAP has been the difficulty in reconciling the needs of the

Table 5.2 Farmers' subsidies as a percentage of gross farm revenue

Country	1986/8	1999/2001
Switzerland	72	70
Norway	65	64
Japan	62	60
EU (15)	42	35
USA	25	22
Poland	4*	12
Australia	9	5
New Zealand	11	1

* 1991/3
(Source: OECD)

Table 5.4 Characteristics of agriculture in the member states of the EU (1999)

Country	A	B	C	D	E	F	G	H	I	J
EU15	7023	−29	79.2	24.9	137.9	213,097	7277	83,770	212,954	119,616
Austria	178	n/a	90.9	8.4	20.4	4,736	86	2,198	442	3,680
Belgium	79	−22	87.4	32.3	30.9	2,633	11	3,184	165	7,436
Denmark	98	−12	60.4	50.8	36.8	9,306	117	1,974	142	12,004
Germany	657	−36	72.0	28.3	56.5	44,580	1277	15,227	2,412	24,795
Greece	597	−30	87.6	7.6	56.1	4,568	605	581	15,394	938
Finland	126	n/a	95.0	12.7	39.9	3,812	33	1,145	109	1,467
France	958	−34	77.0	29.8	98.2	67,706	2087	20,389	11,505	15,430
Italy	1798	−16	85.7	21.0	84.6	20,636	1037	7,328	12,237	8,281
Ireland	202	−20	93.0	33.1	182.3	1,865	4	7,093	5,624	1,801
Luxembourg	5	−25	86.0	36.2	34.8	*	*	*	*	*
Netherlands	209	−11	73.7	44.1	74.2	1,434	4	4,292	1,724	11,438
Portugal	520	−47	82.7	8.0	39.3	1,258	92	1,295	7,115	2,365
Spain	1099	−32	70.8	13.2	191.3	22,196	1279	5,839	27,437	19,346
Sweden	82	n/a	76.8	29.6	46.9	5,697	46	1,706	407	2,309
UK	416	−21	62.3	68.2	488.7	22,670	599	11,519	44,471	8,146

Key: A   Total labour force (000)
     B   % change in total labour force 1987–97
     C   Family labour force as a % of total
     D   Dairy cows – average herd size
     E   Sheep – average flock size
     F   Cereal production (mt)
     G   Oilseed crops production (mt)
     H   Cattle – total numbers
     I   Sheep – total numbers
     J   Pigs – total numbers
     *   combined with Belgium
   n/a not available
(*Source*: Eurostats)

consumers with the intention of providing satisfactory incomes for farmers in countries with very different farm-size structures and specialisms (see Table 5.4; Figures 5.2 and 5.3) (Bowler, 1992d). In the 1980s there were frequent charges that it was farmers who were protected at the expense of the consumers, i.e. food prices were too high whilst farmers were receiving large amounts of support. However, that support has varied considerably between sectors. For example, the initial effects of the UK's entry increased incomes most for cereal-growers (Josling and Hamway, 1976). This has continued to be the case, with a stark contrast developing between the rewards accruing to large cereal producers and those to small livestock producers. Meanwhile UK horticulturalists suffered from competition with Mediterranean producers who had lower overheads and from more heavily subsidised Dutch growers. The UK government attempted to counter this with its own 'modernisation' grants for horticulturalists (Ilbery and Bowler, 1994).

Because the CAP has contributed to pressures on farmers to increase the size of operation needed to ensure viability, it has tended to divide farming businesses into 'haves' and 'have nots' in terms of ability to obtain benefit from these economies. It has encouraged the growth of enterprises in which farmers have made heavy investment in plant and machinery, especially on arable farms. This has produced a highly capital- and energy-intensive sector within farming whilst also fostering the demand for land, as land has remained the essential capital input in farming. It is difficult to

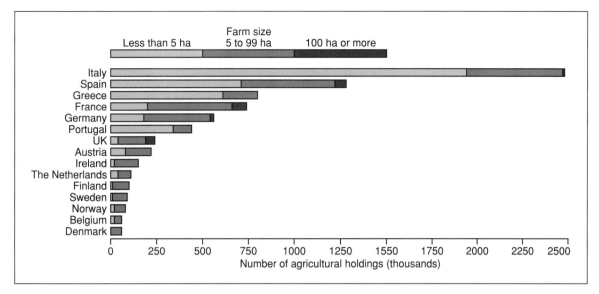

Figure 5.2 Farm size distribution in the EU, 1995 (data source: Eurostat, 1999)

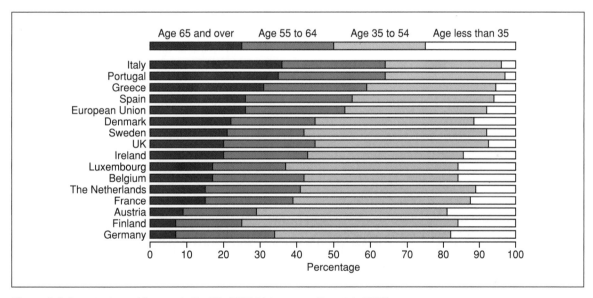

Figure 5.3 Age structure of farmers in the EU, 1995 (data source: Eurostat, 1999)

disentangle impacts directly attributable to the CAP from developments both downstream and upstream of farming activity itself, but, in general, the CAP has been associated with a greater differentiation of farming, on both a sectoral and farm-size basis. Twenty per cent of farmers have accounted for 80 per cent of FEOGA guarantees, with large arable farmers receiving the biggest amount. Other well-documented impacts have been substantially increased output in particular sectors and negative environmental effects.

5.2.2 Increased production

The stimulus of advantageous prices, maintained at artificially high levels by FEOGA, can be seen in

Table 5.5 Agricultural productivity in the EU, 1961–2001

	Gross agricultural output		Cereals		Crops		Food		Livestock	
	2001	*1961*	*2001*	*1961*	*2001*	*1961*	*2001*	*1961*	*2001*	*1961*
EU15	102.5	61.7	106.4	44.7	105.9	64.9	102.4	61.6	102.0	61.4
Austria	104.3	73.8	90.8	41.5	102.0	87.4	104.3	73.8	102.8	69.8
Belgium–Lux.	112.5	64.8	108.9	77.0	136.7	73.6	112.4	64.1	110.6	62.8
Denmark	104.6	71.1	104.2	52.4	92.9	49.6	104.6	71.1	116.7	85.0
Finland	90.4	81.2	95.3	48.2	96.8	57.0	90.4	81.2	91.8	91.6
France	102.4	64.5	104.5	33.1	101.9	59.1	102.4	64.2	103.5	68.8
Germany	98.7	61.5	136.2	42.5	120.6	71.5	98.6	61.4	88.6	64.8
Greece	106.4	54.4	70.4	39.6	108.9	56.7	101.7	54.8	96.7	42.0
Ireland	112.9	51.1	112.7	67.9	112.3	100.0	113.6	50.8	115.2	48.6
Italy	102.6	71.1	110.5	77.0	102.2	80.2	103.3	71.7	106.5	55.8
Netherlands	100.7	41.8	128.8	126.8	113.2	56.8	100.8	41.6	99.0	47.8
Portugal	100.2	74.3	79.8	84.6	90.8	97.8	100.2	73.4	118.0	37.4
Spain	115.7	45.9	96.8	39.4	111.8	53.8	115.3	44.8	130.1	29.4
Sweden	97.2	88.5	101.9	65.3	89.7	73.5	97.2	88.4	101.6	95.0
UK	89.1	64.3	83.4	37.8	88.2	50.3	89.4	64.1	92.1	74.8

Note: 1989–91 = 100
(*Source*: FAOSTAT)

substantial increases in agricultural output. As shown in Table 5.5 the principal production gains occurred prior to 1990 and affected the 'late entrants' to the EU as well as the original six member states. There are numerous examples of major production gains, one being the response of British farmers following the UK's accession in 1973. The output of cereals had risen by 24 per cent by the end of the 1970s, and milk by 33 per cent (Soper, 1983). The UK's self-sufficiency in temperate foodstuffs had increased from around 50 per cent in 1973 to 75 per cent a decade and a half later, and the UK became a net exporter of cereals for the first time since the first half of the nineteenth century. In contrast, UK horticultural producers lost out to producers in east and south Europe who had cheaper production costs and to producers in the Netherlands whose costs were more heavily subsidised (Ilbery, 1985c). Other good examples of a direct response to CAP price supports are milk production and the growing of oilseed rape.

The rise in French milk production between 1960 and 1975 accounted for half the total increase in milk production within the member states. This reflected the stimulus of favourable prices, maintained at artificially high prices by the CAP, coupled with improved production systems that helped raise yields. Indeed, milk and milk products absorbed over 40 per cent of FEOGA's guarantee funds in the 1970s, whilst growth in output led to production outstripping demand. The consequence was the first wholesale sectoral policy reform to the CAP in the form of a milk quota scheme introduced in 1984 (discussed in Section 5.3.4 below).

A good geographical example of the impact of CAP subsidies is that of the distinctively flowered oilseed rape. Given the deficiency in vegetable oils and protein meal in Western Europe in the 1960s, the CAP established attractive target prices to encourage farmers to produce oilseed rape, which yields processed edible oil suitable for salad and cooking oils and margarine. A by-product, rapemeal, is a useful protein concentrate in animal feeding-stuffs. The rise in world commodity prices in the early 1970s favoured rape-growing as did the introduction of new crop varieties making for production of less acidic animal feed. However, the key to its successful introduction lay in favourable prices supported by FEOGA. Initially, this meant an increased value of rape compared with barley, but rape was also a suitable 'entry crop' to grow before wheat, whose price was also sustained at a high level under the CAP. After being grown on a few farms in the south of England in the early 1970s, by 1983 it had become the largest non-cereal arable crop in the UK, being grown on over 11,000 holdings, with an output valued at £200 million and providing 30 per cent of Britain's demand for cooking oil (Wrathall and Moore, 1986). Subsidies to oilseed producers to persuade them to take British rapeseed instead of cheaper oil-yielding crops from abroad also helped in this dramatic expansion.

5.2.3 Environmental disbenefits

The CAP's encouragement of greater farming intensification to obtain maximum advantage from the price supports has had a dramatic impact upon the appearance of the European landscape (Brouwer, 2000). Long-term destruction of hedgerows, woodland, rough grazing, downland, moors and wetlands has been one readily identifiable outcome (Table 5.6). This has been closely linked to the development of mechanisation and to the economics of its use (discussed further in Chapter 10). However, it has also been the economics of the CAP that have driven farmers to put more land under the plough. In particular, the returns upon cereals, whose prices have been artificially supported, have encouraged farmers to plough land that has either never been ploughed before or has only been ploughed when cereal prices were extremely high. This has impacted sharply upon landscape features in lowland areas.

The uplands have also been adversely affected by the CAP, even through the operation of its Guidance section. For example, in the UK there has been a reduction in rough grazing related to the increased numbers of sheep brought about by the headage payments available under the Less Favoured Areas (LFAs) scheme introduced in 1975 to compensate farmers in harsh environments (e.g. upland areas) for the difficult physical environment within which they operated (Wathern *et al.*, 1986; 1988). In designated LFAs farmers have increased stocking rates to increase their headage payments, exceeding the carrying capacity of the semi-natural vegetation of the rough grazings. They have then sought to make grant-aided improvements to their rough grazing under grassland conversion projects, at the expense of dwarf shrub vegetation and open moorland. Loss of rough grazing has been especially prominent in the UK's national parks.

More recently the environmental disbenefits have extended to the quality of produce from industrial farming. Scares over the presence of salmonella in chicken and certain types of cheese (North and Gorman, 1990), and the spread of bovine spongiform encephalopathy (BSE or 'mad cow' disease) and its link to new variant Creutsfeld Jacob disease (nvCJD) in humans have provoked increased calls for a different way of evaluating agriculture beyond the narrow economic viewpoint. However, as will be shown in Chapter 10, the strong support for industrial-style farming within the CAP has fostered the farming methods

Table 5.6 Examples of habitat destruction in the UK, 1945–90

	% lost or damaged		% lost or damaged
Lowland meadows	82	Limestone pavements	43
Chalk downlands	79	Lowland heaths	39
Lowland bogs	60	Upland woodlands	28
Lowland marshes	51	Ancient woodlands	25

(*Source*: O'Riordan, 1987, p. 36)

leading to these problems associated with animal welfare and new systems of livestock husbandry. Indeed, it has been argued that the link between the environmental disbenefits and prevailing government policy and ethos can be extended beyond the CAP to international policies within the GATT/WTO. In particular, this has been through the undermining of existing environmental and consumer protection standards introduced by new modes of regulation in the GATT/WTO as a result of the pursuit of free trade by the world's trading superpowers (e.g. Lang, 1992). Within the CAP itself, though, there has been a growing recognition of the need to address the twin problems of environmental disbenefits and over-production (Ritson and Harvey, 1997). This has resulted in a series of policy reforms, though without dramatically dismantling the ongoing support for production (Winter, 2000).

5.3 Reforming the Common Agricultural Policy

In many parts of the Developed World during the last two decades there have been attempts to deregulate market relations as part of the economic logic of the 'New Right'. This has often involved restructuring of government departments and agencies, as illustrated by the reforms in the USA, UK, Canada, Australia and New Zealand (e.g. Lawrence et al., 2001; Liepins and Bradshaw, 1999; Robinson, 1997b). In particular, though, reform of agricultural policy throughout the Developed World has been considered within a new macro-economic framework in which the economic significance of agriculture continues to decline (it accounts for only 2.5 per cent of GDP and 5.5 per cent of employment in the EU) and there are greater fiscal stringencies and deregulation. The latter was typified by the 1996 US Farm Bill, which aimed to largely phase out commodity programmes early this century, though the protectionism still prevalent in both American and EU farm politics threatens to block these intentions. Governments are under more pressure to limit public expenditure, and hence justification for farm support measures has become harder, not least because external

pressures through the WTO and agri-exporting nations in 'the South' have been exerted upon the United States and the EU to reduce their farm subsidies. Media stories about 'featherbedded' farmers have raised opposition in some quarters to government spending on agriculture, as have lobbies championing free trade in food. This opposition to subsidies has been most vocal in North America, north-west Europe and Australasia, but strong support for the social and environmental dimensions of farming remains in parts of the Mediterranean and Scandinavia (e.g. Lowe et al., 1994; Monke et al., 1998a). Moreover, the current US government is set to reverse America's post-1996 moves towards lower subsidies.

In 1985 the European Commissiom initiated a wide-ranging debate on the future of European farming and its role in economic and social development (CEC, 1985). A year later the Commission stated, 'European agriculture has to accept economic realities and learn to produce for the market, to adapt to commercial demands and to continue to modernise' (CEC, 1986, p. 8). The situation whereby higher farm incomes could be achieved by increasing output at high guaranteed prices could not be sustained; there were real concerns over the escalating costs of the CAP, especially to store surplus food generated by the system, and the environmental consequences of continued agricultural intensification (Ilbery, 1992; Robinson, 1991a; 1991b; 1991c). The cost of storage of surplus production had become a major concern: in 1991 in storage there were 25 million tonnes of cereals, 800,000 tonnes of beef, 700,000 tonnes of dairy produce and 2200 million litres of wine (Robinson and Ilbery, 1993, p. 204). Various proposals for reform were made, with the introduction of agri-environmental measures in the late-1980s to promote environmentally friendly farming, e.g. the Environmentally Sensitive Areas scheme (Whitby, 1994). Subsequently other agri-environment measures were introduced, sitting alongside schemes operated by the member states (these are considered in more detail in Chapter 10). However, the major phase of reforms came under the auspices of the Agriculture Commissioner, Ray MacSharry, in 1992 (Table 5.7) (Kay, 1998).

Table 5.7 Reforms proposed by EC farm ministers, May 1992

Main points

1. 29% cut in cereal support over the next three years
2. £80/t feed wheat intervention price in 1995
3. Compensatory aid payment of £209/ha for cereals harvested in 1995
4. Comensatory aid payment of £300/ha for protein peas and beans harvested in 1995
5. Payment of compensatory aid:
   (a) conditional upon rotational set-aside of 15%
   (b) restricted to the remaining 85%
6. Base area is calculated from average of cereals, peas, beans, rape and permanent set-aside in 1989, 1990 and 1991. This will be based on current exchange rates and estimated average regional yields.
7. Compensation for rotational set-aside of £209/ha in 1995
8. Oilseeds support regime to be incorporated within package
9. Crops not eligible for inclusion in the base area are potatoes, sugar beet, linseed, permanent pasture, fruit and vegetables
10. Co-responsibility levy abolished
11. Additional environmental aid
12. New rotational set-aside to be introduced

(*Source: Crops,* 6.6.92)

The MacSharry reforms of 1992 cut support prices for cereals, with farmers compensated through a flat-rate area payment on each hectare planted (see details and their effects on French agriculture, in Naylor, 1995). To qualify for compensation farmers had to take 15 per cent of their arable land out of production (set-aside). Area payments were also introduced for oilseeds and protein crops thereby creating an Arable Area Payment Scheme (AAPS). The payments were increased under Agenda 2000. Reforms for beef producers reduced support prices but reinforced the system of headage payments (also applicable to sheep). Grants were introduced to promote farm forestry, early retirement schemes and diversification programmes. Smaller cuts in support to beef, dairy and poultry producers were introduced, but the idea of applying production quotas, as followed earlier for milk, was rejected on the grounds of cost and impracticality. Subsequently, further reductions in price supports have been implemented across a number of sectors as the CAP has gradually shifted away from production subsidies. Different aspects of these reforms are now examined: how the EU copied the American policy of set-aside, the American and EU experience of set-aside,

the application of milk quotas, and the more recent reforms in Agenda 2000.

5.3.1 The American experience of set-aside

For many farmers the concept of obtaining income from setting aside land and diversifying into alternative, non-agricultural enterprises on the farm was novel and unattractive, but increasingly diversification has become necessary as incomes from 'traditional' farming activities have fallen. This can be seen in other parts of the Developed World, including the USA where the concept of paying farmers to set aside agricultural land from productive output originated as a means of curbing cereal production and, subsequently, to reduce the risk of soil erosion by encouraging farmers to plant trees on the most erodable farmland. For example, the US Payment-in-Kind (PIK) program, launched in 1983, provided financial compensation to farmers volunteering to take land out of production. This had over a million participants, representing a third of the country's cropland and resulting in a 38 per cent decline in the size of the maize harvest, a substantial Treasury saving in storage costs and an increase in farm incomes (Goodenough, 1984). Subsequently, as part of the 1985 Farm Security Act, a Conservation Reserve Program (CRP) was introduced to set aside 18.2 million ha over a ten-year period. However, despite 31.6 million ha being converted to 'conserving uses' in the 1980s (Bedenbaugh, 1988) and over 375,000 farmers enrolling on the programme, many farmers have looked upon this and set-aside as the antithesis of 'proper' farming (Potter, 1997a). Indeed, farmers have often responded by setting aside their poorest land while intensifying production on the rest of their holding (Ervin, 1988). Hence 'slippage' has occurred whereby the output of cereals has not fallen in proportion to the amount of land retired.

By 1990 in the USA 13.7 million ha had been enrolled in CRP (Heimlick and Kula, 1991), after which it was modified through the 1990 Food, Agriculture, Conservation and Trade Act. This expanded eligibility criteria and aimed to increase the area enrolled (Bjerke, 1991). This effectively added a further one million ha to the CRP (Osborn et al., 1994). Analysis of the CRP by Nellis et al. (1997) highlighted several problems:

- high costs: at around US$1.4 billion per year, the CRP was criticised for its high costs when compared with the amount of erosion prevented (Young et al., 1994). However, after five years of the programme the economic value of these benefits was assessed at US$10 million (Ribaudo et al., 1989);
- declines in output may affect certain parts of the Great Plains, impacting on some agribusinesses (Saltiel, 1994);
- rising prices of the best farmland, as the focus of farming activity is concentrated upon this land;
- increases in pests and diseases, as CPR lands are perceived as a weed source during the time that it takes to establish 'permanent' plant cover (Steiner, 1990);
- expansion of land for hay or grazing, as allowed in later CRP regulations, has concerned stock rearers and graziers;
- weaknesses of its voluntary and impermanent nature, with regulations varied over time.

Additional aims have been added subsequently, including wetland protection and improvement of water quality (Young et al., 1994). Significant environmental benefits have been reported through reductions in use of fertiliser and pesticides.

Farmers' incomes were expected to rise as a result of CRP, through a mixture of decreased production costs and federal rental payments. However, the reality was highly variable, with losses accruing in higher yielding locations, but gains in areas susceptible to erosion, such as south-east Washington (Nellis et al., 1997, p. 23).

The main areas of enrolment have been the Great Plains (60 per cent of CRP-enrolled area), the Corn Belt and the historic Cotton Belt. There is a big concentration in the 'Dust Bowl' region where cropland has been heavily dependent on irrigation post-1945. Similarly, irrigated areas in the Columbia and Snake River Basins are prime CRP concentrations. Especially in marginal areas, where farmers have problems with access to water for irrigation, it has often been profitable to enrol lands in CRP, though there is little evidence that the majority of farmers would pursue conservation aims as opposed to returning land to crops or grazing once the programme ends.

5.3.2 Regulation 1094/88: voluntary set-aside in the European Union

Of the MacSharry reforms perhaps the most controversial was that of set-aside, first discussed in the UK in a Green Paper, A Future for Community Agriculture, published in 1985. This drew upon the set-aside policies already in place in the United States to suggest that set-aside could be used to decrease over-production; assist structural change in agriculture (for example by being linked to cessation of production by some farmers); and help meet environmental/ecological objectives (Haynes, 1992). This regarded set-aside as a control mechanism through restrictions on the amount of land input to the farming system so as to reduce output. It recognised the potential for using incentive payments to promote set-aside. In the same year, in response to EU structural reforms, a set-aside scheme was introduced in Schleswig-Holstein, Germany (CEC, 1986; Jones, 1990), followed by a similar measure in Lower Saxony the following year (Jones, 1992). As a result of these pioneering schemes the German government was an enthusiastic supporter of proposals for introducing set-aside throughout the EU. This was brought about in 1988 via EU regulation 1094/88 for the voluntary set-aside of arable land. The response to this was greatest in Germany, with 170,000 ha set aside in the first year of the policy's operation (Jones, 1991), though there was less enthusiasm in the UK. Overall, by the end of 1993 4.7 million ha had been set aside in the EU (Jackson, 1994).

Regulation 1094/88 offered incentives to farmers to take a portion of their arable land out of production for a period of five years. Eligible land was defined as land used to grow crops such as cereals, oilseed rape, dried peas and beans, linseed, sugar beet and hops. Rates of compensation for farmers participating in the scheme varied according to the type of set-aside practised and whether or not the land was in a Less Favoured Area (LFA). Set-aside land had to be fallowed permanently, or rotated annually, with a suitable green cover crop, or planted with trees or used for a permissable non-agricultural use, such as a golf course or car park for a farm shop. However, partly because the scheme was only voluntary for farmers, there was limited up-take in the UK: affecting only 4 per cent of the arable area compared with twice this percentage in Germany where set-aside was frequently viewed as a financially beneficial preliminary to retirement or scaling down of farm operations (Jones, 1992). In a survey of farmers in Hampshire, southern England, common reasons for non-participation in the voluntary scheme were that it was not economically justifiable, the lack of flexibility associated with the five-year time period, and the common view that setting-aside land was not compatible with the farmer's *raison d'être* (Robinson and Lind, 1999).

5.3.3 The Arable Area Payments Scheme

Reforms to the CAP agreed by farm ministers in May 1992 ushered in a range of measures designed to reduce output, including reductions in price support and compulsory set-aside for certain producers (Robinson, 1993a; Robinson and Ilbery,

1993). Set-aside took the form of the Arable Area Payments Scheme (AAPS). Farmers with an annual production in excess of 92 tonnes of cereals were required to set aside a minimum stated proportion of their base arable area (MAFF, 1993). If farms did not have a sufficient area of eligible crops on which to set aside land, the farmer was still eligible for AAPS on any arable crops.

The AAPS was sub-divided into a Main and a Simplified scheme, though an individual farm could be entered into only one of them:

- The Main Scheme. Farmers claim payments for each eligible crop type covering a minimum of 0.3 ha if a corresponding area of land is set aside. This is a percentage of the total area of land, including that set aside, on which the farmer wishes to claim payments. Rates of payment were dependent on crop type and the proportion of land set aside is dependent on the set-aside option adopted. There is a range of options available (see Table 5.8), though the principal impacts on land use relate to three practices:
  - rotational set-aside, in which land is entered into a six-year rotation;
  - flexible set-aside, in which there is a combination of rotational set-aside and setting aside a parcel of land for a fixed period (often referred to as permanent set-aside);
  - guaranteed set-aside, in which land is taken out of production for a five-year period (permanent set-aside). Each of these three forms had to account for 10 per cent of a defined base area.

  At the outset participants in the Main Scheme have to specify which set-aside option to use.
- The Simplified Scheme. In addition to the requirement to set aside a given proportion of arable land on certain holdings, payments for set-aside may also be claimed for between 0.3 and 15.62 ha. For this area, payment is at the rate for cereals regardless of the type of crop being set aside.

Certain management practices for set-aside land are specified as part of the scheme. As summarised in Table 5.9, these practices refer to the range of

Table 5.8 Summary of types of set-aside

Type of set-aside	Comments
Rotational set-aside (10%)	The set-aside land must be entered into a six-year rotation
Flexible set-aside (10%)	The set-aside can be a combination of static and rotated land. This may be left in the same place or moved. Different parcels of land within/between fields may be treated differently
Guaranteed set-aside (10%)	Payments are guaranteed for five years if the land is permanently set aside for this period. Any land entered into the Countryside Access Scheme must also be entered into Guaranteed set-aside
Voluntary set-aside	This is land set aside in excess of the grower's basic obligation. The total must not exceed the cropped area on which they are claiming. Growers cannot claim on voluntary set-aside if they are in the Simplified Scheme
Additional voluntary set-aside	Only available to farmers who had land in the old Five-Year Scheme
Penalty set-aside	This is the extra uncompensated set-aside that will be imposed if the regional base area is exceeded
Transfer of set-aside obligation	All or part of a grower's set-aside obligation may be transferred to either another farm within a 20 km radius or to a designated environmentally approved area, e.g. Nitrate Sensitive Area. In all cases the basic set-aside obligation increases by 3%
Other	This includes land entered into the Farm Woodland Premium Scheme or the Habitat Scheme and counting towards the set-aside requirement

(*Source*: MAFF, 1995)

Table 5.9 General uses and management practices on set-aside land

Type of use	Management practices
Non-agricultural uses	Various, including grazing family ponies, shooting
Agricultural uses	Production of some non-food usage crops, e.g. peas, spices, millet, mustard seeds. These are all required to have pre-sowing contracts with a processor. Arable Area Payments are made on these crops but no set-aside payments are made
Types of cover crop	Natural regeneration, wild bird cover, grass cover, mustard or a cover mixture of at least two crop groups (such as cereals and brassicas)
Management of set-aside land	During the set-aside period farmers must not damage or remove any of the following features which are sited on or immediately next to land which is set aside: vernacular (traditional) buildings, stone walls, hedges, trees including hedgerow trees, watercourses, ditches, ponds, pools, lakes and archaeological remains
Day-to-day management	Use of herbicides, fertilisers and pesticides is very restricted, as is cutting, mowing and cultivation of the cover crop. Within strict controls farmers have to develop their own management plans

(*Source*: MAFF, 1995)

uses to which the land may be put whilst it is set aside. This use may include production of specified non-food crops as well as natural regeneration and sown grass or clover as a cover crop. For the latter, restrictions are imposed on the cutting, mowing and cultivation of the crop. Use of herbicide, fertilisers and pesticides on set-aside land is also strictly controlled. This and other restrictions on usage mean that set-aside has possessed some potential for contributing to environmental improvement in addition to its role in reducing production (Silson, 1995). However, the extent to which this potential has been realised has been questioned (e.g. Craighill and Goldsmith, 1994), partly because the conservation aspects of set-aside have been an incidental aspect of the policy rather than the prime consideration. Moreover, tensions between the idea of set-aside and the farmers' view of their role in land management and food production have been experienced with respect to compulsory set-aside.

After three years of the operation of set-aside in the EU there were 7.2 million ha set-aside, 1.5 million ha of which were in the UK. In England at this time there were 29,199 claimants for payment under the Main Scheme and 17,509 claimants under the Simplified Scheme (*MAFF news release* 262/96). The former produced 286,442 ha of rotational set-aside and 240,642 ha of flexible set-aside, with 84,676 ha of non-food crops on set-aside. The distribution of set-aside is shown in Map 5.1, where the located proportional circles show that the greatest amounts of set-aside under the Main Scheme were in the east. The eastern counties also tended to have a predominance of rotational set-aside, which accounted for over 60 per cent of the set-aside in several of these counties. This allows land to be utilised as a fallow or for a rest crop. It is also quite versatile, allowing for relatively swift action in the event of policy changes. Western counties and those with a significant urban fringe tended to have higher proportions of flexible set-aside, in which set-aside land can either be rotated or left in one fixed place. This geographical variation seems to reflect the greater attraction of rotational set-aside for farmers who can use it as a means of giving land a temporary break from growing cereals. The flexible option offers a means of combining this rotation with some longer-term removal of land from cereal production. The latter has proved most attractive in areas not traditionally known for growing cereals.

Having decided upon the type of set-aside to adopt, the farmer's next decision is how to locate the area to be set aside within the holding. In north Hampshire, over two-thirds set aside whole fields, some farmers combining this with use of field margins or strips of land to make up the remainder of their requirement (Robinson and Lind, 1999, pp. 302–5). The most popular cover crop on the set-aside land was natural regeneration, which requires no ploughing or sowing. This seems to offer greater conservation benefits than a sown green cover (Wilson and Fuller, 1992) and incurs few additional costs, unlike annual crops or specially planted cover for wild birds. In contrast, the planting of industrial crops such as oilseed rape gives low returns once the costs of the inputs have been considered. However, around 10 per cent of set-aside in the survey was under an annual crop grown under the rotational option. A common reason for this was to use the crop to clear the land of weeds or to prepare the ground for future cereal production. Those farmers setting aside combinations of whole fields and strips or margins were most likely to participate in combinations of cover crops, enabling them to adjust the cover crop to best suit the area and type of set-aside. The potential for set-aside to be used for environmental benefit is discussed further in Chapter 10, though it has been argued that environmental benefits from extensification can be better delivered by other means (e.g. Midmore and Lampkin, 1989). Moreover, Walford's (2002) study of large farms in south-east England found that 'farmers in significant numbers were using set-aside as a management tool to support agricultural production' (p. 255).

5.3.4 Milk quotas

The rising costs of the CAP first led to substantive reforms to the policy in the 1980s, in the form of milk quotas, first imposed upon individual farmers in 1984 (Downs, 1991). In France this reduced output by 2 per cent in its first two years of operation, whilst 113,000 French dairy farmers left

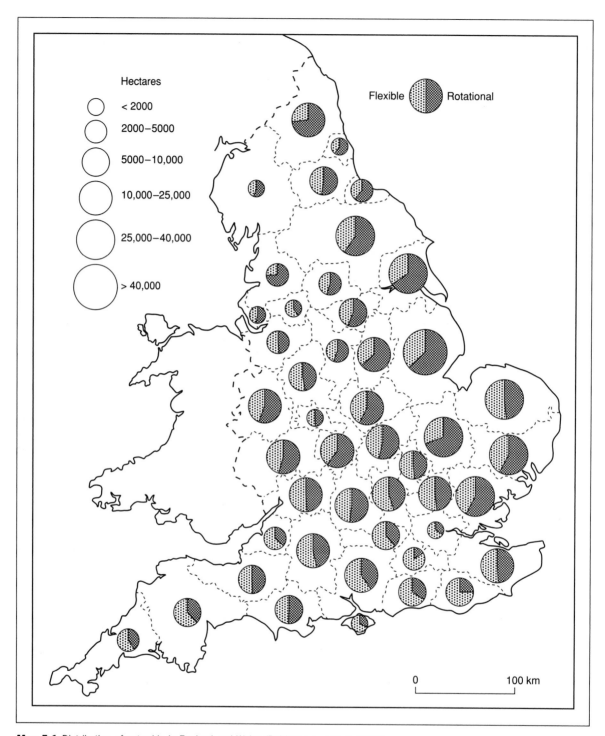

Map 5.1 Distribution of set-aside in England and Wales (Robinson and Lind, 1999)

dairying between 1984 and 1989 through payment of an outgoers grant (Naylor, 1993). Land released from dairying was converted to cash crops or livestock rearing or fattening (Naylor, 1986). In the UK, quotas initially cut the national dairy herd by 15 per cent and milk production by 18 per cent, whilst output per producer rose by 3 per cent. Quotas helped maintain prices paid to producers and hence made quotas a much sought-after commodity, with buying and selling of quotas becoming a multi-million pound business.

In the UK, milk quota in 1984/5 was allocated to all dairy farmers on the basis of their production in 1981 plus 1 per cent. This entailed a cut of 9.4 per cent from the 1983/4 production level irrespective of whether the producers were high- or low-cost producers. However, economic efficiency was increased by virtue of the quotas being tradeable. The original intention was to attach quota to land, but quota has been both leased and sold without land. Hence it has been possible for producers who have ceased production to continue owning quota and lease it out on an annual basis. These are the so-called non-producing quota holders, known popularly in Germany as 'sofa farmers'. In the UK in 1997/8 5284 quota holders (14.2 per cent of the total) leased out all of their quota, accounting for 756 million litres or 5.3 per cent of the total national quota and 59 per cent of all UK leased quota that year (Colman, 2000).

This trading in quota has been instrumental in increasing efficiency of production by enabling the size of production units to be increased. However, Colman (2000) concludes that up to one-quarter of the UK's national quota would need to be transferred to eliminate existing inefficiencies. Moreover, the quota system imposes high costs on expanding and potential new producers, which represent substantial grounds for arguing that abolition of quota would entail significant benefits (and some risks) for the UK milk-producing sector. However, following the Agenda 2000 proposals (see below), which were a wide-ranging set of policy reforms, it appears that quotas will exist at least until 2008.

In terms of the geographical impact of quotas, the number of herds and the number of cows in the Less Favoured Areas (LFAs) of the EU has increased, so that the LFAs now account for 49 per cent of milk producers and 36 per cent of herd numbers. Hence dairying has become more of a mainstay in marginal areas than previously. Three-quarters of the EU's dairy herd is located in Germany, France, the UK, the Netherlands and Italy, representing a combination of the best pastures and an assured income for small farmers in some of the more marginal pastoral areas. A major impact has been the reduction in CAP expenditure on dairying: from 28.5 per cent of the budget in 1984 to 6.5 per cent in 2000, a decrease of nearly 50 per cent in financial terms despite a decline in milk output of only 8.7 per cent (CEC, 2002).

5.3.5 Agenda 2000

The Agenda 2000 reforms to the CAP were agreed in March 1999, extending the MacSharry reforms of 1992 in terms of shifts from price supports to direct income payments to farmers. These further reduced commodity support prices in the beef, dairy and cereals sectors. Compensation to farmers was increased through increases in direct payments from schemes such as the AAPS, but many commentators felt that this round of reforms was, at best, half-hearted and insufficient to achieve its stated aims (Buller and Hoggart, 2001; Lowe and Brouwer, 2000; Tangermann, 1999). However, as part of Agenda 2000 the European Commission (EC, 1997; 1998) defined a number of policy objectives for a 'new' CAP (see Table 5.1) whilst maintaining two of the original objectives, namely stability of incomes and a fair standard of living for the agricultural community. Ramsay (1997) notes that as part of the latter aim, it is implicit that there will continue to be compensation paid for any fall in commodity prices. Moreover, Potter and Goodwin (1998), anticipating the reforms, referred to the desire to attain a 'double dividend': a reduction in prices of farm produce, which will encourage extensification, and the resources previously invested in price support will be re-invested in agri-environment schemes.

The main element in the Agenda 2000 reforms was a reduction of 15 per cent in the intervention price of cereals, with only half of this being compensated for by an increase in area aid payments.

A 20 per cent cut in support prices for beef production was accompanied by elimination of public intervention buying, the institution of a private storage regime, and compensation through higher headage payments. This was intended to increase domestic beef consumption by raising prices, whilst also enabling non-subsidised export of EU beef and veal to third countries. The first of these aims has been seriously compromised by the spread of BSE to member states that had previously been free of the disease, notably Germany where consumption of beef fell sharply in late 2000. Other measures included the targeted increase of milk quotas (favouring certain member states, young farmers and producers in mountain regions).

Under Agenda 2000 a system of 'cross-compliance' has been introduced whereby full receipt of arable area, headage and other payments is conditional upon respect of environmental standards (Beard and Swinbank, 2001). This will allow member states to make direct aid payments conditional on farmers fulfilling nationally defined environmental management standards, which, it has been suggested, farmers will be happy to meet (Russell and Fraser, 1995). Penalties or even complete withholding of payments can be made for non-compliance. A 'modulation' proposal will allow member states to redirect up to 20 per cent of compensation payments per farm, as a function of the level of employment on the individual farm or overall prosperity of the holding through funding shared with the European Commission, for 'accompanying measures' such as agri-environment programmes, afforestation and early retirement schemes (Falconer and Ward, 2000; Lowe and Ward, 1998).

Agenda 2000 also continued beef sector reforms introduced under MacSharry; price supports were reduced by a further 20 per cent, with compensation made through increased direct payments, a new slaughter premium and 'natural envelope' payments differentiated on a national basis. The impacts on farmers' incomes of these changes are disputed. For example, in the case of Ireland, it is claimed that they have led to a drop in on-farm margins, and a rise in cattle farmers reliant on off-farm employment to maintain incomes (Binfield and Hennessy, 2001). However, a more positive outcome is attributed for Scottish farmers (Mitchell and Doyle, 1996).

Although containing fewer reforms than many had wanted, Agenda 2000 did allow greater adaptation of the CAP to national and regional circumstances by incorporating certain elements of national discretion. In particular this has been built into what were termed the first and second pillars of the CAP (Table 5.10).

The pre-eminent role of agriculture within the rural economy has been challenged, as exemplified by the introduction of the Rural Development

---

Table 5.10 The pillars of the Common Agricultural Policy

The First Pillar
- Opportunities to apply environmental conditions where direct community payments are made
- National discretion in the application of a proportion of direct payments (the national envelope) to the beef sector (and eventually the dairy sector)
- National discretion to modulate the total CAP direct payments for individual farmers in order to increase expenditure on the second pillar

The Second Pillar
- The new framework of the Rural Development Regulation and its opportunities to promote the integrated and decentralised planning of agri-environmental, agricultural and rural development measures
- Eco-conditionality – agricultural aid to be made conditional on meeting environmental requirements
- Differentiation of aid paymemts – agricultural policy directed away from Pillar 1 towards rural development whilst limiting the amount of development aid paid to farmers

(*Sources:* Colson and Mathusin, J. 2002; Lowe et al., 2002)

Regulation (RDR) which requires that EU member states produce a Rural Development Plan (RDP). RDPs have as a key objective the delivery of vibrant rural communities which draw their livelihoods from a variety of activities, including farming, forestry, rural industry, services, tourism and recreation, and the provision of environmental qualities for the public good (Gray, 2000). These RDPs are likely to play an increasing role in determining the character of Europe's countryside as they will be used to shape specific policies, as illustrated in the UK's decision in 1999 to increase funding for agri-environment initiatives under the auspices of RDR. However, because of agreed funding arrangements, the proportion of CAP expenditure on rural development and accompanying measures will fall to just below 10 per cent in 2002.

Lowe *et al.* (2002) note that both France and the UK have strongly committed themselves to support of the RDR, in part because of their desire to reduce farmers' direct payments. The French are using modulation to achieve a more equitable distribution of public money to farmers and to develop policy that extends beyond a focus on food production. The outcome will be cuts in payments to one farmer in eleven, though for most their direct payments will drop by less than 5 per cent. Around 1400 farmers, mainly in the rich northern cereal regions, will have their direct payments cut by the maximum 20 per cent (Lowe *et al.*, 2002, p. 9). These measures have been challenged by the main French agricultural union, the FNSEA. French modulation will refocus support on small and medium-sized farms and will transfer funds from agriculturally prosperous to marginal regions. In contrast, the UK has adopted universal modulation, with a flat-rate reduction of all direct payments, rising from 2.5 per cent in 2001 to 4.5 per cent in 2005. This universal approach is likely to bring greater gains to cattle and sheep farmers and will boost resources allocated to agri-environment schemes. In addition, both France and the UK are developing RDPs to benefit farmers, the former using voluntary farm management contracts designed to combine pre-existing and new aid schemes, Contrats Territoriaux d'Exploitation (CTE), with the triple objective of maintaining and improving the economic, social and environmental contribution of farming to rural areas (Buller and Brives, 2000).

5.4 Agriculture and the expansion of the European Union

One of the underlying rationales for Agenda 2000 was the prospect of further enlargement of the EU through the accession of new members, comprised primarily of Central and East Europe countries (CEECs). There are largely countries formerly within the Soviet sphere of influence, with communist regimes in government from the end of the Second World War until the early 1990s. Clearly these countries have had very different agicultural structures and management to that prevailing in the EU for the last 50+ years (see Table 5.11), and hence there are likely to be profound difficulties in bringing CAP rules and regulations to bear across such a varied set of national agricultural profiles (Rabinowicz, 2000).

If all the candidates for EU membership from the CEECs were admitted, the EU's surface area would increase by one-third and its population by 29 per cent (to around 500 million) (Map 5.2), but its GDP would rise by only 5 per cent at current exchange rates. Indeed, in terms of purchasing power, the average GDP per head in the enlarged EU would be 16 per cent less than in the 15-member EU. This compares with a fall of 6 per cent in the EU's average income following the admission of Greece, Portugal and Spain in the 1980s (Banse *et al.*, 2000). Agriculture is one of the key problems to address in any expanded EU as all the aspiring new member states have farm sectors that account for a higher proportion of the work-force but produce smaller output per worker than the current EU average.

Possibilities for transforming farming in the new member states via application of CAP subsidies are being ruled out on the grounds of expense. Instead there are contested political moves to investigate direct payments to the applicants to encourage some agrarian modernisation and structural reforms prior to wholesale reforms of the CAP later this decade. There may also be some

Table 5.11 Characteristics of agriculture in the CEECs, mid-1990s

Country	Agricultural employment as % of total employment	Agricultural trade as % of total imports	Food expenditure as % household income
Bulgaria	21.2	10.6	48
Czech Republic	5.6	9.6	32
Estonia	8.2	16.7	39
Hungary	10.1	7.4	31
Latvia	18.4	na	45
Lithuania	22.4	10.8	58
Poland	25.6	11.1	30
Romania	35.2	9.9	60
Slovakia	8.4	9.3	38
Slovenia	10.7	8.2	28
CEEC – 10	26.7	10.4	41
EU – 15	5.7	9.5	22

(*Data source*: Commission of the European Communities)

potential for the low-technology farming prevalent in several applicant countries to be a key element in the growth of organic production in the EU, though this presupposes sufficient bureaucratic efficiency to develop a certification system capable of dealing with the many smallholders who might be eligible. And the numbers involved are not inconsiderable, e.g. around half of the farmland in Bulgaria has never had chemical fertilisers applied to it (*Economist*, 19–25.5.01, 'A survey of European enlargement', p. 11).

The wealth gap between the current member states and the applicants may have several difficult to manage consequences should enlargement proceed. For example, land prices on either side of the Polish–East German border can vary by as much as a factor of ten. This could result in Polish farmland being purchased by wealthier entrepreneurs from richer parts of the EU. In particular, descendents of ethnic Germans, expelled from areas that are now part of Poland or the Czech Republic may wish to 'reclaim' their heritage. Although enlargement would liberalise agricultural trade across the enlarged entity, this may not be of significant benefit to the new members in the short term as their farm sectors are too inefficient and unmodernised to take advantage of the opportunities afforded. Aid for modernisation will also be restricted by EU rules that no country can receive

more than the equivalent of 4 per cent of its GDP in EU aid, and thereby assuring that, unless the regulation is removed, the poorest countries will receive the least aid in absolute terms.

One of the key concerns emerging from the preliminary discussions between the EU and the CEECs is how to harmonise policies across the enlarged membership. With specific reference to agriculture, the increased variation in agricultural circumstances will be extremely difficult to manage (see Hartell and Swinnen, 2000). However, in the decade following the demise of Communist rule in the CEECs there have been dramatic changes in the structure and management of farmland throughout the region.

5.4.1 Land reforms in the Central and East Europe Countries

The need to modernise the structure of farming has been a feature of agrarian reforms in many parts of the world. For example, in the early years of the twentieth century revolutionary political change in Mexico was accompanied by the transference of land from large estates (*haciendas*) to landless labourers to work in collectives. Similar widespread collectivisation followed the revolutions in both the Soviet Union and China. However, the Developed World has not been immune

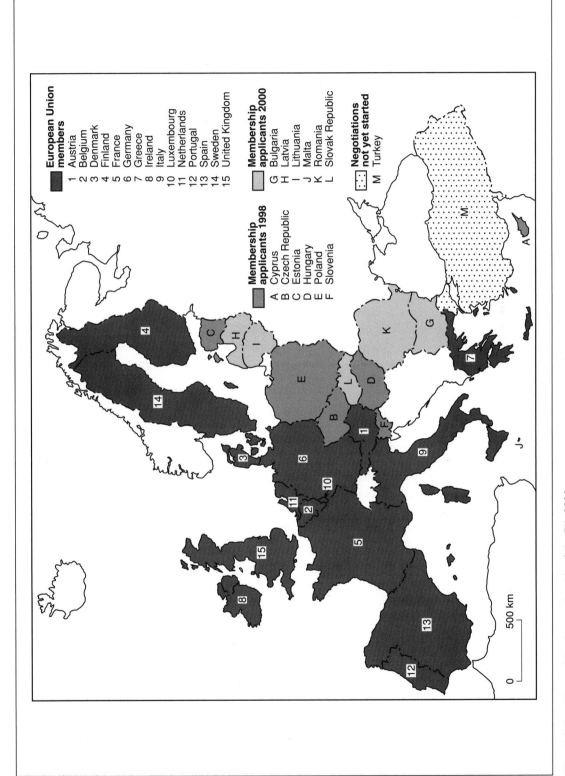

Map 5.2 Countries applying for membership of the EU, 2002

European Union members
1 Austria
2 Belgium
3 Denmark
4 Finland
5 France
6 Germany
7 Greece
8 Ireland
9 Italy
10 Luxembourg
11 Netherlands
12 Portugal
13 Spain
14 Sweden
15 United Kingdom

Membership applicants 1998
A Cyprus
B Czech Republic
C Estonia
D Hungary
E Poland
F Slovenia

Membership applicants 2000
G Bulgaria
H Latvia
I Lithuania
J Malta
K Romania
L Slovak Republic

Negotiations not yet started
M Turkey

from such change, with the enclosure movement in England representing widespread modification to farm structures extending over several centuries. In other parts of Western Europe more modern reforms have had to tackle the disparities between large and small holdings, concentrations of land in very few hands, and fragmentation of holdings associated with generational transfers of land under the Napoleonic Code, notably in France (Clout, 1988; Jones, 1989a; 1989b), southern Italy (the Mezzogiorno) (King, 1987), Spain (Jones, 1984), Greece (Keeler and Skuras, 1990) and Portugal (Unwin, 1987; 1988). Hence there have been programmes of reform in several countries that have endeavoured to tackle these problems. Some schemes have simply encouraged farm consolidation whilst others have promoted more wholesale changes in the arrangement and management of farming (Robinson, 1994a, pp. 189–210). However, in the last two decades a more rapid process of reorganisation has occurred in the CEECs as communist regimes have been replaced by governments pursuing more capitalistic policies and, in many cases, preparing for accession to the EU and the umbrella of the CAP.

The farming system of Eastern Europe had been dominated by production cooperatives and state farms, but with the move to multi-party democracy from 1990/91 centralised planning has been largely replaced by deregulation, liberalisation and privatisation, so that in several countries individual farms and new corporate farms are now the dominant type of operation (Table 5.12) (Hartell and Swinnen, 2000; Swinnen et al., 1997). The persistence of collectives, cooperatives and state farms in some countries reflects the extent to which these systems of production had become the norm and often their workers resisted attempts to alter them. Other forces of inertia have included the lack of expertise in running individual holdings, the lack of a risk-taking culture that is taken for granted by many Western farmers, insufficient capital, and the fact that many individual farms would have been too small to be cultivated efficiently (Berger, 2001; Swinnen and Mathijs, 1997).

Throughout the CEECs and the Soviet Union agricultural policy emphasised collectivisation and agri-industrial integration, controlled prices and margins, and rural industrialisation. In pursuing these policies the state was the dominant actor, with production dominated by either collective farms (*kolhozes*) or state farms (*sovkhozes*). However, small plots of land operated by the households of workers on these farms, or associated with second homes of city dwellers, have been of significance. Caskie (2000) estimates that 60 per cent of Russian households produce a significant proportion of their own food needs, including three-quarters of all fruit and vegetable production. A similar dualism of small private plots and large collectives has been reported elsewhere in eastern Europe, including the Baltic States (Alanen, 1999; Kostov and Lingard, 2002) and in China post-Mao (Veeck and Shaohua, 2000).

The logic behind the collectives, in addition to their furtherance of communist ideology regarding state control and ownership, was that they created scale economies, offered more scope for

Table 5.12 Distribution of farmland by organisation in selected CEECs, 1998

Country	Collective/cooperative farms	State farms	New corporate farms	Individual farms
Albania	0	20	0	80
Bulgaria	42	6	0	52
Czech Republic	43	2	32	23
Slovak Republic	60	15	20	5
Hungary	28	4	14	54
Romania	12	21	0	67

(*Source*: Burger, 2001, p. 262)

Table 5.13 Arguments for land reform in the CEECs

Argument	Rationale	Mechanism	Conflicts
Historic justice	Land should be returned to 'rightful owners'	Restitution	Choice of date for basis of restitution
			Rules of division, exit and provision for workers
			Abandoned land
Economic efficiency		Privatisation, auctions	Division of lots
			Inequalities and social tensions
			Foreign ownership
Food security	Land provides means for self-consumption and food security	Distribution	Basis for distribution
			Transaction costs
			Inefficiency of subsistence farming

(*Source*: Gorton, 2001, p. 270)

mechanisation and modernisation of peasant/feudal production systems and hence also increased productivity. However, it was frequently the case that these potential benefits were simply lost through combinations of overly 'bureaucratic management, prescribed forms of organisation, mistaken central directions, targeted investment and production, lack of interest, and forced farm mergers creating excessively large farms to take advantage of size' (Berger, 2001, p. 261). Gorton's (2001) summary of the arguments favouring reforms of these collectives is shown in Table 5.13.

Decollectivisation throughout the CEECs has involved transfer of property rights from state and collective farms to individual or corporate ownership. This has been based on claims for historic justice, improving agricultural efficiency and ensuring food security. The most popular method of land reform for collective farmland has been through restitution to former owners based on ownership structures from over 50 years ago (Swinnen, 1999). However, this has recreated the fragmentation of pre-collectivisation and has been associated with many legal wrangles over ownership. The fate of collective farmworkers has been highly problematic within the redistributions of land.

The rapid economic transition brought dramatic falls in GDP (up to 40 per cent in some cases) in the early 1990s, followed by recovery in the middle of that decade. The magnitude of the decline and subsequent recovery has varied according to the extent of the previous economic weakness, but a common factor has been a diminished contribution of agriculture to GDP as other sectors have benefited more from the new economic and political conditions. Agriculture has been disrupted by modernising reforms including privatisation and changes to land tenure (Young, 1993). Initially, like the economy as a whole, these reduced output, but in most East European countries there has subsequntly been some recovery, generally based on the creation of family farms alongside remnants of the former collectives and state farms (e.g. Alanen, 1999). Hence there is an ongoing dualism within the farm-size structure. Unwin (1994; 1997) illustrates this for Estonia where the number of private farms rose from zero in 1991 to 10,153 just three years later. Meanwhile the collectives and state farms were reorganised in the same period from 394 units to 1013, with many of the latter run under some form of co-operative arrangement.

The land reforms have taken longer to implement than was forecast in the early 1990s and they have had very mixed results (Swinnen, 1997; 1999; Swinnen *et al.*, 1997; Swinnen and Tangermann, 1999). Gorton (2001) illustrates these problems associated with land reform with respect to Moldova, one of the East European countries most reliant on agriculture at the time of political upheavals in 1990 (50 per cent of the active labour force employed in agriculture, and agriculture and food industries accounting for 40 per cent of GDP). Here a policy of small-scale land privatisation and various de-collectivisation initiatives have been pursued despite political resistance to the latter.

There is insufficient room to examine the reform process in detail, but to illustrate the types of dramatic reforms introduced and their traumatic aftermath, examples are taken from two of the leading aspirants to EU accession, Poland and Hungary.

5.4.2 Land reforms in Poland

In parts of eastern Europe land reforms have shifted radically between calls for collective and private modes of production to enhance perform-ance. For example, in Poland, following libera-tion from German occupation in 1944, land was redistributed from large estates to landless peas-ants and/or new settlers. Family farms became the dominant type of farming despite collectiv-isation programmes, which started in 1948. In contrast to most other countries under Russian influence, private farming remained viable so that by the end of the 1980s the private sector owned 75 per cent of the arable land in Poland, with high concentrations in the central and south-eastern regions. Nevertheless the number of family farms rose and fell in accord with policy shifts, espe-cially between 1948 and 1956 when large and medium-sized farms were often incorporated administratively into collective cooperatives. Some of the latter were then dissolved and new private farms established as 'peasant-worker' farms aver-aging 2–3 ha.

Further reforms in 1974 made it difficult for private farms to expand as, when possible, arable land was transferred to the State Land Fund to be cultivated by collective farms or newly created cooperatives and state farms. This reform was intended to improve the condition and structure of Polish agriculture by increasing crop specialisation.

The area in state farms rose by one-quarter as a result and 300,000 small farms were eliminated (Whitaker, 1984, p. 177). This trend towards fewer farms was maintained through the 1980s because of the flight from the land associated with low incomes from farming. Rationalisation occurred within the private sector as the average farm size in this sector rose from 5.4 ha in 1980 to 7.1 ha in 1990 (Loboda et al., 1998, p. 101).

In the early 1990s neo-liberal farm policies were introduced, with greater emphasis upon the role of the market economy. This contributed to a sharp decline on the state farms where out-put of the staple cereal and potato crops declined dramatically. In many regions land on state and cooperative farms was sold, with a newly estab-lished Agricultural Property Agency, an arm of the State Treasury, taking over the assets of the 1600 state farms, or 3.7 million ha, for sale or lease (one-fifth of the country's arable land). How-ever, lack of available capital and doubts about long-term returns on investment have meant that sales of this land have been limited. Nearly two-thirds are currently being leased by the private sector, though the average size of private farms is still under 10 ha, and Polish agriculture in general lags well behind Western Europe in terms of its use of modern equipment and methods (Loboda et al., 1998).

Despite the reforms, compared with the situation in the EU, Polish agriculture still has a much lower level of production intensity, with low crop yields and lower productivity in the dairy sector (3136 litres per cow per annum in Poland and 5271 litres per cow per annum in the EU). Around 80 per cent of Polish dairy farms keep five cows or less and only two per cent have more than 20 cows (Piskorz, 2000, p. 91). Accession to the EU will be crucial to modernising this and other farm sectors. For example, if Poland receives a favourable milk quota (around 16 million tonnes) milk production could grow by 18 per cent within two years of accession (Piskorz, 2000, p. 92). Poland could also become a major pro-duce of pork and poultry within the EU, though competition with other EU producers would probably lead to the elimination of small Polish producers.

5.4.3 Land reforms in Hungary

State control of agriculture went further in Hungary than in several of its East European neighbours so that by the early 1970s cooperatives administered 80 per cent and state farms 14 per cent of the arable land. The remainder was on very small private farms or plots cultivated by non-agricultural workers. The average size of cooperatives was 2000 ha of agricultural land, with 450 members and employees. State farms averaged 4900 ha, with 900 employees (Suli-Zakar *et al.*, 1998). From the mid-1960s these farming operations ceased to follow targets set by central planning authorities, so there was independent planning of production and development activities. This helped increase investment, credits and subsidies to farming, facilitating technological change based on West European and North American farm technology. A strong degree of state control was maintained through national trusts which dealt with the processing and distribution of agricultural products and which were the sole agents for the sale of agricultural products.

Despite the size of the collectivised farming sector, small-scale individual farmers continued to play a significant role in food production, contributing between one-quarter and one-third of total production despite occupying less than one-seventh of the land operated collectively (Keefe *et al.*, 1987). With restrictions on the amount of farmland that could be farmed individually, this meant that such land tended to be cultivated intensively, and hence the predominance of vineyards, orchards, vegetables and fodder crops for household animal husbandry. Much of this production supplemented low rural incomes whilst diversifying food supplies in the country. The juxtaposition of small- and large-scale farming helped retain population in some rural areas. Where conditions for farming were less suitable, however, depopulation occurred.

During the 1980s it proved difficult to sustain both the technological advancement of agriculture and its profitability. Capital for investment was sought from exports. However, as production costs remained high in Hungary compared with the falling prices of agricultural products on the world market, it became harder for Hungary to export its agri-commodities. Cooperatives sought to diversify into non-agricultural sectors in order to maintain income, and this detracted from the attention devoted to agriculture. Meanwhile, legal restrictions on business and land acquisition restricted the emergence of private entrepreneurship.

Hungary's agricultural sector pre-1989 did enjoy some measure of reform not found in most CEECs (Mathijs and Meszaros, 1997). This included partial liberalisation of prices, abolition of central plan directives and significant independence of decision-making for farms. Most new cooperatives received state support and this helped to mechanise arable production between 1965 and 1975. Moreover, from the early 1970s there was greater promotion of the use of household plots and contracting out of grape and fruit cultivation to individuals and families (Symes, 1993). By the late 1980s, most production cooperatives pursued a mix of production, procurement, sales, services and processing. This helped a significant number of them to survive the reforms of the 1990s.

Transformation to a market economy began in 1989 as trade between the Comecon countries collapsed. There was a chaotic period as the state withdrew much of its support for and involvement in agriculture. Some processing operations were purchased by TNCs, but often these switched their sourcing to foreign suppliers. Loss of traditional markets, notably the shrinkage of the Soviet market, and the limited buying power of the domestic market adversely affected farming, as did land reforms based initially on ownership of property held in 1947. Cooperatives were compelled to put up some land for auction, either for purchase by the former proprietors or by cooperative members and employees. As a result some very small holdings were created, but it was primarily the cooperative members who obtained land in order to continue farming. The majority of the former proprietors had ceased to maintain direct ties to farming and had no desire to return to it. Many of these former proprietors leased their land to permit re-formation of cooperatives. This re-establishment of cooperatives was not successful in all regions, with problems occurring in peripheral and marginal areas (Agocs and Agocs, 1994). However, the practice of re-forming or creating new cooperatives was widespread, encompassing around 90 per cent of the original land area in cooperatives. The transformation effected in property relations is summarised in Table 5.14.

The economic decline and disruption of the reforms came to an end in 1994 when agriculture accounted for 6.4 per cent of GDP, as compared with 15.6 per cent in 1989 (OECD, 1996, p. 35). However, recovery was rapid, with exports to the EU increasing rapidly in the mid-1990s and trade within Comecon growing nearly as fast. The former was hastened by the 1994 Agreement between Hungary and the EU, which had the objective of integrating the Hungarian economy into the European market through tariff reductions and increased import quotas. This has helped stimulate Hungarian agriculture, which is now responding under the unique combination of land-leasing, re-formed cooperatives and a network of small producers.

Overall, one-third of the land of production cooperatives has been sold at compulsory auctions; one-third has remained in the name of members of cooperatives; and one-third has been redistributed to those cooperative members who did not own land earlier. This has effectively created millions of scattered parcels of land, only capable of becoming part of a modern mechnaised farming system at great expense and which cannot be productively cultivated. Thousands of new landowners have also been created, though many are retired and/or do not live in rural areas. In the weak land market many of these new owners have simply held onto their land whilst not utilising it productively (Berger, 1998).

As Berger (2001) notes, land tenure is much more concentrated than ownership, and some of

---

Table 5.14 Changes in Hungarian agriculture following reforms in the mid-1990s

Former collective agricultural land	m ha	%
Total	5.6	100.0
Allocated to private owners during communist period	1.9	33.9
Awarded as compensation	1.8	32.2
Divided between farm workers, cooperative members or their descendants	1.9	33.9

State farmland	ha	%
Total	925,925	100
Set aside for compensation	370,370	40
Distributed to employees	55,555	6
Retained in state ownership	500,000	54

- Compensation vouchers and land auctions have created over 600,000 new landowners, each acquiring on average 3.5 ha
- There are now 1.8 million owners of agricultural land; their proportion of total land increased from 22% in 1993 to 46% in 1996 to 54% in 1998
- 90% of agricultural land is in private hands
- 80% of land owned by individuals is leased out
- In 1995 farms over 100 ha accounted for 58% of agricultural land; farms of 5–100 ha for 20%; farms under 5 ha for 22%
- Individual landowners own on average a land area of 4.4 ha, but 11% own less than 1 ha and 60% less than 10 ha
- An additional one million people cultivate small household farms or gardens
- 45% of the land is cultivated by corporate farms, 96% of which are larger than 300 ha

(*Sources*: Burger, 2001, p. 263; OECD, 1996; Suli-Zaka et al., 1998, pp. 138–9)

the largest farms have survived either as renamed and restructured cooperatives or as different companies organised from the former cooperatives and state farms. This sector now cultivates 45 per cent of the land in the CEECs and a large part of the livestock. A consolidation of land has also occurred in the small-farm sector. A sample survey in 1998 of 11 counties (of the 19 in total) showed that 60 to 70 per cent of the land of individual farms is cultivated by farms larger than 50 ha. Farms under 50 ha are mostly run part-time or belong to the retired and unemployed who do not run this land for productive purposes. Hence it is the larger cooperatives and reconstituted state farms that stand to benefit most from the opportunities afforded by the umbrella of the CAP.

5.5 Macro-level change

5.5.1 Challenges to traditional farmer–state relations

Traditionally there has been a strong 'farm lobby' in many Developed Countries, having a significant influence on the policies of national governments, especially in many European countries. For example, it has had a direct influence on EU officials, and an indirect one by enlisting the support of affiliated EU lobbying organisations. There are close relationships between most national farm ministers and domestic farm organisations, and there is a long history of farm organisations delivering political legitimacy and political support for governments. In return, farm organisations have been allowed to participate in policy-making. There has been public support for this in some cases, notably France, via the 'agrarian myth' in which rural life is regarded as a vital ingredient in European culture, and the continuation of inefficient, small ('peasant') producers is seen as a national priority. For example in the mid-1990s the motto of the chief group organising rural protests, Coordination Rurale, was 'Tuer les paysans, c'est tuer la France' (Scargill, 1994).

The decline of the farm lobbies' political power, partly related to the diminishing contribution of

farming to national economies (Grigg, 1992b), can be seen in the advancement of reforms to the CAP in the 1990s, and also in the fragmentation of agricultural lobbies both in individual states and supra-nationally, e.g. the farm lobby in the EU was unable to block the conclusion of the Uruguay Round negotiations of the GATT (McMichael, 1993). Perhaps the main achievement of this Round was to establish agriculture as part of the agenda of multilateral negotiations and to promote a move away from open-ended price support in agriculture (Vanzetti, 1996). A negotiation framework covering four areas was established: internal support, border protection, export competition, and sanitary and phytosanitary regulations.

External pressures through these negotiations also played a role by promoting 'fair trade', thereby legitimising controls over marketing and wages. In turn this increased the power of the non-farm sector of the agri-food chain, particularly retailers and processors. However, as government regulation has waned so private interest regulation has grown, for example with retailers assuming greater responsibility with respect to food hygiene (Marsden and Wrigley, 1996).

In general terms the views of the EU farm lobby can be contrasted with those of most TNCs. The farmers have been largely opposed to further deregulation, fearing the greater competitive conditions that deregulation may bring. With wholesale deregulation, industrial countries of the EU, North America and Japan could be freed from a traditional but expensive alliance with national farm sectors, whilst also freeing controls on the investment and sourcing activities of transnational agri-food capital. Hence there is the prospect of agriculture in the Developed World being driven increasingly by a new international structure, with TNCs growing in importance and a smaller number of farm operators surviving. It is also likely that trade liberalisation will enhance the farm commodity exports of the USA and aid efficient producers of particular commodities (Ufkes, 1993).

Set against this interpretation is the argument by Atkins and Bowler (2001, p. 52) that the global influence of TNCs has been exaggerated. They contend that multi-domestics (characterised by international operations that are an extension of national operations and with central coordination and ownership of relatively autonomous overseas affiliates) are more dominant in the agri-food sector. In part, this reflects the fact that food is still largely supplied/sourced domestically and is processed and distributed by businesses operating independently of their parent companies. A good example of this is Outspan International, operating as a monopoly citrus producer in South Africa from where it has developed a global market strategy (Mather, 1999). Nevertheless, TNCs are making gains through the assistance of supragovernment organisations such as the WTO. For example, TNCs can appeal to the WTO over national restrictions placed on their commercial activities, and can then be rewarded as the WTO is constituted to uphold liberal trading principles. This has meant that the WTO may override the concerns of particular social groups and democratic institutions within nations and regions. In recent years this can be seen with respect to a range of issues, including the incorporation of genetically modified (GM) organisms into foods, animal welfare interests regarding the long-distance transport of livestock, and use of artificial growth stimulants in livestock. Furthermore, the gradual dismantling of national regulatory policies has given TNCs greater scope to develop their operations whilst removing food policy further from public scrutiny (Robinson, 1997a, p. 46).

5.5.2 Coloured boxes and multifunctionality

Despite the Agreement on Agriculture reached in 1994 at the conclusion of the Uruguay Round of the GATT negotiations, import barriers to and export subsidies on agricultural produce in the Developed World remain high, and indeed it can be argued that the agricultural trade policies throughout the industrialised countries remain very similar (Grant, 1993). From January 1995 it was agreed that tariffs would be reduced by 36 per cent over a six-year period, possibly boosting global agricultural trade by 50 per cent. However, the policies implemented by the leading economies remain subject to different degrees of discipline/ exemption according to their degree of trade distortion. These distortions are classified into different 'boxes'. Green box subsidies have no or minimal trade-distorting effects and so are exempted from any reduction. Blue box subsidies are linked to production factors but not to price or the volume of output, and are implemented under production-limiting programmes. The reforms of the CAP in 1992 moved from market price supports towards blue box payments. Amber box measures impact on trade and are subject to a 20 per cent reduction over the six years based on price and volume of output. However, these arrangements have not resulted in significant decreases in support for domestic agriculture in the USA and EU, where use of green or blue box area payments have risen. Meanwhile the EU and Japan have made extensive use of amber box measures. Overall, the rate of domestic agricultural support in the Developed World rose by US$361 billion from 1997 to 1999 (Landau, 2001).

In terms of the extension of exemptions the EU has argued that agriculture is 'multifunctional' in that, as well as having a food producing role, it also has major economic, environmental and social roles in rural areas (Skogstad, 1998). Included in this term are the generation and management of rural landscapes and ecological features, and the social role of supporting population in peripheral areas. Reforms to the CAP and to trade policy therefore have to accommodate this multifunctional role of agriculture (Potter and Burney, 2002; Redclift et al., 1998). This role may seen as another pillar of agricultural policy, following those for general market support and rural development policy. Hence the EU argues that multifunctionality is a specific item to be considered in negotiations and, specifically, an exemptable item, and there is a desire to have this recognised within the WTO as one of the agricultural policy objectives to be pursued as a non-trade distorting set of measures. Meanwhile, the USA and the Cairns group of agri-export nations contend that the multifunctionality argument allows the EU to circumvent the subsidies discipline.

Removal of agricultural support in the EU has been proposed for some time, though with various forms of initial compensation suggested to shield agriculture from the immediate losses that would result (e.g. Nash, 1965; Marsh, 1982). These compensation payments would be 'decoupled' from production decisions of those wishing to remain in agriculture, though other suggestions such as the producer entitlement guarantee advocated by Blandford et al. (1989) are not fully decoupled. Decoupling was a central part of the Uruguay Round, referring to support payments not linked to production and which, in consequence, had no impact on trade and would not be subject to the new GATT (later WTO) disciplines. The Agreement on Agriculture of April 1994 specified reductions in the Aggregate Measurement of Support – of 20 per cent over a six-year period for Developed Countries and a smaller reduction and ten-year implementation period for Developing Countries. However, various exemptions were identified under different headings (so called green box and blue box payments) primarily to encompass the USA's deficiency payments scheme and the EU's MacSharry compensation payments. However, in the USA the Federal Agricultural Improvement and Reform (FAIR) Act of 1996 prompted a substantial decoupling of payments from production for wheat, maize, grain sorghum, barley, rice and upland cotton. With FAIR expiring in 2002 and a reversion to stronger price and income supports looking likely the green and blue box exemptions remain significant.

5.5.3 Some comparisons with US policy

The reforms to the CAP carried out in 1992 and 2000 have some parallels to changes in American agricultural policy. Agenda 2000 bears some resemblance to the FAIR Act of 1996 (Guyomard et al., 2000), in which the US government proposed to progressively reduce production supports over a seven-year period (four years for dairying) and terminate set-aside programmes. This held the prospect of reinforcing American pressure upon the EU in future negotiations regarding the levels of support for farmers within the EU. Meanwhile, the doubts about the viability of existing EU agricultural policies in the face of new members from the CEECs are prompting suggestions for more radical reforms to the CAP than those previously pursued. Indeed, 'decoupling' of government from supporting farmers may be pursued by the EU as a

means of encouraging a polarisation between highly intensive production, where conditions are optimal, and other areas, where different objectives are followed, perhaps with issues such as sustainability and the social capital of farming playing a more significant role. In this scenario, most of the farmers in upland areas of Europe 'would become entirely dependent on increasingly sparse welfare payments for environmental or local economic activities' (Marsden, 1998b, p. 269).

Described by Potter (1998, p. 161) as signalling that the US intended to pursue a strong line with respect to the definition of green box subsidies and the paring down of agricultural support, the FAIR Act comprised three dominant features:

1 Decoupling of income support policies for major crops, in which the link between income support payments and farm prices was largely removed. In place of target price/ deficiency payment provisions, producers receive payments (over a seven-year period during which payments gradually decline) that do not depend on type or amount of crop produced nor on the level of the market price. Furthermore, the Acreage Reduction Program (ARP), which restricted the acreage that participants could plant to any single programme crop, was abolished and voluntary set-aside provisions were ended. However, this decoupling did not apply to all products, notably it excluded dairy products, sugar, peanuts and tobacco. Certain loan deficiency payments were also not decoupled from production and market conditions.

2 Reduction of export subsidies. Export subsidy programmes remained largely unchanged, though their funding was reduced and some programmes were removed, e.g. the Sunflower Oil Assistance Program and the Cotton Seed Oil Assistance Program. Export Credit Guarantee programmes were maintained, but with changes to target emerging, but riskier, markets; to reduce domestic content requirements, allowing for a broader range of products to be exported; and to enhance the potential for exporting US high-value and value-added products.

3 Existing conservation programmes have been maintained and new ones introduced. These include efforts to improve the targeting of the CRP, to focus on areas most susceptible to erosion. The maximum CRP area was fixed at 14.7 million ha (compared with 15.4 million ha in 1990). New conservation programmes addressed limitations within the CRP (Lant and Roberts, 1990) and included the Environmental Quality Incentive Program (EQIP), with initial funding of US$1.3 billion over seven years, providing technical, educational and cost-share assistance to reduce soil, water and related resource problems. Half the funding must be used for environmental concerns associated with livestock production.

Guyomard *et al.* (2000) argue that the FAIR Act should force EU agricultural reforms in the form of including in the green box compensatory payments for support price cuts, or at least making them more decoupled than at present. Hence one future possibility would be an increase in US exports of several agricultural products and strong US pressure for reduced EU trade barriers. However, Agenda 2000 did not decouple direct aid payments sufficiently, and has been followed by measures both in the EU and the USA that are likely to diminish rather than increase decoupling.

The new US Farm Bill (2002) could add US$5 billion per annum for six years to market supports for grain, cotton and milk producers whilst extending so-called 'safety net' payments to growers of peanuts, pulses, onions, honey, wool and mohair. One estimate is that production subsidies will total $190 billion over ten years (*The Economist*, 13.7.02, p. 36). It will partly repudiate the market-oriented direction in US farm policy as undertaken in the 1996 farm legislation (the 'Freedom to Farm' Act) prompted by the Uruguay Round GATT agreements and promising a series of benefits (e.g. Beach *et al.*, 1995). The level of farm spending will be increased by about half, focusing most of the payments on large-scale producers of cotton, rice, wheat, maize, soybeans and minor grazing and oilseeds. Additional payments will be made that supplement existing guaranteed annual per-unit payments to growers and to 'loan deficiency payments' which are made when market prices fall below the level at which USDA crop financing loans are extended. For the first time regular deficiency payments to dairy farmers will be applied for a three-year period. However, there will also be substantial increases in funding for agri-environmental and rural development projects and a reduction in the ceiling on individual farm subsidy payments (from US$460,000 to US$360,000). In addition, a new protectionist element will be added, through the enforcement (in 2004) of rules to require labelling to show the country of origin of red meat (but not poultry), fish, peanuts, fruits and vegetables. The approval of the Bill by both Republican and Democrats in Congress indicates that neither party wanted to risk being labelled an 'anti-farm party' in advance of November 2002 elections to decide which party controls Congress.

Meanwhile, in July 2002 proposals for further CAP reform from the European Agriculture Commissioner, Franz Fischler, have met with a strong rebuttal from the member states. In line with the thinking in the US FAIR Act, Fischler proposed to decouple subsidies and production and to redirect more CAP spending towards environmental and rural development projects. There were five key elements in these proposals:

- removal of the link between subsidies and production for beef, sheep, oilseed and cereal farmers;
- introduction of a ceiling of €300,000 for subsidies accruing to a single farm;
- linking of aid to farmers to environmental and food safety standards;
- introduction of a staged 21 per cent cut in direct payments to farmers, with the smallest farms exempt from cuts;
- retention of the overall farm budget at €40 billion per annum.

These proposals were rejected immediately by France, the biggest beneficiaries from the subsidies, and, in an agreement between France and Germany in October 2002, the reforms were blocked, with the crucial exception of point (5). As a result, from 2007 the amount of money going into farm subsidies will be pegged at 2006 levels. From 2007 spending will be capped and will not increase beyond the rate of inflation up to 2013. However, this limit on spending may hasten the accession of new member states, who will have access to a share of the funds: probably one-quarter of the direct subsidies initially.

Other critical elements in ongoing agricultural policy in both the USA and the EU have been linked to external tariffs. The USA has consistently utilised measures aimed at reducing direct competition with home-grown produce whilst the EU has operated a system of import tariffs and export subsidies. Although during the 1990s EU farm export subsidies fell from 30 per cent of the CAP budget to 8 per cent, recent EU proposals to reform these measures have been strongly attacked as 'cosmetic', e.g. reductions in export subsidies which will not take effect until 2013, excluding direct payments to farmers from consideration when discussing reductions to trade-distorting subsidies, and excluding key exports such as sugar, dairy products and beef from consideration. Hence, recent policy trends in both the USA and the EU promise little scope for both the elimination of subsidies for production and elimination of protective tariffs. Criticism of these 'retrenchments' has been strong from Developing Countries and overseas aid agencies, though some 'green lobbies have been partially placated by the increased spending promised on both sides of the Atlantic for environmental measures associated with agriculture. These are considered in more detail in Chapter 10.

6 Specialisation and diversification

North American agriculture in particular has pioneered the development of an industrial structured production model to such good effect that the contemporary global agri-food system, comprised of agri-technology industries and heavy government involvement in agriculture markets, is central in the USA and Canada. As illustrated in Chapter 3, key characteristics of this industrialisation of agriculture, especially in the Developed World, have been various aspects of intensification, concentration and specialisation. These can be contrasted with the opposite set of processes – extensification, dispersion and diversification respectively – that some contend is indicative of post-productivism (Ilbery and Bowler, 1998). Elements of all these processes may be present in any given region and hence the industrial character of North American agriculture does not dominate all sectors or regions, and indeed its progress is being slowed in places by responses to growing consumer mistrust of the quality of industrially produced food products.

Whilst the polar extremes are an over-simplification of the range of processes operating within contemporary farming systems, there is a large geographical literature that has examined aspects of one of these dichotomies in particular, namely that represented by specialisation versus diversification. In many respects it is this dialectic that has been used most frequently to illustrate, first, the transformation of farming activity during the second half of the twentieth century, and second, the response of farmers to the growing 'crisis' in their industry in recent decades. Given the emphasis upon this particular dichotomy, it is worth exploring it further by means of two sets

of examples: one dealing with the development of specialisation par excellence, primarily in the farming systems of North America, and the second with the growth of farm diversification in Western Europe. In both cases examination of the processes involved illustrates how an inter-weaving of different economic, social, political and environmental factors has shaped the nature of agricultural change and, in turn, how geographers have responded to this in their studies.

Underlying the two processes is the fact that agriculture is no longer at the heart of the economy in the Developed World. So, whilst much primary produce, such as wheat and meat, is still in high demand, occupies large amounts of land, and is produced in large volume, it directly employs only a tiny fraction of the modern labour force (Grigg, 1992a, pp. 24–31). Instead it is activity 'downstream' from the farm that generates employment and wealth, both in an ever more highly differentiated manufacturing sector and in the service sector. Both downstream and upstream activities have altered dramatically in the last half-century, but work on farms has also been transformed, with major impacts upon the composition of the labour force, the size of farms, structure of ownership, types-of-farming, and the nature of farm work.

Whilst these impacts have permeated farming systems in most parts of the world post-1945, key differentiating features remain in the form of variations in farm structure, organisation and type-of-farming. Thus the impacts of change have varied substantially even within the Developed World. For example, despite the massive shedding of farm labour that has occurred, the Mediterranean countries still have a larger farm labour force with a greater predominance of smallholdings and fragmented farms than in other parts of Europe. The structure of farming also varies widely, especially between the monocultural grain-producing regions of the New World and mixed farming regions in Western Europe. These variations will be examined below, specifically with respect to the characteristics of specialisation and diversification.

6.1 Specialisation

At the heart of changes to commercial agricultural operations post-1945 has been increased output through specialisation. This pre-dates World War Two, but it has accelerated in recent decades, largely through applications of capital to promote technological change. This in turn has favoured certain types of farming, for example irrigated agriculture in parts of the American Great Plains, reducing the influence of market access upon location (Visser, 1980). The growth of specialised production systems has diminished the extent of mixed farming, with its harmonisation of crop and livestock production, thereby restricting the operation of joint economies and incurring higher environmental costs (e.g. Lawrence, 1987). This can be seen in both North America and Western Europe: in the Paris Basin and East Anglia, for example, intensification has been signified by heavy investment in machinery, which has only proved profitable through use of larger areas under cereals and accompanying break crops such as sugar beet and oilseed crops (Robinson, 1988a, pp. 185–205).

6.1.1 Crop production

Specialisation has had profound spatial consequences, as shown in the USA where specialist production of cotton, tobacco, peanuts, rice and sugar cane has helped concentrate these crops in environmentally favoured 'islands' in the southern states, eclipsing less competitive areas which have poorer quality crops (Hart, 1978). This has led to substantial areas passing out of farming altogether: nearly 40 million ha in the Piedmont area of

Georgia and the Carolinas between 1939 and 1974 (Hart, 1980).

The drive towards increased specialisation can be seen in the Corn Belt of the USA where what was once a mixed farming area has been transformed into one specialising in production of grain for sale, usually by combining maize and soybeans in rotation. The latter crop has now become the second crop of the Corn Belt and is especially important in north-central Iowa, the Grand Prairie and the Maumee Plain. The traditional reliance on livestock has been replaced by specialisation in corn and soybeans for direct sale as cash crops, as the Corn Belt has become a specialised feed-producing region for new livestock-producing areas emerging elsewhere. Family farming has remained the norm, but the size of holdings has doubled during the last four decades whilst the landscape has been altered radically to accommodate modern machinery. The number of tractors and combine-harvesters in this area increased by 50 per cent between 1960 and 1990 whilst the size of individual machines also increased and yields doubled. Hart (1986) dates the transformation of the Corn Belt to 1933 and the introduction of hybrid maize seed, but 'take-off' did not occur until the 1950s and the application of agricultural chemicals and machinery to raise yields (Kenny et al., 1989).

The process of increased holding sizes in the Corn Belt has followed trends similar to those for other parts of the Developed World. Small family farms have become uneconomic and have either been purchased by neighbours or added to other farms via leasing arrangements. Family farming has survived with fewer holdings and through use of estate tax laws, which have encouraged incorporation and growth of more elaborate corporate structures. However, these have largely been devices to facilitate inter-generational transfers of assets within a family. Indeed, it is often claimed that farming is an exception to the general tendency for large-scale corporate operations to increasingly dominate production. This is at the heart of the so-called 'agrarian question' introduced in Chapter 2.

The increased output associated with specialisation in the Corn Belt has necessitated greater reliance on overseas markets, with over one-third of both the maize and soybean crops being exported. This has made farmers much more vulnerable to global recession and to fluctuations in the world grain market. Hence a direct product of technological success and innovation has been to render farmers more vulnerable to global economic forces despite protective measures adopted by the US government (Munton, 1988). Another consequence of the advance of both specialist production systems and the use of sophisticated technology has been negative environmental consequences, including soil loss, pollution of watercourses and loss of wildlife habitats.

Specialisation has also contributed to dramatic growth in other sectors of arable farming, notably in the highly labour- and capital-intensive sectors of horticultural production. For example, Lee (2000) refers to the worldwide expansion in the production of hardy nursery stock (hns). In the UK, for example, the output of hns in 1968/9 was £17 million, based on 6325 ha. By 1994 the value of output had increased by almost 1500 per cent, to £269 million (at current prices) based on a land area that had increased by only 32 per cent. The level of output is now comparable with that of sugar beet in the UK, but more than that of commercial fruit production in the country. This expansion has been stimulated by growing consumer demand for hns and other horticultural produce, accompanied by the advance of new forms of retailing, notably the growth of large-scale garden centres and do-it-yourself (DIY) stores with substantial gardening sections, e.g. B&Q, Focus and Homebase. For hns and cut flowers, mail-order business has also been important, e.g. Flying Flowers, the Jersey-based mail order and retail services group specialising in hns (Lee, 2000, p. 143).

There has been a consumer-led growth in the hns industry throughout the Developed World, based primarily on the emergence and rapid development of container-grown stock. Hence the growers/farmers have specialised to take advantage of the changing market conditions, but have largely been subsumed into a marketing system in which price control is dominated by the large retailers and wholesalers, a common factor in many sectors of modern agricultural production.

One of the most highly specialised areas of crop production in North America has been Canada's Prairie provinces. Here the combination of the physical environment (especially on the Dark Brown and Black chernozem soils) and favourable government support produced one of the world's major grain producing areas from the late nineteenth century. The key legislation was the establishment of the Crow Rate (the 'Holy Crow'!), a concession made by the Canadian Pacific Railway in 1897 to reduce rail freight rates on specific products, especially cereal crops such as wheat transported to Thunder Bay on Lake Superior. Subsequently the concession was extended to more routes and to other railroads and was made statutory on grain and flour in 1925. It had the effect of shortening the distance between cereal growers and their markets, and gave rise to a dense network of railroads and grain storage depots (elevators) across the Prairies (Carlyle, 1991). In the early 1950s there were over 5000 grain elevators and over 2000 grain delivery points. The Prairies, and especially southern Saskatchewan, became known as Canada's breadbasket, with large farms, strong cooperative farming organisations and advanced methods of production typified by rapid mechanisation and use of purchased additives post-1945. In the early 1980s 36 million tonnes of Prairie grain were marketed, of which wheat accounted for just over two-thirds. Specialisation on grain was reinforced by agribusiness development and economies of scale, though mixed farming and ranching were present in more northerly and drier areas respectively (Map 6.1a).

Despite remaining as a major grain producing region, the Prairies has responded to globalisation and other changes in the last 20 years by some dramatic production changes (Carlyle, 2002; Seaborne, 2001), generally advancing industrial models of production (Gertler, 1999). A crucial stimulus has come from the elimination of the Crow Rate, partially through gradual reductions in subsidy after the 1983 Western Grain Transportation Act, and finally in 1996 when farmers received a one-off compensation payment (worth Cn$550 million overall) based on acreage and land quality (Ramsey and Everitt, 2001). Large stretches of rail track and elevators had already closed as

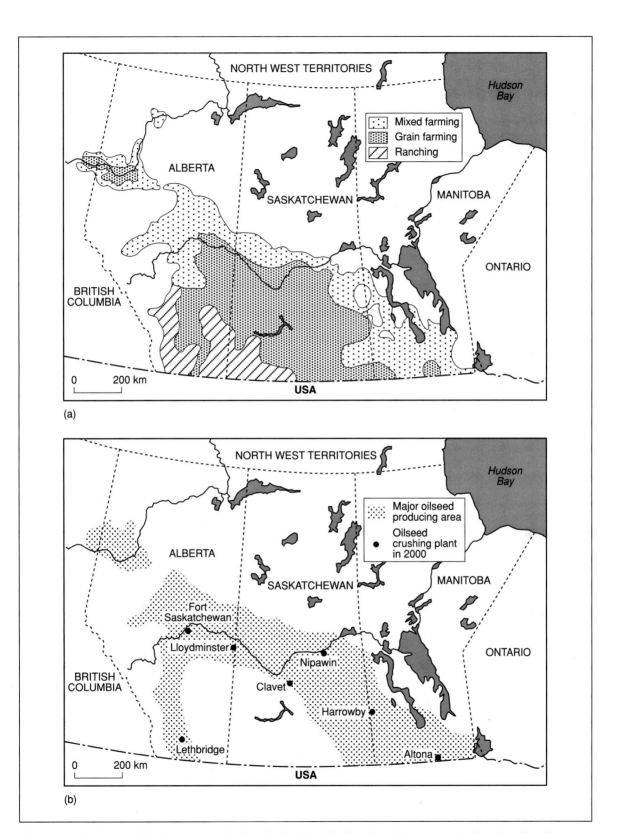

Map 6.1 (a) Distribution of farming types on the Canadian Prairies; (b) oilseed producing areas and oilseed crushing plants on the Canadian Prairies (based on Seaborne, 2001 and Carlyle, 2002)

farmers started to use road haulage, but the loss of the rail subsidy, coupled with falling prices and rising production costs, has produced an exodus of farmers and helped promote diversification (Epps and Whitson, 2001).

In some cases there has been diversification into speciality crops such as field peas, lentils, mustard seed and canary seed, with Saskatchewan dominating Canadian production of all four crops (69%, over 95%, 82% and 86% respectively of the national total) (Seaborne, 2001, p. 155). Peas are a replacement for soybean meal in livestock feed as a protein supplement (especially for pigs and poultry). Canada is now the world's leading producer; it is also the largest exporter of lentils, competing strongly with North Africa. The growing of mustard and canary seed has been helped by the development of 300 specialist processing firms in Western Canada, extending the local agri-food network in a way not seen under grain production. However, the largest area converted from grain has been to the oilseed crops, canola (oilseed rape) and flax, although their yields are lower than wheat. New varieties have been developed in Canada, well adapted to local soils and climate. The price of both crops has risen steadily against that of wheat over the last two decades, and the region has exported the oil from these crops (as salad and cooking oils, margarines) and a high protein livestock feed by-product from the late 1980s. Oilseed crops now occupy over 10 per cent of Prairie cropland (Table 6.1), having expanded steadily from the 1970s. This development has introduced new TNCs into the region in the form of oilseed crushing plants, of which there are now seven (Map 6.1b).

Despite these innovations and changes in farming practice (e.g. Carlyle, 1997) grain still occupies two-thirds of cropland on the Prairies, with wheat accounting for just under half (Table 6.1). The long-term mainstay, spring wheat, grown for pan breads and noodles, is gradually declining, though it remains a major export crop. Durum wheat (for pasta) and barley have been direct grain alternatives, the latter being grown primarily as livestock feed in Alberta where there are slightly cooler summers. Alfalfa and hay for fodder have also grown in importance as part of the expanding livestock industry in southern Alberta (Broadway, 2001; Carlyle, 1999), including beef feedlots and industrial-style pig production (Gertler, 1999).

Elsewhere in Canada new cropping specialisations have also occurred, notably in southern Ontario, where first wheat and then soybeans have become mainstays on arable holdings. By the mid-1990s soybeans had become the leading cash crop in Ontario, occupying just under one million ha (one-quarter of all the province's cropland) on one-third of the holdings (Keddie and Wandel, 2001). As with oilseed rape on the Prairies, it was the development of improved cultivars (higher yield, less time to mature, cold tolerance, lodging resistance, resistance to root rot and herbicide resistance) that helped the expansion of soybeans in Ontario, allied to developments in chemical weed control, inocculants and field equipment. It was the expansion of maize (grain corn) from the 1950s (Joseph and Keddie, 1981; 1985; Keddie, 1983) that paved the way for soybeans by developing a cash-cropping culture in an area where it was previously absent. Now soybeans are frequently grown in rotation with winter wheat and maize. This type of change, and that of oilseed crops on the Prairies, indicate how the combination of local factors (e.g. location-specific cultivars) and external factors (e.g. government policy, market forces) can radically alter patterns of specialisation.

6.1.2 Livestock production

The trend towards specialisation and greater economies of scale is well illustrated by livestock production in the USA where small numbers of very large producers account for significant proportions of the country's livestock products (Royer and Rogers, 1998). Livestock farmers have intensified traditional extensive activities by applying new systems of stock rearing and fattening. Hart and Mayda (1998) note that in 1949 three of every four farms in the USA had a flock of barnyard hens and a few cattle, two of every three had a milk cow or two, and more than half slaughtered a few pigs. In 1992 less than one farm in ten had any pigs, dairy cows or chickens, though

Table 6.1 Cropping patterns on the Canadian Prairies

	1961–6		1976–81		1991–96	
	000ha	%	000ha	%	000ha	%
Spring wheat	10,222	50.2	9,862	42.3	10,836	38.6
Durum wheat	590	2.9	1,539	6.6	2,021	7.2
Winter wheat	81	0.4	116	0.5	84	0.3
All wheat	10,893	53.5	11,517	49.4	12,941	46.1
Oats	2,932	14.4	1,888	8.1	1,432	5.1
Barley	2,504	12.3	4,592	19.7	4,463	15.9
Rye	224	1.1	303	1.3	197	0.7
Mixed grains	285	1.4	187	0.8	168	0.6
All grains	16,838	82.7	18,487	79.3	19,201	68.4
Alfalfa	977	4.8	1,702	7.3	2,414	8.6
Other hay/fodder	1,100	5.4	1,212	5.2	1,348	4.8
All alfalfa/hay/fodder	2,077	10.2	2,914	12.5	3,762	13.4
Oilseed rape/canola	468	2.3	1,026	4.4	3,285	11.7
Flax	794	3.9	396	1.7	533	1.9
All oilseeds	1,262	6.2	1,422	6.1	3,818	13.6
Speciality crops	143	0.7	303	1.3	1,151	4.1
Others	41	0.2	187	0.8	140	0.5
Total	20,361	100.0	23,313	100.0	28,072	100.0

(*Source*: Carlyle, 2002, p. 99)

over half kept beef cattle, which have become the favoured livestock of small farmers because of their low labour, feed and land demands.

The trend towards large-scale production has been dictated not only by farm-based economics but also by the demands of consumers and the processing sector. Consumers increasingly desire 'leaner cuts of meat in convenient ready-to-use packages of predictably uniform quality' (Hart and Mayda, 1998, p. 60). In order to meet this demand, the processors require a steady supply of animals of nearly identical size, shape and quality. They also prefer to have dealings with a limited number of producers capable of regularly delivering large numbers of 'standard' animals rather than a larger number of small producers whose supply is of more variable quality. In addition, at the processors' behest, animal geneticists have developed more prolific breeding stock that will grow faster and produce leaner meat on less feed. Such animals tend to be better suited to large holdings because they require a greater capital investment and more skilful management.

A good example of this tendency towards large-scale specialist livestock production is feed-lot production of beef, originating in the western USA, with capacities in the tens of thousands and reliance primarily on purchased feed (Hart, 1975, pp. 92–4). Developed initially in Colorado, and with major concentrations today on the Colorado Piedmont and the Platte River Valley, the current main foci are in the Denver–Omaha–Lubbock triangle, especially in south-west Kansas and the 'panhandles' of Oklahoma and Texas. This area has over seven million cattle, or one-quarter of the national total. In part, this reflects its position with respect to suppliers of feeder cattle from small Southern farmers and from Western ranchers. However, the development of centre-pivot sprinkler irrigation systems fed from deep wells has also enabled the farmers to increase their own production of feed crops such as maize, sorghum and alfalfa. The rapid development of feedlots in the area has also attracted the meatpacking industry, as the costs of shipping feed grains and cold processed beef have shifted in favour of the latter. So new plants, designed to process one type of meat highly efficiently, were established near the feedlots.

In 1959 there were 2618 meatpacking plants and 197,000 employees, with concentrations in major Mid-West cities, especially Chicago. In 1985 there were 1588 plants and 128,000 employees, with a move to small towns in the heart of the beef feedlot areas, especially in Texas and Kansas (Broadway and Ward, 1990). The major urban-based processing firms, Swift and Montfort, were being replaced by the new 'giants' of this industry, IBP, ConAgra, National Beef and Excel (who slaughter 84 per cent of the nation's cattle); and McDonalds had become the country's largest purchaser of beef. In 1968 McDonalds purchased its beef from 175 local suppliers; by the mid-1970s it relied on just five (Schlosser, 2002, pp. 137–8). ConAgra and Excel operate their own huge feedlots whilst IBP relies on contract arrangements with the country's largest ranchers and cattle feeders ConAgra has the largest meatpacking plant in the USA – in Greeley, Colorado – which is supplied from two feedlots, each of which can hold up to one hundred thousand head of grain-fed cattle. A high proportion of the workforce at the Greeley plant, and at many of the other new large plants, are recent immigrants from Latin America who are low-paid and non-unionised. A similar process of reliance on feedlots and large new processing plants has occurred in Canada, primarily in the province of Alberta (Broadway, 1997; 2000).

There have been related developments in the processing of milk in New Zealand (Robinson, 1988b; Willis, 2001) and Ireland. For the latter, Breathnach (2000; Breathnach and Kenny, 1997) distinguishes six characteristics of the growing involvement of TNCs in the dairy processing sector:

- elimination of privately owned creameries;
- amalgamation of dairy cooperatives into larger units;
- internal rationalisation of processing in these units;
- the latter includes the closure of branch creameries;
- internationalisation and a movement towards privatisation of the largest dairy processing cooperatives.

These changes have also helped to reduce the numbers of dairy farmers and converted many of those remaining into contract suppliers to large agribusinesses whose primary responsibility is to private shareholders and overseas operations.

Another related development occurring in similar areas has been drylot dairying. This originated near major cities in sub-tropical and tropical areas where cultivable land was in short supply, especially in irrigated oases, e.g. the Greater Los Angeles area in the 1950s (Gregor, 1963, p. 312). This form of dairy production relies on feeding the cattle hay, grain and concentrates, and high-volume throughput to offset costs of mechanised milking and feed. This system has been displaced from urban fringes by building development, with relocations concentrating in the San Joaquin and King Valleys.

Although both beef and dairy cattle are increasingly part of 'industrial'-type farming systems in the Developed World, the pig and poultry sectors have been the ones most frequently associated with the growing dominance of industrial methods applied to agricultural production. The term 'factory farming' has generally been associated with pig and poultry production, and 'battery' hens and 'factory' pigs have been symbols of this industrialisation. Yet in terms of labour process post-1945 pig farming lagged behind other livestock sectors. Whereas the production of most meat products shifted from independent producers to a vertically integrated system with either contract growers or corporate production, pig farming remained largely in the hands of independent producers until recently (Heffernan and Constance, 1994; Qualman, 2001).

Furuseth (1997) points out that in the USA most pig producers supplied high proportions of feed from crops grown on their own farm: in 1975 almost 80 per cent of the grain fed to pigs was grown on the same farm. Moreover 80 per cent of pig production, from farrow crates to finished animals for slaughter, occurred on the same enterprise as opposed to rearing on one farm and fattening/slaughter on another. Hence the majority of production was concentrated in the Midwest Corn-Belt states. This was essentially low-density production with a small number of pigs per unit, on average 650 head per annum produced in four groups around the cropping schedule.

The small-scale system has been transformed in the last two decades as high-density specialised production systems, pioneered earlier by the broiler industry, have taken over. In recent years operations with over 3000 pigs have become almost commonplace (Rhodes, 1998; Zering, 1998). However, many of these are now corporate-owned pig farms with much larger numbers of pigs per farm: Furuseth (1997, p. 299) refers to a farm in Utah planning a herd of 100,000 sows capable of producing two million pigs per year for the west coast market. The geographical shift in production accompanying these changes has seen North Carolina account for 37 per cent of the increase in the national herd on farms with over 2000 pigs between 1992 and 1996. Most of the expansion has occurred in states not known traditionally for pig production, e.g. Utah, Arkansas, Mississippi. Close coordination between packers and producers has enabled growth of both pig production and construction of new large packing plants in these states, e.g. the country's largest plant, at Tar Heel, North Carolina, owned by Smithfield Foods (Zering, 1998, p. 213). Larger contractors have tended to depend more on contract finishing than smaller contractors. Control of the industry by small numbers of producers is indicated by the fact that about 10 per cent of the producers market over 80 per cent of all commercial slaughter (Rhodes, 1998, p. 236).

Large-scale broiler farms have purpose-built factory-like structures housing thousands of birds, and have served as a model for other industrial farming systems. Broilers and six- to ten-week old chickens are raised in large 'houses', which are little more than oversize hen-coops. This type of broiler production has evolved from a multitude of small poultry farms and small processors, with industrial-style methods contributing to a concentration of both the raising of the birds and their processing (Benson and Witzig, 1977; Bowler, 1994). This has had the effect of transforming both the production and consumption of chickens. In the 1940s chicken tended to be a 'luxury' food eaten for Sunday dinner and on special occasions. Now it is the cheapest and most popular meat in the Developed World. Per capita consumption per annum of chicken in the US has risen from 20 pounds in the 1930s to 70 pounds in the 1990s. In part, though, this reflects the rise of the fast-food industry, with Kentucky Fried Chicken and McDonalds' chicken McNuggets respectively the first and second largest purchasers of chicken in the USA. In 1992 American consumption of chicken surpassed that of beef for the first time (Schlosser, 2002, p. 140).

Broiler chickens, turkeys and pig production have all become examples of vertical integration in which a single decision-making unit controls successive stages of production, either by direct ownership or contract. For broilers this has involved control over three stages: manufacturing the feed, feeding the birds until they are ready for slaughter, and their slaughter and subsequent marketing (Figure 6.1). In developing an integrated system of production, the example of Sun Valley Foods (SVF) in the English Midlands is fairly typical (Robinson, 1988a, pp. 244–7), though the Tyson broiler business of northwestern Arkansas, with its familiar brand names such as Holly Gold Kist, Jewell, Perdue and Tyson, is a larger American equivalent (half of the McNuggets in the USA are manufactured by Tyson Foods).

SVF began as a farmers' cooperative in 1960/61 in Herefordshire in the west Midlands. The individual farmers originally each produced poultry as a subsidiary enterprise, but were able to increase output through their cooperative venture. Both vertical and horizontal integration were introduced by construction of a factory for broiler production. Fourteen farmers in the cooperative produced hatchlings whilst there were 17 intensive production units fattening the young chicks ready for transfer to the factory where the birds were prepared for market, largely being sold to major supermarkets. Over time the number of participating farms increased, as did throughput, with sales of poultrymeat trebling between 1965 and 1973. New purpose-built accommodation for feeding birds was established via a subsidiary company, SV Chickens Ltd, and more farmers became involved through the growing of cereals for feed. This in turn stimulated diversification, with feed also being used for turkey and pig production. The cooperative built its own hatchery, feed mill, farm

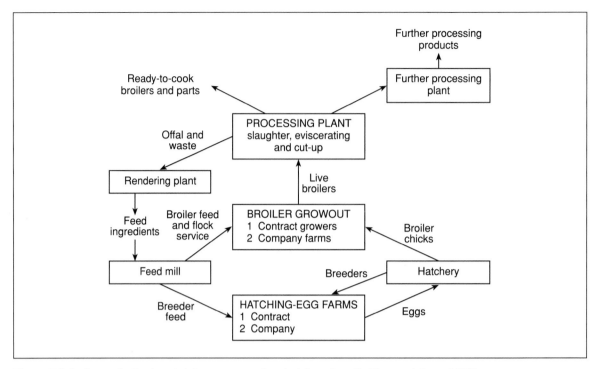

Figure 6.1 Broiler production from hatchery to processing plant (based on Chul-Kyoo and Curry, 1993)

maintenance service and poultry-growing farms, before becoming part of the American-based Union International Group (UIG) of companies with factories elsewhere in the UK to tap new markets. The evolving complexity of the operation is summarised in Figure 6.2, covering developments prior to the takeover by UIG. This is illustrative of the growing emergence of complex linkages between farms, suppliers and markets that have been characteristic of 'industrial-style' specialised production. This arrangement differs from that of Tyson Foods. The chicken 'growers' with Tyson contracts are supplied one-day-old chicks for rearing, but the birds belong to Tyson not the grower and it is Tyson which supplies the feed, veterinary services, technical support, and determines feeding schedules, required equipment upgrades and employs flock supervisors. In this system the chicken grower provides the land, labour, the poultry houses and the fuel, but it is a precarious arrangement, and Schlosser (2002, p. 141) claims that 'about half of the nation's chicken growers leave the business after just three years'.

These brief examples of specialist crop and livestock production systems highlight the steadily increasing interlinkage between different elements of the agri-food system in the Developed World. As a consequence there are new pressures placed upon the family farmers who have long been the backbone of farm-based production. These farmers are frequently locked in a stranglehold between the large supermarkets and wholesalers on the one hand and the suppliers of inputs on the other. This 'squeeze', in which costs of inputs rise but returns on sales are static or diminish, is an integral part of the international farm crisis. Hence a central element in the crisis is a question mark over the very survival of family farming businesses.

6.2 The survival of family farming

6.2.1 The agrarian question revisited

Within the context of increased specialisation allied to rising levels of production there has been one

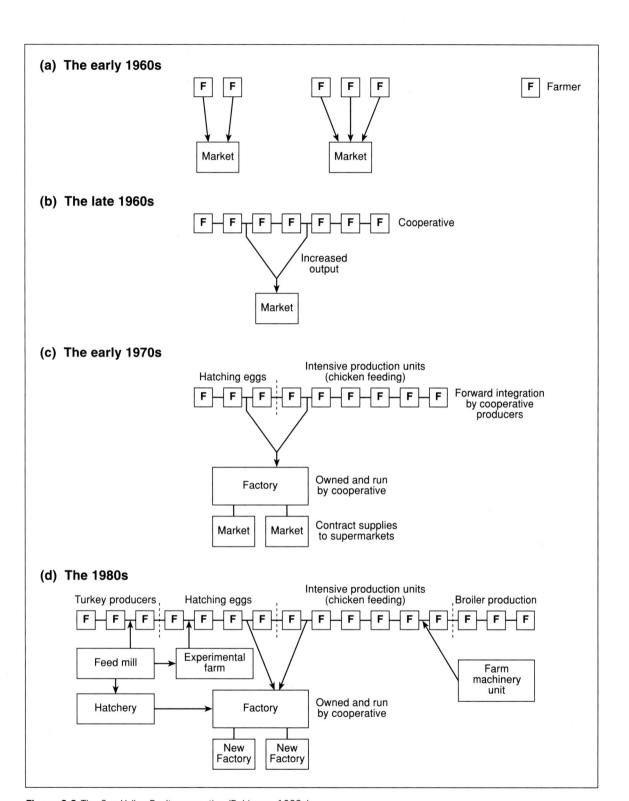

Figure 6.2 The Sun Valley Poultry operation (Robinson, 1988a)

issue that has provoked particular academic debate, namely the impact of the manifold changes upon the family farm – the bedrock of agrarian capitalism in the New World and of rising productivity in Western Europe since the Second Agricultural Revolution. Part of this debate has been concerned with the apparent contradictions between the continuing survival of the family farm and theories about the most profitable forms of business arrangement (Gasson *et al.*, 1988; Gonzalez and Benito, 2001; Marsden, 1985).

According to Gasson and Errington (1993, p. 18), the farm family household displays six distinctive characteristics:

- business ownership is combined with managerial control in the hands of the business principals;
- these principals are related by kinship or marriage;
- family members (including these principals) provide capital to the business;
- family members, including business principals, do farm work;
- business ownership and managerial control are transferred between the generations with the passage of time;
- the family lives on the farm.

Under these criteria, family farms are the dominant agrarian business structure in many parts of the world. For example, in the UK over 90 per cent of farm businesses are family businesses, and 75 per cent are run by members of just one family (Gasson *et al.*, 1988; Hill, 1993). Yet, such farms are seen as being threatened by the changing economic circumstances associated with, first, the growth of specialisation, and then the emergence of growing pressures upon farm businesses associated with the international farm crisis (Hanson, 2001; Hill, 1999). Indeed, the numbers of farms, and family farms in particular, have been falling sharply in the Developed World during the last thirty years. For example, Gale Jr.'s (1996) projections show that the number of farms in the USA will have fallen from 1.93 million in 1992 to 1.29 million in 2007, and with a steadily rising average age of farm operator.

There have been three main sets of explanations for the survival of family farming within the broader context of modern capitalist economies:

- The coincidence it represents between household and enterprise (Friedman, 1978a; 1978b; 1980; Lem, 1988).
- The non-coincidence between labour-time and production-time (Mann and Dickinson, 1978; Mann, 1990). This is discussed below.
- The ability of family farmers to resist conditions tending to subsume them into fully capitalist production (Buttel and Newby, 1980; Mann, 1990; Mooney, 1987; Whatmore *et al.*, 1987a; 1987b); in part related to the high degree of flexibility required for production in systems dependent on local physical conditions. Other factors are the ability of farmers to extend their resistance to the full costs of capitalist market relations beyond the farm gate via exchange of labour and machinery (Lem, 1988), membership of cooperatives and belonging to political lobbies (Winter, 1996b).

The classic investigation of the maintenance of the family farm in the USA was made in the late 1970s by the rural sociologists, Susan Mann and James Dickinson (1978). They referred to the maintenance of non-wage-labour on farms as a contrast to the social (and wage) relations prevailing in most industries, including those agribusinesses which had appropriated operations once performed on the farm (Friedmann, 1978a; 1980). This special characteristic of agriculture, based on the physical properties of land, was noted by both Marx and Kautsky, who argued that land posed a problem for capitalists by virtue of its 'natural' properties and its control by a landholding class (Marx, 1999). So it has often been argued that agriculture is distinctive, differing in kind from other economic activities by virtue of the fact that it has to contend with the vagaries of 'fractious' nature, and, because of its special character, agriculture has not been transformed in a linear, uniform fashion into an arena suitable for large, capital-rich corporations.

The Mann–Dickinson thesis was based on an investigation of why some branches of agriculture become capitalist more rapidly than others (Mann, 1990, p. 32). They argued that the key was turn-over time of capital, and the problem that value was not created when capital spent longer in production (production time) than it did in labour (working or labour time). This problem occurred especially in crop farming because the time during which crops mature in the ground involves little or no application of labour, thus proving unattractive as a source of capital investment and more likely to be left in the hands of 'petty producers' (family farmers). Other problems include the lack of flexible production schedules in agriculture and the presence of slack periods for both labour and machinery. Another set of problems is posed by the lengthy periods during which no return is generated by an agricultural commodity because it can sit for a long time in storage or in transit or only be sold over a long period of time.

The deterrent effect of the lengthy capital circulation time in agriculture has been questioned by a number of researchers (e.g. Davis, 1980; Kulikoff, 1992; Mooney, 1982; 1987), and recently by Henderson (1998; 2000) who reaches different conclusions to Mann and Dickinson, following his analysis of the use of credit in Californian agriculture in the early twentieth century. He argues that, in this period, creditors made money by fuelling farm expansion and by 'floating' farmers during periods of contraction, with the credit system overcoming some of the limitations of the agriculture–nature conjunction: termed by Henderson (2000) as 'fictitious' capital.

Henderson demonstrates that the obstacles arising from the differing temporal needs of plants, animals, people and capital also represent opportunities. For example, whilst crops mature in the fields this provides an opportunity for the selling of credit, the provisioning of temporary labour or the investment in innovation. Hence farmers can move from 'exploiter' to 'exploited'. Henderson's account is based on examining these changes through notions of class position and formation (Mitchell, 2000, p. 94). In his research the different circulation times of capital interact with these

shifting class locations and the spatial systems governing the contingent development of particular places and regions. This involves examination of the role of credit, the significance of marketing cooperatives for fruit and vegetable producers, and the exploitation of cheap labour. However, the nature of the latter was highly variable and included migratory labour, share-cropping, family labour, the substitution of machinery for labour, and labour for machinery (Wells, 1996). Moreover, none of these changes occurred in a simple orderly fashion and so there was much spatial and temporal variation, directly attributable to the need for farmers to control waged labour.

Other studies have shown that there are numerous examples of large-scale penetration of agriculture by corporate enterprise (and often with a long history to this), none better than in parts of California where, for example, Carey McWilliams (1939) highlighted the impact of the 'seven-league' purchases of land by San Francisco businessmen for the establishment of wheat farms and cattle ranches in the late nineteenth century. The continuation of similar industrial-style operations in Californian fruit and vegetable production in the twentieth century has led Walker (1996) to argue for the primacy of industrial capital in Californian agriculture, the result being 'singularly large landholdings, massive investment in water systems, rapid mechanisation wherever possible, harnessing of science, mass proletarianisation, and a choke-hold on the government in its domain' (Walker, 1997, p. 278). He contends that this demonstrates the existence of a profitable, workable agri-industrial model that is being mimicked elsewhere around the globe, with barriers to agriculture's 'exceptionalism' being broken down (see also Pincetl, 1999).

There are some cases, though, that support the maintenance of exceptionalism. For example, in Tasmania a rising level of external penetration can be observed in terms of potato growers' dependence on technology, marketing and institutional finance. However, the farmers still exert a controlling influence across a range of key decision-making. Some control has been forfeited as a result of the contract into which the farm families enter

.th the contracting company (Fulton and Clark, 1996). Subsumption levels appear high, but in reality farm families have not lost control over production (Whatmore *et al.*, 1996). Elsewhere there are examples of family farmers increasingly losing control over farm-based production processes as they have been bound by contracts secured in order to guarantee a market outlet (Hart, P. W. E., 1992, pp. 194–9). This can be seen for numerous types of production, for example broiler chickens for broiler 'integrators' like Tyson Foods (Chul-Kyoo and Curry, 1993), growers of iceberg lettuces contracted to grower-shippers, and tomato growers contracted to processing firms like H. J. Heinz (McKenna *et al.*, 1999a). These contracts have become institutional supports for family farming whilst suborning the farmers to the needs of their contractural overlords. The contracted farmers have been termed 'peasant-workers' by Boyd and Watts (1997) and 'propertied labourers' by Davis (1980) to reflect the fact that, although they own the means of production, they control few of the input factors or the methods of production, which are usually specified by the contractor.

An additional loss of control has also occurred downstream, where the overall impact of farming on the rural economy has often been considerably reduced. For example, Argent (1997) has showed how downturns in the farm economy on Kangaroo Island, South Australia, affected the nature of the linkages between farming and the local rural economy. He referred to the advance of 'uncoupling' whereby the prime position of agriculture in the rural economy has been consistently undermined in recent decades by increased rural economic diversity (Stayner and Reeve, 1990), and especially through the growth of post-productivist use of farmland, in this case for rural tourism. Hence, during the 1990s, there was a shift of firms, from those dependent directly on the farm sector towards those dependent on tourism, demonstrating the power of the new tourist-dominated economy to offer alternative opportunites, as those associated with traditional farm activities diminishes.

The political ramifications of this reduced economic impact of farming are widespread as part of the diminishing power of the so-called farm lobby. As the number of farms in the Developed World declines so farming has less political leverage with government, whilst the growth of subservient relationships between farmers and retailers, wholesalers and processors also weakens the farmers' position. However, an additional aspect of this general situation is the overall decline in the agricultural population as the size of the farm labour force falls.

6.2.2 The decline of the farm workforce

Table 6.2 shows how the agricultural workforce has declined throughout the Developed World and that the contrast between the Developed and the Developing Worlds is stark: the latter has continued to apply more labour to the land whilst the former has cut its labour force by two-thirds within half a century. The most dramatic fall has occurred in the EU, where there have been rapid declines in both the numbers of agricultural holdings and the overall farm labour force during the second half of the twentieth century (Hoggart *et al.*, 1995). These trends have been sharpest in the Mediterranean countries where the proportion of the population employed in agriculture has fallen from around 50 per cent in the 1960s to 10 per cent today (Monke *et al.*, 1998a; 1998b) (Table 6.3). However, the continuing significance of the agricultural labour force within the overall labour force in Greece and Portugal is apparent as are the absolute numbers still on the land in France, Germany, Italy and Spain, which between them have 4.75 million farmers and farm labourers, over half the EU total.

In the last three decades the decline in the farm labour force has been especially marked in southern Europe. For example, the number of farm holdings in Spain fell by over 40 per cent between 1976 and 1999, whilst the agricultural population was reduced from comprising 22 per cent of total population in 1976 to just 7.4 per cent in 2000. As elsewhere in the Developed World, the contribution of non-farm-family labour has been greatly diminished – cut in half since 1976 (Vinas, 2000). Moreover, in those remaining labour-intensive parts of Spanish agriculture labour costs are being kept to a minimum by utilising cheap workers from North Africa (Hoggart and Mendoza, 1999).

Table 6.2 Total agricultural labour force, 1950–2000 ('000s)

Region	1950	1970	1990	2000	Percentage change 1950–2000
Developed World	137,382	88,596	62,296	47,866	−65.2
Developing World	672,087	840,136	1,158,929	1,270,762	+89.1
Africa	86,759	117,909	165,621	197,116	+127.2
Asia	571,643	693,722	953,454	1,040,036	+81.9
Central America/Caribbean	11,639	13,604	16,494	17,233	+48.1
Far East	531,670	657,568	920,488	996,068	+87.3
Near East	31,111	37,663	42,742	49,107	+57.8
Oceania	1,743	1,905	2,507	2,835	+62.7
South America	21,074	26,681	28,160	26,898	+27.6
South Asia	175,177	227,235	297,667	341,909	+95.2
Canada/USA	9,385	4,532	4,134	3,417	−63.6
Eastern Europe	27,869	21,134	13,000	9,674	−65.3
European Union (15)	37,562	19,237	10,858	7,607	−79.7

(*Source*: FAOSTAT)

Table 6.3 The decline in the agricultural labour force in the EU, 1950–2000

Country	Agricultural workforce ('000s)			Agricultural workforce as a % of all labour force	
	1950	2000	−%	1950	2000
Austria	1,186	191	83.9	34.2	5.1
Belgium	415	77	81.4	11.9	1.8
Denmark	534	111	79.2	25.7	3.8
Finland	703	143	79.7	35.0	5.5
France	5,960	899	84.9	30.9	3.3
Germany	7,484	1013	86.5	23.0	2.5
Greece	1,709	775	54.7	55.3	16.8
Ireland	516	163	62.4	40.2	10.2
Italy	9,041	1352	85.0	44.0	5.3
Luxembourg	33	4	87.9	24.1	2.2
Netherlands	709	248	65.0	17.7	3.4
Portugal	1,721	650	62.2	49.8	12.7
Spain	5,618	1293	77.0	51.6	7.4
Sweden	639	151	76.4	20.8	3.2
UK	1,294	537	59.5	5.5	1.8
EU15	37,562	7607	79.7	28.3	4.3

(*Source*: FAOSTAT)

As a consequence of the growing flight from the land, over half of the farmers in the European Union are over 55 years of age, and they are especially dominant on the smaller holdings and in the Mediterranean countries where between one-quarter and one-third of farmers are aged over 65 (Laws and Harper, 1992). For example, in Spain, elderly and 'pre-retired' farmers own 60 per cent of farms and 72 per cent of holdings smaller than 10 ha (Mazorra, 2000, p. 113). In France elderly farmers have been viewed as potential suppliers of land for amalgamation and restructuring (Naylor, 1982). This view has been echoed in the CAP reform process, for example in Regulation 2079/92, which encouraged the restructuring and amalgamation of farms and the retirement of mature and elderly farmers. Other member states have also instituted their own early retirement schemes for farmers, for example Spain incorporated EU guidelines on early retirement for farmers in policy from 1989 onwards in various royal decrees, as a means of modernising agricultural structures. However, the French and Spanish experience has been that farmers have responded reluctantly to the structural dimensions of such policies, thereby prompting policy-makers to introduce 'social' dimensions that have permitted 'the gradual incorporation of the traditional transfer processes to close family members and offspring' (Mazorra, 2000, p. 120). The diversity of factors affecting the decisions of farmers to retire has meant that there has been no clear systematic pattern of take-up of the various schemes in either a geographic or farm-size distribution sense. This also reflects the continuing gap between the nature of early retirement schemes and the concerns of farmers, especially regarding farmers' views of their own best interests and specifically those associated with inheritance issues. The latter are highly significant, given that, for example, in the EU at least two-thirds of farmers inherited their farm from relatives (Blanc and Perrier-Cornet, 1993).

The broad figures in Tables 6.2 and 6.3 hide the fact that many of those enumerated in official statistics as farmers and farm labourers actually hold other jobs and derive income from sources other than agricultural activity. These sources may include farm-based processing (such as the traditional practices of butter and cheese making, jam making and farmhouse crafts), or more recently introduced farm-based work such as running a farm shop or a farm bed-and-breakfast. They may also include off-farm work, such as agricultural contracting, or activities not associated with farming at all and which may extend to several members of the farm household. These activities may constitute various forms of farm diversification, multiple job holding and pluriactivity, all of which have tended to become more prominent in recent years throughout the Developed World as part of multi-faceted responses by farm households to changing economic and social conditions (see Shucksmith and Herrmann, 2002).

6.3 Farm diversification and pluriactivity

6.3.1 Defining farm diversification

Farm diversification includes a variety of on-farm activities through which the farm household attempts to raise income, largely from non-traditional farm enterprises. Because of the wide range of activities involved it has proved difficult to produce a simple all-embracing definition, but several classification schemes have been developed which provide useful guides to the types of activities involved (Table 6.4; Figure 6.3). These show that farm diversification should not be equated with enterprise diversification, in which there is a move to a more diversified system of mixed farming (e.g. adding a 'conventional' farming activity to the farm enterprise). Instead, literature on farm diversification places emphasis upon the non-traditional aspects of agricultural diversification, such as growing 'exotic' crops (e.g. Spellman and Field, 2002; Wood, 1994) or energy crops (Collins, 1999), and includes structural diversification, in which the use of farm resources is directed towards markets beyond the productive farming system. Therefore the activities included within the term 'farm diversification' are generally not based on the traditional or conventional production of food and fibre, and, within the EU, fall outside the price support mechanism of the CAP (Ilbery, 1987a, p. 23). The diversified activities include farm-based enterprises related to tourism and recreation (e.g. bed-and-breakfast, gîtes) (e.g. Ilbery, 1989; Muller, 1992), unconventional enterprises (e.g. linseed, rare breeds), adding value to conventional products (e.g. pick-your-own schemes), use of ancillary resources

Table 6.4 Types of farm diversification in the UK

Structural diversification	Agricultural diversification
Tourism	**Unconventional enterprises**
• *Accommodation*	• *Crop products*
• Bed-and-breakfast	• Linseed
• Self-catering	• Teaseed
• Camping and caravan sites	• Evening primrose
• *Recreation*	• Borage
• Farmhouse teas/cafés	• Triticale
• Demonstration/open days	• Fennel
• Farm zoo/childrens' farm	• Durum wheat
• Water-/land-based sports	• Vineyards
• War games	• Lupins
• Horticulture	
• Craft centre	• *Animal products*
• Nature trails/reserves	• Fish
• Country/wildlife parks	• Deer
• Farm museum	• Goats
• *Combined*	• Horses
• Activity holidays	• Llama/ostriches
	• Sheep (for milk)
Adding value to farm enterprises	• Rare breeds
• *By direct marketing*	• Rabbits
• Farm gate sales	
• Farm shops	• *Organic farming*
• Delivery round	
• Pick-your-own scheme	**Farm woodland**
	• Energy forestry
• *By processing*	• Amenity/recreation
• Butter/cheese	• Wildlife conservation
• Ice cream/yoghurt	• For timber
• Cider/wine/juice	
• Jam/preserves	**Agricultural contracting**
• Potato packing	• For other farmers
• Flour milling	• For non-agricultural organisations
• *By selling skins, hides, wool*	
Passive diversification	
• Leasing of land	
• Leasing of buildings	

(*Source*: based on Ilbery, 1991)

(e.g. farm forestry, game birds, use of redundant buildings as industrial premises), and public goods (e.g. agri-environment payments).

Farm diversification may also be considered as part of a broader process of rural diversification, which includes rural industrialisation (Fothergill *et al.*, 1985). The latter has been termed the 'ruralisation' of industry and arises from both indigenous growth of rural industry and a shift of industry from urban to rural locations (Healy and Ilbery, 1985; Keeble *et al.*, 1983; Owens *et al.*, 1986; Townsend, 1991). Government policy has

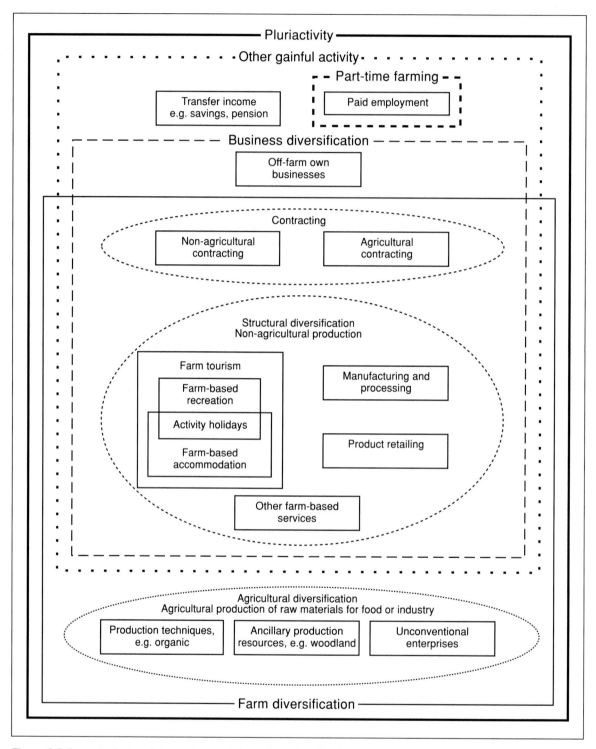

Figure 6.3 Conceptualisation of the relationship between farm diversification and pluriactivity (based on Higginbotham, 1997)

often contributed to the process of rural industrialisation through instigation of area-specific and non-area-specific measures. In the UK the latter include the Redundant Building Grant, available from 1982, offering assistance for re-utilisation of redundant rural assets for new and non-agricultural purposes, with commercial usage stipulations. Various specially designated areas have been established for receipt of infrastructural improvements, labour subsidies, low-cost premises and tax incentives. For example, in the UK, the Rural Development Commission was active in the 1980s in the provision of premises for small businesses (Martin and Tricker, 1989), representing opportunities for farm and non-farm rural diversification.

In their conceptualisation of rural diversification, Kelly and Ilbery (1996) saw it as a response to changes occurring within both 'internal' and 'external' environments. Significant broad changes in the external environment have been:

- policy emphases moving away from a productivist philosophy and in favour of supply control and reduced price support (Robinson and Ilbery, 1993);
- a growing impetus within policy encouraging farmers to diversify and seek alternative means of generating income from the farm resource base. This has been accompanied by a shift from the primary to the secondary and tertiary sectors within the rural economy, providing additional opportunities for the diversified usage of land, labour and capital (e.g. McInerney et al., 1989);
- the growth of farm diversification and ruralisation of industry have often been mutually self-reinforcing, contributing to new employment opportunities in the secondary and tertiary sectors at a time when agriculture is still shedding labour (Townsend, 1991);
- the 'international farm crisis' in which farmers affected by the 'cost–price squeeze' have had to invest more to maintain income levels. This has encouraged diversification as a viable option and promoted linkages between farming and both the secondary and tertiary sectors. In turn this has introduced

non-farm capital to the process of diversification.

The development of farm diversification is primarily related to farmers and the farming household becoming pluriactive, whereby a range of new sources of income generation is obtained both on- and off-farm (MacKinnon et al., 1991). These activities include farm diversification but, in addition, also embrace redeployment of human capital into other gainful activities (OGAs) off the farm as an employee or on a self-employed basis (Fuller, 1990). Therefore pluriactivity includes all of the various aspects of farm diversification referred to in Table 6.4 as well as such activities as employment as hired labour on other farms and off-farm waged labour. Moran et al. (1993) also refer to pluriactivity as including mutual labour exchanges and informal arrangements, inclusive of waged and unwaged labour and farm family members being self-employed on and off the holding (Evans and Ilbery, 1993) (Figure 6.4).

Different definitions of diversification and pluriactivity have been employed in different countries. For example, three types of pluriactivity have been recognised in France (Campagne et al., 1990):

- Business pluriactivity. In Picardie, household members are using agricultural resources to increase non-agricultural activities as part of an enterprise culture involving farm wives working outside agriculture.
- Pluriactivity for maintaining farming. In Languedoc, a wine-producing region, there is a long history of pluriactivity, with income from off-farm sources generated mainly by farm wives and children, for farm modernisation and maintenance.
- Pluriactivity for survival. In Savoy, south-east France, where farming is marginal, there is a combination of income-generating activities on- and off-farm.

Whilst pluriactivity has a long history in France and some parts of Europe (e.g. Franklin, 1969), there is evidence for increasing pluriactivity throughout the Developed World in recent decades. In France, the proportion of farm household

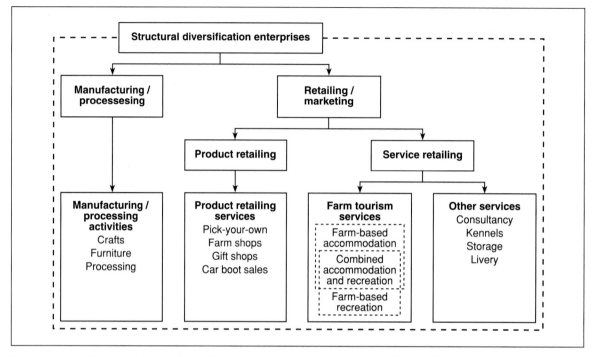

Figure 6.4 Classification of structural diversification enterprises (based on Chaplin, 2000, p. 10)

income from pluriactivity increased from 15 per cent in 1956 to 42 per cent in 1988 (Benjamin, 1994). In Ireland one estimate is that 60 per cent of farm operatives and/or their spouses will derive income from off-farm activity by 2010, and that in 2000 only 53 per cent of Irish farm household income came directly from farming (Kinsella *et al.*, 2000).

The relationship between farm diversification and pluriactivity may be extended further by referring to the concept of business diversification, which occurs when members of the farm household set up new on- or off-farm businesses, the latter representing an entrepreneurial element of OGAs. Indeed farm diversification may be a significant factor in the maintenance of a vibrant rural economy (Mayfield, 1996). The off-farm business may be related to agriculture, as in the case of a contractor for agricultural machinery, whilst the on-farm business may be only loosely related to agriculture, e.g. farm-based tourism. Moreover, increasingly, the productive assets of a farm can also be seen as multifaceted capital assets, which

can either be sold at high value or else exploited by non-traditional means. The latter has been part of the farm diversification process but has also involved various innovative forms of property relations such as divisible rights, short-term leases, and options.

6.3.2 Motives for and impacts of diversification

The different development paths open to the farm family are summarised in Table 6.5. Pathways 3 and 4 constitute the development of pluriactivity and farm diversification, though number 6 may also include some element of diversified income sources. Farm businesses may combine two or more of these paths so that the paths can be broken down into three principal strategies (Bowler and Ilbery, 1992, p. 307; Ilbery *et al.*, 1996):

- maintaining full-time, profitable food production as the farm business (paths 1 and 2);

Table 6.5 Pathways of farm business development

I	Continuation of agricultural production
	1. Extension of industrial model of farming: traditional products
	2. Redeployment of resources into new agricultural products
II	Diversification of the income base
	3. Redeployment of resources into new non-agricultural products
	4. Redeployment of resources into off-farm OGAs
III	Marginalisation of farming
	5. Traditional farm production with lower income/inputs
	6. Hobby or part-time (semi-retired) farming

(*Sources:* Bowler, 1992b; Ilbery *et al.*, 1997)

- diversifying the income base of the farm business by restructuring resources into non-farm enterprises and occupations (paths 3 and 4);
- surviving as a marginalised farm business at a low level of income, perhaps supported by investment income, pensions or other direct state payments (paths 5 and 6).

This categorisation can be compared with the pathways for farm business development recognised by Damianos *et al.* (1998; Damianos and Skuras, 1996) for 'lagging regions' in Greece, in which economic and social factors were found to exert independent effects upon farmer decision-making:

- maintaining conventional (mainstream) farming: accounting for 29 per cent of farm businesses;
- alternative farm enterprises, primarily introduction of new industrial crops, such as tobacco (41%);
- Conventional farming combined with off-farm employment (27%).

For Greek agriculture as a whole, Daskelopug-lou and Petrou (2002) translated pursuit of these strategies into a three-fold categorisation:

- Subsistence producers. These were primarily small farmers (some farming under 1 ha) for whom off-farm income was the primary source of funds. Farming was not commercially viable and many of these farmers were pursuing an exit strategy from farming.
- Survivalist. These farmers tended to farm between 5 and 50 ha, some using modern machinery. They were generally pluriactive but farming was the chief source of income.
- Productivist. These were termed modernisers. They generally had more than 50 ha and were market responsive, relying far more on their farm incomes than those in the other two categories.

The predominance of particular paths and strategies in different localities is suggested by Bowler and Ilbery (1992): for example paths 3 and 4 may be dominant in urban fringe areas whereas marginal upland areas may be dominated by paths 3 and 5. They also consider the possibilities of the transition of farm businesses from one path to another, though once 'marginalised' it is difficult for a business to move up the pathways to an agri-industry mode.

The development of pluriactivity and farm diversification has attracted a substantial amount of research by geographers and other social scientists, attempting to understand the reasons for such developments, the motivations and personal characteristics driving the diversifications, and the resulting impacts upon rural economy and society. The fundamental motive for diversification lies in farmers' need for additional sources of income in the face of falling revenue from traditional agricultural sources. However, the reasons for reduced profits are complex and there are also non-agricultural factors at work, including the growth of new business opportunities associated with changing public demands on rural resources and varying government responses to agriculture, rural industry and services (Gasson, 1987; 1990; 1991; Halliday, 1989).

Common motives include the external imposition of limits on traditional farming, changes in the family life-cycle (Errington and Gasson, 1996;

Gasson and Errington, 1993), personal interest and the utilisation of available resources (Evans and Ilbery, 1989; Ilbery, 1988; Kelly *et al.*, 1992). Personal aspirations are closely linked to the motivations of diversifiers. For example, those attaching a low priority to generation of large profits are less likely to seek business expansion by pursuing new enterprises. However, aspirations are strongly conditioned by personal characteristics and experiences. Youthful, well-educated and relatively affluent farmers have tended to be those most likely to diversify, but there is no clear guide as to the 'essential rural diversifier' (Kelly and Ilbery, 1995). The nature of the diversified activity largely determines the additional level of training and experience required, but the activity is highly likely to reflect the existing qualifications of the farmer and/or members of the farm household (Townroe, 1992).

Farm diversification may be favoured by the presence of exploitable opportunities relating to a particular configuration of land, labour and capital resources. Hence a key factor may be farm type. For example, Evans and Ilbery (1993) observed that arable production can enable farm operators to devote a proportion of their labour to off-farm activities in addition to any farm-based enterprises. Ilbery (1987a) reported that OGAs were most associated with farms with lower labour requirements (cereals, permanent crops, livestock grazing) and less with intensive dairy, pigs, poultry and horticultural enterprises. Bateman and Ray (1994) suggest that lower levels of pluriactivity on dairy farms might relect the capital-intensive nature and time-consuming work practices of dairying. Alternatively, recent relative profitability may mean an absence of financial pressure to be pluriactive (Ilbery, 1991).

Larger and more industrialised farm businesses have been better placed to pursue profitable full-time production of food and fibre. However, larger farms have also been most likely to develop new on-farm enterprises. In contrast, smaller farms have been more likely to be associated with farm family members developing OGAs off-farm (Gasson, 1986; 1988a; 1988b; Shucksmith *et al.*, 1989). Gasson's (1998) survey in England and Wales found that off-farm pluriactivity was most pre-

valent on small farms (< 50 ha) whereas on-farm OGAs were better represented on large farms (> 200+ ha). Ilbery *et al.* (1996) reported that on-farm business diversification was concentrated amongst both very small (< 40 ha) and very large farms (> 400 ha). On-farm activities were most common on medium-sized farms (120–200 ha), with leisure and recreational enterprises favouring the larger farms. Ilbery (1996) reported that a majority of farm attractions had invested under £5000 to establish their recreational activities and less than 30 per cent had spent over £15,000. This scale of investment is small in comparison with the capital required for new agricultural machinery and buildings.

In England and Wales, research has suggested that farm diversification is most associated with larger farm businesses, higher net incomes and higher levels of net indebtedness, especially if farmers are in younger age categories and have continued their formal education beyond school. The presence of children to whom the farm may be passed can also be important. Although larger farms favour on-farm diversification, OGAs dominate farm households on small farms (Gasson, 1988a).

In the mid-1990s in the northern Pennines in England, Bowler *et al.* (1996) found that one-third of farm households had off-farm OGAs and 29 per cent had on-farm OGAs. There was, though, a certain degree of resistance by farmers to diversification, so that OGAs were only considered once 'traditional' farming activities were revealed to be inadequate to address the income needs of the farm household.

In the late 1980s the Arkleton Trust made similar findings in a survey of 24 regions of Europe (Arkleton Trust, 1992; Fuller, 1990), though with considerable inter-regional variation. Nearly 60 per cent of farm households surveyed were pluriactive, with half of the households having off-farm OGAs compared with 10 per cent having on-farm OGAs. OGAs were more numerous than on-farm diversification and were dominated by farm wives, often reflecting social reasons affecting female labour force participation. In all regions surveyed, farm operators were more prevalent than their spouses in off-farm work, though in Spain, Italy, Austria and Ireland operators' children played a bigger role in off-farm work. The Arkleton Trust project revealed that one-third of farm households obtained over half their total income from off-farm sources. Only 17 per cent of farm households derived all of their income from farming whilst 43 per cent obtained under 30 per cent in this fashion.

Position in the household life-cycle has been shown to have significant consequences for the farm business and pluriactivity (Potter and Lobley, 1996a). For example, Shucksmith (1993) found that the largest reallocation of labour from agriculture and into off-farm work had been amongst households with farmer, spouse and other family members present. These households were most likely to reallocate labour in favour of off-farm work, especially where the farmer was in his 40s and when the household could be expected to include teenage or young adult offspring. Non-diversifiers were more likely to be single farmers or couples with no children. Halliday's (1989) study of Devon also suggests that in the later stages of the family life-cycle, new possibilities for engaging in alternative sources of income may exist. On-farm pluriactivity appears to be favoured by farmers in mid-life-cycle with older children.

The Arkleton Trust project and more recent studies show that the spatial patterns of pluriactivity and farm diversification are complex, reflecting the interaction of several factors external and internal to the farm business (e.g. Shucksmith et al., 1989; Shucksmith and Smith, 1991). Regional socio-economic conditions are important as these help determine the number of opportunities available. So regions with a good farm-size structure and diverse labour markets are likely to favour pluriactivity (Marsden, 1990). However, this can be modified by the presence of local cultural factors and the pressure of specific landscape designations, such as national parks (Bateman and Ray, 1994). Personal preferences and household characteristics are other key factors (Edmond and Crabtree, 1994; Edmond et al., 1993; Eikeland, 1999; Eikeland and Lie, 1999; Ilbery, 1988).

One crucial factor affecting both farm diversification and rural industrialisation has been the role of the state in attempting to increase rural employment, for example through provision of small business premises (Chisholm, 1985) and assistance in starting a new business in the form of information, advice, training and cheap loans.

The establishment of new buildings has involved both the use of converted redundant farm buildings (RFBs) and special purpose-built units (PBUs) on rural industrial estates. The former has enabled traditional and often obsolete buildings to be turned into assets rather than falling into disrepair.

Financial support for farm diversification has been available in the EU in areas designated as Objective 5b, initially under the structural fund programme (1994–9). In England and Wales measures introduced in 2000, except for EU Objective 1 areas, offer support for various rural development activities, including diversification. The UK's Farm Diversification Grant Scheme (FDGS), introduced in 1988, offered a 25 per cent grant for diversifying the farm business (Ilbery and Stiell, 1991).

6.3.3 The role of women in farm diversification

One persistent strand of agricultural geography has been work on the impacts of agricultural change, whether dominated by increased specialisation or diversification, upon the lives of farming people. There have been studies of how the nature of farm work was changing on farms where mechanisation had greatly reduced the size of the labour force or restricted it solely to members of the farm family (e.g. Errington and Gasson, 1996). However, a more substantial amount of research has concentrated recently on the gendered arrangements of farm families (Darque, 1988; Gray et al., 1993, Kothari, 1990; Phillips and Gray, 1995; Repassy, 1991; Whatmore, 1991a; 1991b), and also on gender and the cultural politics of agriculture (Brandth, 1995; Cloke, 1996; Liepins, 1996; Mackenzie, 1992a; 1992b), focusing on discourses to show how meanings are formed and power is articulated (Liepins, 1998a; 1998b; 1998c).

With respect to the latter, for example, Liepins (1998c) has shown how specific meanings of masculinity and femininity are articulated through farming organisations and rural print media in Australasia. These meanings have encompassed various images of men and women working in farming. For example, the image of 'tough farm men' can be seen as part of hegemonic masculinity on the farm and in industrial affairs. This contrasts with 'relatives and carers' for traditional farming femininity or 'the real woman farmer' as an alternative occupational femininity on farms. New constructions are now appearing in terms of references to sensitive new-age farmers, implying new masculinities in farming and the 'alternative agricultural activist', which recognises the changing position of women in agriculture (see Brandth and Hauger, 1997; Fink, 1991; Oldrup, 1999).

Although the presence of women as farm labourers and in a variety of other roles on farms and in rural areas in general was ignored or barely remarked upon in academic study, both in the Developed and the Developing Worlds (Barrett, 1995), the diversity of women's roles within agriculture has been acknowledged and studied in much greater detail during the last decade or so (e.g. Dempsey, 1992; Whatmore, 1991a). This has revealed a broad spectrum of identities occupied by rural women, from the traditional supportive homemaker image of a farmer's wife to the woman working alone or with other partners. Furthermore, women often occupy more than one position, frequently combining a domestic role with responsibility for certain types of activity related more directly to commercial return, e.g. managing a farm-based accommodation outlet, or taking responsibility for a particular sector of the farming enterprise (Bouquet, 1982). This has been demonstrated in a series of studies by Alston of the changing roles of rural women in Australia (Alston, 1991; 1995a; 1995b; Alston and Wilkinson, 1998; for European comparisons see Burg and Endeveld, 1994; Haugen, 1990).

Liepins (1995) distinguishes different theoretical perspectives in the literature about farming women, of which two have been most significant:

- Socialist feminism. This treats the family farm as rich examples of capitalism and patriarchy, combined in an uneasy alliance in which

home and work combine in one group, that is an interdependence of family and enterprise (Whatmore, 1991a, p. 43). In applying theories of patriarchal gender relations with a broader political economy analysis of women on family farms, women are viewed as unequal members of a domestic political economy. This approach is useful in helping to understand women's experience of patriarchal family labour and property relations within family farm production (Blanc and MacKinnon, 1990). It enables women's roles on the farm to be viewed differently from the standard 'perceived capitalist patriarchal organisation of agriculture where women have been confined to the sphere of domestic, reproductive and consumption activities, and released from this only when other "service" is required, e.g. in terms of additional labour, farm administration and paperwork or "gofer-ing"' (Liepins, 1993, p. 185). This approach also employs class analysis to consider how class informs the motivations of women working in agriculture.

- Post-structural feminism (e.g. Mackenzie, 1992). This approach stresses the need to study the diversity of farm women's self images and their social and political positions (Haney and Miller, 1991). It emphasises the socially constructed subjectivities of farm women, in which there is a plurality of identities both personally held and socially prescribed. Thus women can variously, and often simultaneously, constitute themselves as family and community carers and nurturers, working women, and individuals seeking equal opportunity. The approach can incorporate some of the ideas of the French philosopher, Foucault, and his writings on the relationships between power and resistance. For example, changes in family farming in Australia have been considered in terms of the exercise of power and resistance to economic, social and political developments affecting agriculture (Alston, 1991). Consideration of 'resistance' can incorporate analysis of social movements and organisations, as in Mackenzie's (1992) work in Canada on the Women for the Survival of Agriculture. In particular, distinctions can be drawn between mass movements, typified by a relatively homogeneous membership, hierarchical internal structure and powerful leadership, and postmodern social movements which are of grass-roots character, voluntary input and fluid and diverse memberships (Rosenau, 1992).

In terms of particular emphases within research on women working in agriculture, from the late 1980s, work on farm diversification considered the roles of farmers' wives and their children in generating both on- and off-farm income for the farm household (Evans and Ilbery, 1992a; 1992b; Gasson, 1992). More work has been done explicitly on the contribution of women to the overall functioning of farms, initially by Gasson (1980; 1987; Berlan-Darque and Gasson, 1991), with the recognition that women's input was one of the reasons why family farms had been able to survive in times of declining profit margins (Jervell, 1999; Whatmore, 1991a). This acknowledged the importance of gender relations in affecting farm production, but most of the work on this, and especially utilising distinctive feminist perspectives, has been confined to work in rural societies in Developing Countries where women frequently take the prime role in agricultural work. This is dealt with in the following chapter.

The variations between regions in the extent and nature of farm diversification reflect the complex interplay of farming, household and labour market characteristics as well as cultural factors (Edmond and Crabtree, 1994; Fuller, 1990, p. 368). Within this interplay studies have revealed the importance of the increasing participation of women in the labour force (e.g. Benjamin, 1994), but strongly influenced by 'internal' factors such as farm size, family life-cycle, succession, age and education (Gasson, 1998; Ilbery and Bowler, 1993). For example, in a study of regional patterns of pluriactivity in Greece, Efstratoglou-Todoulou (1990) found that off-farm opportunities for women were significant in promoting OGAs in less

favoured areas, in which case OGAs were largely a necessity. In areas with better agricultural structures and higher farm incomes, OGAs were the result of choice by farm households possessing alternative opportunities.

Pluriactivity may increase the economic freedom enjoyed by the farmer's wife or it may change her position in the business and household decision-making process (e.g. Shortall, 1994; 2002). However, it may also increase the amount of work that farm wives have to perform, with no corresponding gain in independent income (Evans and Ilbery, 1996; Gasson and Winter, 1992). The development of pluriactivity over time may have the potential to alter the distribution of power between the spouses involved in the farm business and household. This may mean a fundamental change in traditional patriarchal gender relations within the farm, though not necessarily in terms of increased power for women in rural society (e.g. Pini, 2002).

6.3.4 Diversification into farm-based tourism and recreation

One of the significant growth elements within farm diversification throughout the Developed World has been in the number of farms on which additional income is gained from enterprises associated with tourism and recreation (Evans and Ilbery, 1989). This has given rise to the term 'vacation farm' to describe a working farm operation which derives additional income from a periodic tourist clientele (Fennell and Weaver, 1997). Some official definitions of this type of farm have stipulated a maximum number of rooms beyond which the operation ceases to be regarded as a 'farm', e.g. a limit of five bedrooms is applied for France's *gîtes rureaux privés*, whilst the Austrian equivalent is ten bedrooms. Various estimates utilising these definitions in the mid-1980s gave between 20,000 and 30,000 vacation farms in the main European concentration, which includes Austria, France, Germany and the UK (Murphy, 1985), compared with 700 in Canada (Shaw and Williams, 1994). In the UK there is evidence that cooperative marketing of attractions has attracted more farm-based tourists (Ilbery, 1996).

Walford (2001) compiled a database of 6665 farms in England and Wales offering accommodation in 1996. This included serviced accommodation as well as self-catering facilities in both fixed and temporary structures. Of these farms 94 per cent had just one type of accommodation available for tourists: camping (52%), self-catering (26%) and bed-and-breakfast (16%). Farms under 100 ha accounted for 70 per cent of the farms in the database. Larger farms (> 100 ha) were more likely to have serviced accommodation, with camping and caravan sites more typical on smaller holdings. Over one-quarter of farms had sheep and cattle enterprises whilst 18 per cent were mixed farms and a similar proportion were dairy farms. In part, this association between farm-based tourism and livestock enterprises reflects the preponderance of farm-based tourism in western Britain, with its pastoral economy (Ilbery et al., 1998).

The typical provision of accommodation was small scale (Walford, 2001, p. 337):

- serviced accommodation on larger farms with six adult bed-spaces in three rooms;
- self-catering for six adults, usually in two rental properties;
- camping and caravan sites on smaller farms, with five pitches and an adult capacity of at least ten persons.

There was a tendency for more provision in areas immediately adjacent to designated areas, such as National Parks, Heritage Coasts and Areas of Outstanding Natural Beauty, rather than in the areas themselves. Evans and Ilbery (1992a; 1992b), also investigating farm-based accommodation in England and Wales, reported that 65 per cent of farm households with accommodation gained less than 10 per cent of their total income from this enterprise (see also Evans, 1992).

In a detailed analysis of farm-based tourism and recreation Chaplin (2000) focused on further classification of structural diversification to distinguish between manufacturing/processing and retailing/marketing distinctions. Within the latter, two further groups were recognised, namely product retailing and service retailing (Figure 6.4). These categories enable any definitional overlap between farm-based recreation and product retailing to be eliminated. They also enable farm tourism to be clearly recognised as any activity attracting people onto a farm to 'consume' accommodation and/or recreational services, rather than retailing or other services. Farm-based recreation itself was defined as 'all those activities on a working farm which involve people using a recreational/leisure service provided by the farm other than, or in addition to, accommodation which can only be "consumed" on the farm. It does not, in itself, involve the purchase of products or other services' (p. 12). Therefore this includes recreational activities that may generate little or no direct income, such as fox hunting or pigeon shooting.

Table 6.6 represents a classification of farm-based recreation, distinguishing between agricultural types, which can be identified as involving a distinctive farm or agricultural experience, and non-agricultural types, which encompass all those recreational activities that do not involve an agricultural element. Further, residential, educational and specific resource subgroups could be identified. Chaplin (2000) argues that this type of classification is most appropriate for investigating the spatial distribution of farm-based recreational activities at regional or national level, but is insufficient for work on identifying the reasons why farmers adopt different types of recreational activity on their farm. For this, different categories of provision need to be considered, in particular the following:

- activities that are open to the general public without booking;
- activities that are available to the general public, but require booking or prior arrangement;
- activities that are only available to members of private groups, syndicates or clubs;
- activities that are only available for personal use (family, friends and employees);
- short-term recreational events, occurring for less than 28 days a year.

This enables consideration of farms providing recreation to the wider market whilst also including those that supply recreation to family members at the expense of this market.

Table 6.6 Leading components of farm-based recreation

Agricultural	Non-Agricultural
Museums	Game fishing
Farm trails	Access agreements
Farm visitor centre	Coarse fishing
Demonstrations	Horse/pony riding
Educational facilities	Horse/pony riding competitions
Working farm tours/tractor rides	Shooting
Rare breeds/wildlife park/pets corner	Motor sports/motor sports competitions
Farm open days	Indoor sports facilities
Farm shows	Water sports
Lambing/shearing days	War games/paintballing
Ploughing matches	Facilities for model clubs
Sheep dog trials	Village sports pitches
Traction engine/tractor/vintage machinery rallies	Golf courses
Birthday parties	Adventure play areas
Children's farm	Picnic sites
	Restaurant/teas/coffee shop
	Nature reserve/nature trails
	Historic battle re-enactments

(*Source*: Chaplin, 2000, p. 14)

6.4 Conclusion

This chapter has examined different aspects of contrasting processes that are shaping the character of modern agriculture. In highlighting the conflicting trends of specialisation and diversification it is clear that there are distinct geographical outcomes, with some large areas increasingly becoming devoted to specialist production systems, often tied to corporate retail and wholesale concerns. In contrast, diversification is often referred to in terms of its role in the survival of the small family-run farm. In this conceptualisation it may be seen as a process operating in areas that are spatially removed from areas of optimum production and specialisation. However, such spatial differentiation is frequently far from clear-cut, with many regions combining a complex mixture of specialisation and diversification, often on neighbouring farms. It is the various 'situational' factors of farming that contribute to this situation – stage in family life-cycle; other personal, sociological and psychological characteristics of the farmer and the farm family; proximity to an urban area; the flexibility allowed by the underlying physical environment; the impress of government policy. However, a strong factor in determining the extent of farm diversification, and especially the pursuit of OGAs in tourism or work off the farm, has been the role played by the farmer's wife. In examining the changing role of farm wives and women farm workers in general, the great significance of their role in agriculture in the Developing World was suggested. This therefore provides a direct link and introduction to the next two chapters, which endeavour to outline some of the key features of geographical work dealing with the agricultural geography of Developing Countries.

7 The 'other side' of globalisation: farming in Developing Countries

7.1 Differential impacts of globalisation

Most governments in the Western world subscribe to the primacy of economic growth, the need for free trade to stimulate the growth, an unrestricted 'free' market, limitations on government regulation, and consumerism allied to a Western corporate vision. It is this model that the West has promoted throughout the world, producing tendencies leading towards global homogenisation of culture, lifestyle and level of technological development. Within the ongoing arguments over the limits to the future extent of this globalisation are different views over who will benefit. In examining the experience of globalisation to date, there have been strong contentions that benefits are highly confined to those at the hub of the process or to those retaining particular connections to it. Elsewhere the overwhelming mass of humanity has either not experienced any significant GATT benefits or are actually being impoverished (Gallup *et al.*, 1999). These disbenefits have furthered the world divide between the West and the rest of the world, sometimes refashioned into a divide between North (Developed) and South (Developing) and, within the South, between the small wealthy urban elites and the countryside (Lipton, 1982). Moreover, in recent times trading relations between the rich North and the poor South have not changed sufficiently to enable Developing Countries to trade themselves out of poverty (Watkins, 1997). Indeed, recent analyses suggest that the growing openness to trade following the GATT Agreement on Agri-

culture in 1994 has had a negative impact on the income growth of the poorest 40 per cent of people in Developing Countries (FAO, 1999; Lundberg and Milanovic, 2000; Madeley, 2000). The tendency has been to accelerate food imports into the South, especially to meet the demands of small wealthy elites pursuing food consumption trends similar to those in the Western world.

There are differing views about the future impacts of globalisation and global capitalism on the South (e.g. Langhorne, 2001, pp. 133–50). Amongst the myriad of different projections is a simple contrast typified by the following ideas. On the one hand it can be argued that Developing Countries will increasingly be brought within the 'global orbit', so that not only will food produced by Western TNCs be sold in larger volumes in the South, but also the farmers and food processors of the South will play a significant role in this process. In effect, there may be an agri-food parallel of the 'economic miracle' associated with the production of electro-technical goods by Japan in the 1960s and 1970s and the south-east Asian 'tigers' in the 1980s. One scenario allied to this is that, through the increased development of free trade, many Developing Countries will be able to take advantage of the markets of the North for both primary produce and processed foods. Hence there will be, say, a Chinese- or Vietnamese-based agri-food company to challenge Nestlé or Heinz in the same way that Toyota and Hyundai have challenged established Western car manufacturers. On the other hand, the 'anti-globalisation' demonstrators at the Seattle and Genoa 'World Summits' have argued that freer trade simply plays into the hands of the rich North, permitting existing TNCs to further dominate the markets of the South. Protectionism practised by the USA and the EU, and the economic control exerted by the IMF, undermines the competitive abilities of farmers in the South, and maintains them in thrall to the economic interests of the rich North and a minority of rich and powerful individuals in the South. Hence farmers in the South do not receive fair reward for their produce and are often held in neo-colonial relationships with Western TNCs who control the processing and marketing sectors of the agri-food industry (Eicher and Staatz, 1990).

These two sets of opposing views have been aired as part of the Uruguay Round of the GATT negotiations, in the ongoing work of the WTO and at major meetings of the world's political leaders. Each view can draw upon support from different constituencies that are not necessarily readily divided between North and South, but with some key Northern interest groups and class fractions identifying with anti-globalisation ideas, and some farming and political interests in the South identifying with the advocation of free trade, e.g. the role of the Cairns group of primary producers of agri-commodities in the GATT negotiations (Robinson, 1993b; 1995; 1996a; 1996b).

It is clear that these multi-faceted differences of views about the trajectory of globalisation and its impacts on the South are tied closely to how the nature of the relationships between North and South are regarded. Much of the work under the broad umbrella of 'development studies' addresses this either directly or indirectly, and from various philosophical positions (Corbridge, 1995). The scope of this work is too broad and complex to tackle here. However, it is possible to take a narrower agricultural perspective that can illustrate key issues that agricultural geographers and researchers in closely related areas have addressed. In particular, this research has focused on the long-term impacts of colonialism (and neo-colonialism) upon agriculture in the South, and it is this focus that is pursued in this chapter. This entails consideration of the creation of a dual economy, a substitution of commercial production along capitalist lines for indigenous systems of production, the development of a distinctive 'agricultural transition', and an ongoing struggle for control over large-scale production between interests based in the North and the South.

7.2 The dual economy

A crucial differentiating factor between the world's farming systems is the use of capital. In the Developed World high capital investment in machinery and purchased inputs, such as chemical fertilisers, has reduced the need for farm labour whilst substantially raising output. In poorer countries, especially within the tropics, large-scale capital investment has not been widespread, and production has tended to rely on labour-intensive methods. This is not to imply, though, that such methods have failed to bring some increases in output. For example, in China, where population numbers have risen by over 278 million since 1980 (+27.7 per cent), food production has grown at a comparable rate. Here, retention of traditional hand-tool technology, increases in labour input associated with redistribution of the means of production, and improved marketing systems have made it possible to raise output sufficiently to keep pace with the growing demand, largely without recourse to substantial imports of food. Similar increases in output have occurred in a handful of tropical countries, frequently accompanied by the dissemination of improved crop varieties: contributing to increases in the global production of padi rice and maize of 55 per cent and 80 per cent respectively since 1985. However, even with a continuing reliance upon a large input of human labour on farms, agriculture has been unable to provide jobs for all of the rapidly growing rural population in the tropics, and large-scale rural to urban migration has been occurring for several decades. As the numbers of city dwellers have grown, so too has the need for farmers to produce food for the urban market. This has helped bring about the demise of subsistence farming systems in many parts of the tropics.

Subsistence farming is essentially food production solely for the purpose of consumption by farmers and their families. It is inexorably being displaced by production for sale, so that even remote parts of the tropics are beginning to farm on more commercial lines. This can have the effect of promoting increased ouput through the direct incentive of generating more income, but it also has had negative consequences. These have included the abandonment of ecologically beneficial traditional farming practices, thereby contributing to soil erosion and land degradation; the shedding of farm labour, which adds to rural unemployment or swells the numbers migrating to rapidly growing towns; and the production of non-food crops like cotton, coffee and tobacco. In certain respects the last development has had the most far-reaching consequences because it has placed farmers in poor tropical countries at the mercy of fluctuating world markets, exposing them to forces well beyond their own control.

In many parts of the tropics, though, and in some cases for over three centuries, subsistence farming systems have continued to exist alongside an imported system. The 'import' dates to the colonisation of the tropics by the European powers, notably Britain, France, Spain, Portugal and the Netherlands, and the creation of a dual economy in the agricultural sector of the colonies. On the one hand there were indigenous, often largely subsistence, farming systems, generally using available resources efficiently, but tending to achieve relatively low levels of output commensurate with their limited inputs. Alongside these indigenous systems the colonists installed new farming systems that were run along commercial capitalist lines. These systems introduced by colonists were often a type of 'plantation agriculture' in which there was large-scale production of tropical crops by a uniform system of cultivation under central management (Courtenay, 1981; Jackson, 1969). Under this form of production the growing of crops such as rubber, tea, coffee, bananas, sugar cane, cocoa and oil palm was undertaken throughout the tropics (e.g. Harrison, 2001). The demand for plantation crops was related solely to the world's industrial economies. The products themselves usually had little local value, but provided raw materials, basic foodstuffs and beverages for the industrial nations who were the colonial powers. During colonial rule the division between the plantation and indigenous subsistence sectors was usually clear, but the former developed at the expense of the latter, thereby restricting food production for local consumption. Furthermore, during the twentieth century, plantation crops increasingly displaced food crops on smallholdings within the indigenous farming sector (e.g. Robinson, 1989; 1998c).

Small farmers have been attracted by the cash returns for growing plantation crops, and many governments in tropical countries have encouraged their farmers to grow 'plantation' crops for export, e.g. as part of state-sponsored village regrouping schemes in Malaysia (Courtenay, 1990). Indeed, the demise of colonialism has often produced an increase in the production of plantation crops, as their export can provide countries with a vital source of revenue. However, this revenue can be highly unstable as it is reliant upon fluctuating world market prices. It has helped place both farmers and whole economies at the mercy of these markets (Drummond and Marsden, 1995b; Harrison, 2001; Middleton, 1999a). For example, more than three-quarters of the exports from tropical countries consist of primary produce, whilst two-thirds of their imports are manufactured goods. There are numerous examples of economies tied to just one crop and with little recent economic diversification, for example Uganda – coffee, Ghana – cocoa, Cuba – sugar. Ghana is an excellent example of the blurring of the indigenous and plantation sectors. It is the world's largest exporter of cocoa, but the majority of this crop is grown on farms of less than 5 ha.

Dualism (subsistence production versus an 'imported' system producing essentially for an overseas market) is also apparent in the contemporary colonisation of frontier lands for agriculture in the Developing World, especially in south-east Asia and south and central America, where it is not only the expansion of subsistence production that is occurring, but also development of ranching, large-scale plantation production, commercial logging and mining activity (e.g. Macmillan, 1993; 1994). Indeed, such colonisation is often mediated by the expansion of export agri-commodity production. This connects small farmers to the global market through the development of market institutions or through the farmers' own initiative

(Yarrington, 1997). Hence a central feature of the agricultural geography of the South, and especially the tropical belt, is the ongoing retreat of subsistence production in the face of an 'advancing wave' of commercialisation. This is now investigated by considering five different aspects of this process: the retreat of subsistence production, the impacts of colonialism on sub-Saharan farming systems, the changing character of women's farm labour in Developing Countries, the agricultural transition in Asia (especially southeast Asia), and the growth of the 'new' plantations.

7.3 The 'advancing wave' of commercialisation

7.3.1 The retreat of subsistence production

Traditionally the simplest forms of subsistence production, shifting cultivation, have been practised in situations where land is plentiful and there has been no need to maximise output. Natural processes of soil recovery after cultivation are utilised involving the land being left fallow, in some cases for several years. Given that this is likely to necessitate quite large areas lying fallow for long periods of time, the farmers' activities are moved from one piece of land to another. This may be achieved from either a fixed settlement base or, less commonly today, by moving that base when a new location for cultivation is required. This movement of both settlement and cultivation has long been practised in tropical rainforests, the term 'forest fallow' being applied to the system of fallowing in which land is left fallow for over ten years between short periods of cultivation (Figure 7.1). Shorter periods, of between five to ten years around a fixed settlement are termed 'bush fallow' and involve less effort in re-establishing cultivation. Even less is involved for grass fallows of up to five years' duration during which time only grasses and weeds are likely to invade the abandoned land. These generic names have distinctive regional variants, e.g. *milpa* in Central America (Ewell and Merrill-Sands, 1987), and *swidden* in Amazonia (Beckerman, 1987).

Analysis of these systems of fallowing by the Danish economist Ester Boserup (1965; 1981) led her to argue that increases in the intensity of cultivation do not take place in pre-industrial societies unless there is population pressure. It is this pressure that prompts farmers to produce a surplus beyond basic subsistence requirements. It provides an incentive for the farmer to intensify the farming

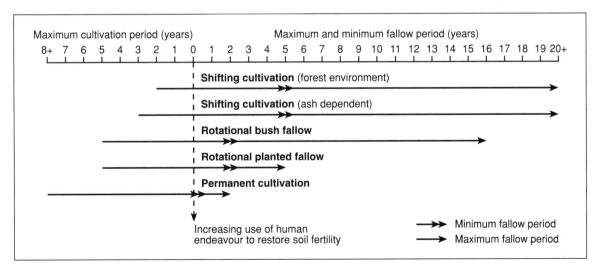

Figure 7.1 Some systems of land rotation in tropical Africa (based on Grove and Klein, 1979, p. 19)

system in various ways, usually by adding more labour input as, at least initially, there is no income with which to purchase inputs such as fertilisers, pesticides or mechanical equipment. Population increases are the catalyst for reducing the length of the fallow period, as land must be cultivated more frequently to feed the growing numbers. Therefore, in farming systems reliant on hand-tool technology, more labour input is added to the land, with more emphasis placed on mulching and weeding in order to maintain soil fertility.

Increased population pressure tends to be associated with fragmentation of land holdings and insufficient land for the rotation of fields with long fallow periods. There is also pressure for change arising from the desire of farmers and their families to participate in the market economy, which involves generating cash by selling farm produce or taking off-farm employment (Jokisch, 1997). Moreover, in many countries, governments have promoted the introduction of cash crops, with profound impacts on traditional farmers (e.g. Chaleard, 1996; Ganganapan, 1997). The growing diversity of income sources for farmers helps drive shifting cultivation out of existence (Adams and Mortimore, 1997; Coomes and Barham, 1997), often as part of the 'dynamic adaptability' exhibited by small farmers (Adams and Mortimore, 1996).

These debates have been refocused since Boserup's arguments in the 1960s that increases in rural population density lead to greater inputs of labour and technology per unit of cropland. Therefore, restrictions to the availability of land relative to labour may stimulate increased investment to improve land productivity (Boserup, 1965). Boserup's thesis involved population pressure as the engine for driving a move from shifting cultivation with low productivity, extensive land use and a long fallow period, towards permanent or semi-permanent cultivation with high productivity, a short fallow period, and the use of inputs to maintain and increase soil productivity (Rothenberg, 1980). Similar, though modified, arguments were advanced by Brookfield (1972; 1984; Brookfield and Hart, 1971) with respect to agricultural intensification in Melanesia.

It should be noted, though, that Boserup's interpretation is not shared by all researchers who have examined the link between population growth and economic development. Blaikie (1985, p. 24), for example, regarded Boserup's assertions of internal or 'autochthonous' innovations under conditions of population pressure as 'remarkably fragile' because they almost completely ignored consideration of the role of the state, relations of production and patterns of surplus extraction. Blaikie himself argued that adaptation in the form of new technology is rarely sufficiently far-reaching or fast enough to eliminate non-sustainable use of the natural environment. Moreover, there are a number of processes that reproduce themselves over time in such a way as to enforce this and to keep poor farmers in perpetual poverty.

Brookfield (2001) has recently added to the critique of Boserup, labelling her model as reductionist and unilinear, and arguing that to see intensification solely as a response to population pressure is a very partial explanation. He adds a wider range of factors, which include diversification of production and livelihood opportunities, investment and finding new ways of using and managing resources. Nevertheless, Allen and Ballard (2001), in re-evaluating Brookfield's own work, stress that both Boesrup and Brookfield had shown that significant associations do exist between agricultural intensity and population density, and their inter-relationship is a key element in the variability in non-industrialised agriculture. This argument is taken further by Kates (1993) and by Stone (2001). The former distinguishes between work by Boserup and Chayanov (1966), which relates production to household needs and wants under subsistence conditions (usually population-driven), and theories that relate production to demands from the market, sometimes referred to as theories of induced innovation (Hayami and Ruttan, 1985).

Aspects of both the population-driven and the induced innovation theories can be applied to change in agricultural systems in sub-Saharan Africa where there has been both autonomous intensification, associated with long-term adaptation to changing population pressures, and 'overnight' change brought about by responses to new national policies. Birch-Thomsen and Fog (1996) summarise the changes that have occurred by

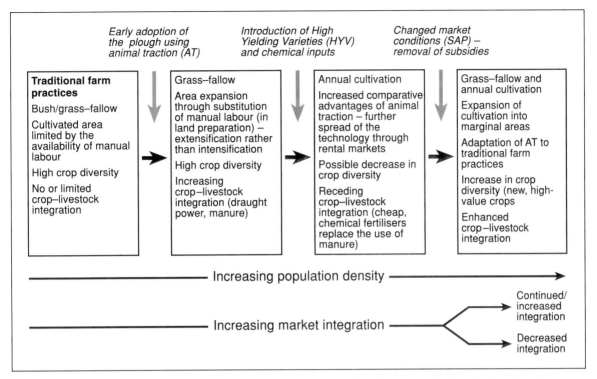

Figure 7.2 Change within agricultural systems in sub-Saharan Africa where the adoption of animal traction has taken place (based on Birch-Thomsen and Fog, 1996) SAP = Structural Adjustment Programme

means of the diagram shown in Figure 7.2, focusing on those systems where the adoption of animal traction has taken place, though each different scenario shown in the boxes will not necessarily occur at any given location.

Stone contends that the Boserupian model cannot be expected to account for all of the complexity in intensification, but that it should be seen as a basic starting point in considering how farming societies have intensified. He shows how more recent work has enhanced our understanding of this process, especially Netting's (1993) analysis of the cultural ecology of smallholders, with its nine characteristics of smallholder agriculture (Table 7.1). Netting's work shows how similar cultural and agricultural features can be found in diverse economic and ecological settings. He argues that 'subsistence versus commercial' is a misleading division that is effectively crossed by the notion of 'smallholder'; the smallholders are driven primarily by local factors, including labour, markets and the land, and that often the state and/

or development agencies, who have limited perspectives on this sector, interfere to the detriment of the production of 'intensive sustainable agriculture' that smallholders can achieve (see also Scott, 1999). Several good illustrations of this are provided in the case studies collected by Reij and Water-Bayer (2001), largely reporting work by African researchers and extension specialists. In these, small-scale farmers are recognised as having the ability to experiment and innovate to overcome adverse conditions and lack of appropriate support associated with the conventional 'transfer of technology' paradigm (see also Reij et al., 1996). This ability lies at the heart of Elliott's (2002) generally optimistic assessment of recent developments in sub-Saharan rural resource development, though she stresses the significance of community-based natural resource management as the key for future stability and greater sustainability (Elliott and Campbell, 2002).

Stone's summary of research on agricultural intensification by small-scale producers in the

Table 7.1 The characteristics of smallholder agriculture

- Smallholders live under conditions of scarcity
- Smallholders are not economically isolated. It is not necessary to hold constant the augmentation of farm income by wage work or the purchase of manufactured goods in order to see the pattern
- The smallholder economy is not devoted to subsistence production and use-value or to profit maximisation and exchange-value. It crosses this divide and shows it to be a poorly informed distinction. Smallholders may sell their goods and/or labour, but are never wholly dependent on the market economy; they always provide for a substantial portion of their own subsistence.
- Smallholder agriculture is sustainable as defined in energy terms: production is predictable and sufficient to feed the producers, and stable over the long run
- Smallholders exhibit a wide range of agricultural technology. Markets, population and other factors favouring higher production concentration do not necessarily result in technological change
- Smallholders farm intensively: they tend to have high rates of production concentration at the expense of low output per worker and per unit of input; they work harder
- Smallholder agriculture demands individual discipline, social co-ordination, physical skill and expert knowledge to carry out tasks such as hand-weeding, transplanting and fertilising, and the making and maintaining of dykes, terraces, ridged fields and irrigation canals
- Smallholders have great incentive for tenure security on their intensively used land, often with sellable, rentable, heritable rights (possibly co-existing with communally managed resources)
- Smallholder agriculture is run by households in which there is a need for high levels of skilled and co-ordinated labour, sustained use and improvement of resources, establishment and transmission of property rights, and multi-year storage and management of resources to minimise risk

(*Source*: Stone, 2001, summarising Netting, 1993)

tropics emphasises the way in which the social system, the economic system and the agri-ecological system interact. The apparent simplicity of the Boserup argument can be subverted by different characteristics emerging from any of these systems. Thus, social and cultural institutions may favour intensification or else encourage migration off the land; market incentives can encourage intensification and introduce variables that override the effects of local population pressure, injecting both higher risk and greater returns; and physical characteristics of the land can alter the nature of the interplay between these systems. Two examples from south and south-east Asia now illustrate the different outcomes within the overall movement towards greater commercialisation of production on smallholdings.

In Sabah, Malaysia, Lim and Douglas (1998) noted how shifting cultivation and a barter trade have tended to survive only in the most inaccessible areas, remote from newly constructed or improved rural roads. Elsewhere shifting cultivators have been increasingly tied to the market economy, with expansions in cash cropping and reduced fallow periods, with many parcels of land being continually used for hill rice for up to four years. The government has promoted this commercialisation of the rural sector, especially by encouraging cash cropping. Key outcomes have included abandonment of traditional practices such as community labour cooperation (*gotong-royong*), and methods to monitor soil fertility, such as maintaining fallow periods and using special hill rice varieties. These have resulted in reductions in labour input, land shortages (despite continuous production), increased use of pesticides and a general failure to maintain sustainable agricultural practices. These mirror findings by research elsewhere in south-east Asia (Hill, 1998), Sri Lanka (Yapa, 1998) and India, where the the incorporation of small farmers into the market economy has given rise to a new class of 'bullock capitalists' (Bentall and Corbridge, 1996).

A striking consequence of the retreat of shifting cultivation has been a widening gap between rich and poor. In the case of Sabah, this gap is between the remaining shifting cultivators, generally operating on the poorerst land in the remotest areas, and those farmers who have converted from rubber and hill rice to pineapples. Hence agrarian intensification and technical change may lead to increased food output, but it may also have unexpected disbenefits. A similar example is provided in a different context by Carney and Watts (1991) who showed that, in Gambia, intensification worsened the situation of rural women by helping to remove many of their traditional claims to land and placing upon them extra work burdens in newly irrigated fields.

Recent extensions of commercial production into areas long dominated by subsistence production have frequently created a dramatic change in the nature of economic, social and environmental problems faced in these areas. New environmental stresses have been especially apparent in fragile mountainous areas where commercialisation has occurred, as in the Himalalyas where land has come under more pressure to provide food, fuel, fodder and grazing. Marginal areas with very steep slopes have been required to provide food and fibre, leading to deforestation and substantial disruption of the hydrological cycle of major river basins (Rao, 1997). Effects have been felt on the adjoining Indo-Gangetic plains in the form of recurrent floods and decreased irrigation potential (Tiwari, 2000), though the relationship between processes in the mountains and on the plains is complex and is disputed (Ives, 1991).

Subsistence production in the Himalayas has traditionally depended solely upon supply of biomass energy from forests and pastures, with fertiliser supplied by a large cattle population. However, as cattle and population numbers have grown, carrying capacity has been exceeded in many areas, destroying young vegetation and furrowing fragile hill slopes. Increased road building has further encouraged this by enabling new areas to be more readily exploited (Tiwari, 1995). The impacts of greater ease of access of farmers to local markets are seen in a shift away from the traditional crop farming and animal husbandry system to production of vegetables, fruit, flowers and milk for sale. In nearly all villages situated along or near the roads, subsistence agriculture has been

virtually abandoned (Monench, 1989). In effect, market access has promoted the introduction of unsustainable resource development practices as evidenced most pertinently in the increased erosion levels in the Himalayas: five times higher than the rate prevailing in the past 40 million years (Valdiya, 1985).

Perhaps an even more pronounced demise of a shifting form of existence has been the sedenterisation of nomadic pastoralists in the arid and semi-arid zones of the world (e.g. Arkell, 1991). Here, for over a millennium, the traditional form of life was that of small groups travelling with their animals seeking available pasture and trading livestock and other products (Salzman, 1980). The variety of animals raised under this system was small: sheep, goats, cattle, horses, camels and donkeys. Some nomads, such as the Masai of East Africa, practised wet season agriculture but were mobile in the dry season. However, the establishment of international borders by colonial powers across the Sahara, the Sahel and the Middle East, and their enforcement by modern states, has severely curtailed the movements of many nomadic groups. Although this has often increased pressure upon the environment, other outcomes have been more positive, e.g. increased staus of women in society, and more diversified production (e.g. Turner, 1999). The institution of collectivisation and land tenure controls in other parts of the world have also been critical in eroding traditional ways of life and systems of production, especially in former Soviet Central Asia.

7.3.2 The impacts of colonialism on indigenous farming systems in sub-Saharan Africa

Colonial administrations in Africa frequently promoted mixed farming as a replacement for prevailing systems of shifting cultivation and nomadic pastoralism. Indigenous farming practices were criticised on a number of grounds, but primarily for failing to provide adequate levels of nutrition for the increasing population, causing unacceptably high levels of environmental damage, and for being incompatible with 'advanced' social and economic development (Sumberg, 1998). A fairly typical view is that cited by Sumberg, and contained in the following quote from an agricultural officer:

> In general, Kenyan Africans are not good agriculturalists...and in almost any area we feel we can vastly improve their methods. In fact, if one were to make an objective assessment of existing methods it would be found that only about 5 per cent of the total used and inhabited areas of the African land units in Kenya is under beneficial occupation; the rest is being slowly but steadily destroyed, or at best is static, even today (Brown, 1957, p. 69).

These ideas and the associated desire for radical reforms of native agriculture were part of widespread beliefs and assumptions about the local cultures, societies and economies. These included a desire to promote changes in land tenure arrangements in favour of individual ownership (e.g. Hall, 1936) and consolidation of small fragmented plots. Many of these prevailing colonial views failed to appreciate the diversity and complexity of African environments and farming systems. Thus, it has been argued that, by focusing on particular ecological and socio-economic conditions in semi-arid areas of West Africa, regional variations were ignored (Powell *et al.*, 1995).

Mixed farming was advocated for various reasons but chiefly because it was regarded as being more productive and efficient than existing systems and it could act as a means of reorganising rural society (e.g. King, 1939). By linking mixed farming to other reforms it was intended to create a new class of farmers, perhaps with direct parallels to the emergence of the so-called yeomen farmers in nineteenth-century Britain (Tiffen, 1976).

The essential characteristic of mixed farming is the integration of crop and livestock production, although this can take many forms and includes numerous different farming systems. The key elements within mixed farming are the positive interactions between crops and livestock. In particular, this involves the availability of livestock manure to fertilise crop production and animal power for tillage. A key aspect is replacement of the soil regeneration capabilities of the fallow period, which was central to shifting cultivation. This replacement may be practised in various ways, for example

involving the housing or penning of animals and the transfer of manure from such enclosures to the fields. Alternatives have included the use of composting, green manures, cover crops and trees, as applied in the case of green manure crops in south-west Nigeria in the 1920s (Vine, 1953).

These ideas on the efficiency of a mixed farming model were based essentially on perceptions of advances in European, and especially UK, agricultural productivity through implementation of the Norfolk four-course rotation and its variants. This crop rotation system and related systems was widely adopted in the UK during the eighteenth and nineteenth centuries and was closely associated with both institutional changes and the widespread use of manure, legumes and other fodder crops within mixed farming systems (Overton, 1996: Robinson, 1988a, pp. 60–1). It is clear that colonial administrators were strongly convinced that it was possible to transfer the apparent merits of these systems to Africa to attain similar results. Sumberg (1998, p. 302) feels that the special attraction of mixed farming was its ability to create a framework through which greater order and control could be brought to the 'chaos' of the African countryside, by replacing small, scattered plots, apparently haphazard crop mixtures, deforestation, malnutrition and free-roaming livestock.

Although the implementation of mixed farming schemes in Africa was not necessarily accompanied without concerns or identification of limitations (e.g. Rounce, 1937), nevertheless numerous such projects were launched, almost all neglecting to incorporate input from local farmers. Moreover, there were many problems pertaining to the adoption of mixed farming that needed to be addressed before projects could be implemented with realistic hopes of meeting their objectives. Problems included the specification of the amount of manure needed to maintain soil fertility, the appropriate ratio of livestock to arable land and of arable to pasture. Research programmes endeavoured to deal with these problems in the 1920s and 1930s, with widespread use of model farms as the context within which mixed farming technologies could be evaluated (e.g. Pedder, 1940). In practical terms, mixed farming systems were introduced in two ways: via resettlement schemes (accompanied by adoption of a completely new farming system) or through gradual integration of elements of mixed farming into existing farming practice, a division of approaches roughly comparable with current debates contrasting radical solutions with evolutionary ones. The resettlement scheme approach was popular in east Africa (O'Connor, 1978, pp. 30–3) whilst the more gradualist philosophy was more common in west Africa (e.g. Benneh, 1972). Yet few of the various schemes met their stated objectives. In the case of the resettlement schemes, the complexity of linking farming practice to other widespread changes could be self-defeating. In the incremental approach, farmers were often interested in only certain elements of the mixed farming model (e.g. Corby, 1941), and, whilst some successes were claimed (Faulkner and Mackie, 1936), these were not widespread, especially in traditional pastoral areas.

Sumberg (1998) asserts that a re-evaluation of African farming systems by 'experts' in the West has been forthcoming since the publication of Allan's *The African Husbandman* in 1965. This emphasised the complexity and variability of African farming systems, including shifting cultivation practices, now more appropriately termed rotational or bush fallow, and regarded as an efficient and sustainable system of crop production if sufficient land is available and population pressure is limited (Fairhead and Leach, 1996; Kleinman *et al.*, 1995). This changing evaluation has accompanied a similar reconsideration of the efficacy of mixed farming in the context of continuing problems affecting African agriculture, notably rapidly rising population numbers that require more productive local agricultural systems, and environmental degradation associated with deforestation, desertification and soil erosion. However, this has helped to return mixed farming to the centre of debates regarding the future direction of African agriculture.

These debates have moved on since Boserup's work in the mid-1960s, and it has been argued specifically that Boserup's model is invalid for Africa because of the continent's fragile soils, the presence of environmental deterioration following reductions in the fallow period, and intensification of land use without addition of supplementary inputs such as chemical fertilisers. Lele and Stone (1989) contend that there has not been a lasting transition of African farming systems from areas with a low population density to intensive systems in areas with high population density. Support for this latter view is given in recent work by Goosens (1997) in a study of cassava production in Zaire.

The growing of cassava dominates Zairian farming systems, being grown typically on smallholdings reliant on family labour, with 1–1.5 ha under the crop. Multi-cropping with maize and groundnuts is typical, with long fallow periods utilised to allow restoration of soil fertility. Cassava has been a traditional subsistence crop, though it has gradually become popular in growing urban markets, thereby stimulating increased production for cash sales. The response to this increased demand has been met almost entirely by traditional systems of production. Without ready access to improved technology or high-yield varieties, farmers have continued to employ slash-and-burn methods followed by hand ploughing. Table 7.2 lists the various factors that have helped prevent modernisation of these farming systems to meet rising demand.

Per capita output of fresh cassava roots has remained remarkably stable during the post-war period, though improved roads have helped widen the supply zone of the main urban centre, Kinshasa. The response to increased urban demand has been met by using more land and labour. Hence, the number of farms has grown, but at the expense of reducing the amount of land available for fallow. Without changes in technology, the result has been to degrade soil fertility and reduce yields. The reliance on a high family labour input has limited growth in physical output per farm. Goosen concludes that the farm population is only able to feed itself and a non-agricultural population of the same size, and hence the traditional farming system has reached its limits.

This theme of the link between farming, population numbers and environmental pressure has been a recurrent one (see for e.g. Cromwell, 1996). For example, much of the research conducted at a regional scale in Africa (e.g. using remote-sensing techniques) has noted environmental degradation, which has then been linked to population increase.

Table 7.2 Factors contributing to lack of innovation amongst Zairian cassava growers

- Low level of knowledge of modern farm inputs amongst rural people; highly traditional attitudes and actions associated with illiteracy and poor education, e.g. belief that soil fertility comes from one's ancestors, and traditional farm practices must be used to preserve it
- Inadequate distribution system for the farm inputs necessary to support desired technological evolution
- Imported inputs are too expensive and there is a low rural investment capacity
- Unstable socio-economic climate reinforces risk aversion
- Insufficient government budget for agricultural extension, resulting in low pay, lack of motivation, low status of the extension service, multiple tasks (e.g. tax collection, agricultural statistics), distrust from farmers, lack of logistical support, lack of overall co-ordination, lack of a record of farmer-proved and tested varieties and recommended cultural practices
- Low market prices because of inefficient marketing

(*Source*: based on Goosens, 1997, p. 38)

However, this contradicts work at a household or even plot level which reveals that farmers are improving the quality of their land through manuring, tree planting and various intensification measures (Gray, 1999; Kelly, 2001; Scoones, 2001). The complexity of the mosaic of land improvement and degradation requires an understanding of the factors determining patterns of land use, but in particular in examining the way in which different land managers (households, governments, corporations) gain access to resources and for what purposes. In tackling this issue, geographers have focused on key themes, as elaborated by Batterbury and Bebbington (1999): relationships between formal and informal institutions, the links between social control of institutions governing access and the dominant ideas regarding how resources should be used, and the political and economic dimensions structuring institutional and discursive influences on resource use and access.

The overarching structural control is exerted by the way in which state institutions determine who is able to use and benefit from land-based resources. However, there are also less formal institutions, relating to kinship, gender, group dynamics and connections to government that determine access to labour and land. Both sets of institutions are critical in determining land quality and how it varies across social groups, with different institutions exerting the dominant influence in different areas, for example government, community, and market and property institutions.

Each of these institutions may place a different demand upon the land resource, in part reflecting variable conceptions of what the resource is for and what may constitute appropriate management. Clearly, the prevailing institution will determine the long-term outlook for land use development and soil quality.

In seeking to understand the relationship between farming and environmental deterioration, Blaikie (1989a; 1989b) provided a chain of explanation regarding land degradation in which the 'higher' levels of the chain encompassed the relationships between the balance of power in a society and the global political economy. These relationships cascade down to the individual community and landholders, thereby ultimately exerting influence over land use. He argued that the relationships between the local level, the national and the global can provide both constraints and opportunities for local people. Blaikie himself stressed the power of colonial policy to alter traditional farming systems, for example through land appropriation by colonists. More recently a lasting legacy of the colonial period can be seen in the retention of plantation crops by smallholders in many parts of the Developing World. However, work in several countries is revealing that this has had a mixture of both positive and negative impacts on local economies and environments (e.g. Brierley, 1992; Gray, 1999).

Given increased pressures upon land resources, policy-makers have refocused their efforts on the economics of animal traction in both sub-humid and semi-arid areas (e.g. Jaeger, 1986; Pingali *et al.*, 1987). However, the outcome has not always been as hoped, with farmers using animal traction to raise labour productivity rather than output per unit area of land. Moreover, it was often applied to cash crop production rather than being applied to food production. Shortages of either labour or capital could undermine key elements in the mixed farming system by limiting the use of compost and manure or using animal power to farm larger areas less intensively. However, new dimensions to the mixed farming debate have been added in the last decade through concerns over the sustainability of African farming systems. There have also been some arguments relating to the way in which mixed farming systems themselves evolve. For example, the so-called 'U' model has been influential, with its four-stage sequence (Powell *et al.*, 1995): specialisation, intensification/integration (i.e. mixed farming), income diversification, and re-specialisation.

With respect to development of the second stage, attention has focused on the nature of crop and livestock interactions and nutrient cycling within the farming system (e.g. Powell *et al.*, 1995). There is renewed interest in the notion that the use of livestock can foster intensification of crop production, but with a recognition that the key to beneficial integration of crops and livestock depends on how animal power and manure are utilised. Hence McIntire and Gryseels (1987) distinguish between interaction and integration between crops and livestock, with the latter ideally involving a single peasant farmer controlling inputs of draft power, feeding of crop residues or fodder crops and recycling manure for soil fertility or for fuel. Alternatives to this reliance on manure are more intensive systems of fodder production and zero-grazing management systems, though neither is widespread in Africa. There has also been concern over the most appropriate ratio of ley to arable land. Generally, though, farmers have continued the practice of grazing fallow land, forest or rangeland even if animals are penned at night on cropland. This use of 'outside' grazing may impoverish the grazed areas, but there is little agreement on the area of rangeland needed to supply sufficient manure to support sustainable crop production. One possibility is to utilise inorganic fertilisers as supplements to manure and management of crop residues. However, the economics of this are usually prohibitive and availability of such fertilisers is often limited.

Sumberg (1998) concludes that, despite taking various forms and even through a move away from the dominance of Eurocentric thinking, mixed farming systems continue to be proposed for farming in sub-Saharan Africa. He comments that 'there is considerable irony in the fact that today's researchers and development workers who are searching for sustainability, end up with essentially the same model as the colonial administrators and agriculturalists who were searching primarily to rationalise and gain some degree of control over the African countryside' (p. 312). Moreover, it is clear that certain components of mixed farming are now much more common than at the beginning of the twentieth century, notably the use of animals to provide power and the use of manure for crop production. This reflects the ways in which farmers have adapted certain ideas from mixed farming to their own particular local conditions (Bourn and Wint, 1994). It is this active role of African farmers in assimilating some of the basic principles of mixed farming systems that lends some weight to the validity of long-term arguments in favour of such systems, though it has taken a long time for the importance of the knowledge and views of local people to be acknowledged by external researchers.

In concluding this consideration of the continuing advance of the 'frontier' of commercial agricultural production in sub-Saharan Africa it must be stressed that the progression of commercialisation has been far from smooth, especially as economic decline in some regions in recent decades has led to an overall decrease in food production per head. In Tanzania, for example, economic deterioration has presaged the withdrawal of peasant producers from the cash economy and their reversion either to production for subsistence or for informal or highly localised markets (Bryceson, 1990). This has led to many rural areas and small towns and villages in the country having insufficient local food supply whilst domestic agriculture's ability to meet the needs of larger urban food markets has been reduced.

Problems in the supply of horticultural produce to Tanzania's capital Dar-es-Salaam have been highlighted by recent research by Lynch (1994; 1995; 1999), with evidence of complex patterns of change in supply, partly related to the government's abandonment of its policy of socialist transformation. The research reveals the continuing importance of gender as a factor in affecting the nature and extent of agricultural commercialisation. In the case of horticulture in Tanzania, in the past the growing of fruit and vegetables was seen as 'women's work' as it was primarily for subsistence purposes, and so was re-productive rather than productive. However, the cash-earning potential of horticulture has been exploited by men, thereby producing a gender division in horticultural activities. Maintenance of commercial production has been dependent on factors affecting both existing and potential male commercial producers and those relating to the changing role of women in rural society. These factors have not been purely macro-economic or associated with government policy, but reflect the importance of local and individual economic, social and environmental contexts (Lynch, 1999).

7.3.3 Women's farm labour in Developing Countries

Women in sub-Saharan Africa produce and market over three-quarters of food grown locally. In terms of individual tasks, they are responsible for over 80 per cent of domestic food processing and storage, 70 per cent of hoeing and weeding, 60 per cent of harvesting and marketing, 50 per cent of livestock care and crop planting, and 30 per cent of ploughing (Atkins and Bowler, 2001, p. 315). Proportions are lower in other parts of the Developing World, but they are still highly significant allied to their responsibilities for the majority of housework and collection of water and fuel wood. Thus it is female labour that underpins the whole food system in much of the Tropics, an underpinning whose importance has grown in many areas as men have migrated from the countryside in search of urban-based job opportunities.

In part, this accounts for one-third of households in sub-Saharan Africa being female-headed. Yet, these vital roles of women in farming and within the household were often overlooked in development schemes implemented in the 1950s and 1960s.

Before the 1980s many externally funded development projects in LDCs failed to appreciate the role of women in rural society. It was frequently assumed that women were primarily confined to the private reproductive or domestic sphere (Marchand and Parpart, 1995; Mohanty, 1991). Hence, this denied women access to credit, rural extension services, land ownership and farm technology, and yet there are many societies in which women are either the principal producers of crops grown for household consumption or they play a significant role in providing farm labour and management. This 'invisibility' of women to development planners has extended to their roles in the management of rural communities through their contribution to the organisation of political, social and economic structures and the broader management of the environment and natural resources (Barrett and Browne, 1991; 1996; Shiva and Moser, 1995: Sittirak, 1998).

Moreover, when women were targeted in more recent schemes, constraints were revealed that limit their activities and hamper attempts to involve women in increasing agricultural productivity. For example, it is usually men who control common property resources and land rights (Quisumbing et al., 1995). Women have limited access to the credit facilities that might enable them to purchase modern technology. They have lower levels of literacy than men, and most of the advisors in agricultural extension services are men.

In India around 84 per cent of all economically active women are involved in agriculture. Here there is a positive correlation between agricultural growth rates and employment of female agricultural labour (World Bank, 1991). However, the extent of female labour in agriculture varies tremendously in the Developing World, and direct wage-labour for females is very low. In general, women in LDCs, and especially those from poorer households, balance the multiple demands of their labour time between the economic (wage earning and income-replacing work like collection of fuel-wood and water, care of livestock) and the domestic (cooking, cleaning and child care). In so doing they work longer hours than men (Kabeer, 1994).

In several parts of the world the agri-export economy has been associated with increased feminisation and seasonality of agricultural production. Bee (2000) gives the example of fruit exporting regions of Chile where female seasonal workers (*temporeras*) are employed for a few months each year, especially during harvest time when they work primarily in packing plants. Women dominate the labour force of the plants but also work in the fields where there is a much less marked gendered labour market. This work is highly irregular and relatively poorly paid, but it is linked to potential for greater empowerment and re-working of household relations (Meier, 1999). For those working in the most highly skilled fruit packing work, wages may rise above those earned by male workers, though traditional gender stereotypes have been slow to change. The outcomes of women's involvement in the agri-export sector are by no means clear cut, but are dependent on a range of factors, including type of production processes, working and pay conditions, women's family and household situation, position in the lifecycle, and alternative employment opportunities. In a comparable example, Kritzinger and Vorster (1996) refer to the growing control that women have been able to exert over their work lives on South African fruit farms. Key factors in this have been the development of neo-paternalistic management, improved labour legislation and farmers' changing perceptions of female labour.

Wickramasinghe (1997, p. 14) describes women in Sri Lanka as 'the silent managers of the environmental resources of the village ecosystem'. They work long hours in the fields, attending multiple tasks. If household tasks are included, they work 50 per cent longer than the men whose work is less complex and diverse, and characterised by clear peak and slack seasons. In a study of the dry zone of Sri Lanka, it was shown that greater reliance upon women's work on farms was occurring during farming's off-season when over half the men migrated to service centres for off-farm employment (Wickramasinghe, 1997). This has been a common phenomenon in parts of sub-Saharan Africa, with women playing a major role in producing and gathering crops for family consumption. Yet, the norm has been for women to be unable to make independent decisions about farming activities because they do not control the resources. One consequence of this has been the marginalisation of their contribution in the eyes of agricultural extension services and development agencies, which have been the source of information on new technologies, new crop varieties and their management needs. This information has tended to be directed at men, and if new practices have subsequently been adopted, it has been men who have had the responsibility for managing them. In the Sri Lankan dry zone this has helped push women's farming activities towards common lands and forests. Their views have often been ignored by the outside agencies, who have tended to dismiss their farming practices as traditional, less technical, undocumented and not scientifically tested. This has reinforced the existing patriarchal social system in which women's roles and opinions are considered subordinate to men's.

7.3.4 Asian 'agricultural transition'

In every Asian country, with the exceptions of Japan, South Korea, Taiwan and Singapore, there are now more farmers than there were in 1960, though significant proportions of these farmers also obtain income from off-farm activity. Excluding China, for the countries of east and south-east Asia, this meant an increase of 18 million farmers between 1960 and 1992 (the figures for China show a doubling in the agricultural workforce over this period but data are unreliable). Hill (1997a) refers to one aspect of this as the 'piling up' of rural poor because of rapid population growth and limited structural change to the economy, the latter being most readily recognised in China and the Philippines. The countries with the slowest structural changes tended to have the most substantial growth in the number of farmers.

Hill (1997a) proposes the model of agricultural transition shown in Figure 7.3. This is a simple stage model in which the trajectory of change is

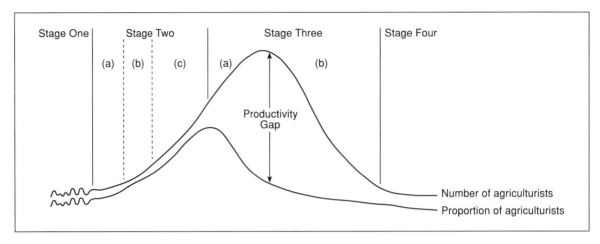

Figure 7.3 Agricultural economies: the move to modernity (based on Hill, 1997a)

Table 7.3 An agricultural transition model for south-east Asia

1 High proportions employed in agriculture but the agricultural workforce, like the population as a whole, is small and fluctuates. This reflects the classical Malthusian controls. Agriculture is fairly extensive because of relatively little pressure on the land

2 (a) A sustained but slow rise in population. Imported manufactures displace some craft industry, maintaining high levels of employment in agricultrure. Agricultural intensification begins
  (b) Imperialistic control accelerates trade, agricultural intensification and economic diversification. The proportion of agriculturalists falls but not their numbers. In areas of high population numbers there may be agricultural involution, i.e. a decline in labour productivity in agriculture (Geertz, 1963)
  (c) Trends in the previous stage continue, but in Asia this continuation occurs within a post-colonial society. Production grows, agricultural colonisation ceases and the proportion of workers in agriculture begins to fall, though numbers increase. The productivity of agriculture in relation to other sectors starts to fall and a 'productivity gap' starts to open

3 (a) The numbers of agricultural workers in the rural areas increases
  (b) A fall in the numbers of agricultural workers begins as the rural poor are drawn increasingly into the urban labour market. The curves for the numbers and proportion of agricultural workers start to converge. Land consolidation may occur and population growth rates may fall

4 The number and proportion of farm labourers in the workforce is small as is the contribution of agriculture to total production. There may be a dualistic structure in which small, 'uneconomic' family farms co-exist with larger fully commercial units. The size and proportion of the agricultural labour force is stable.

(*Source*: Hill, 1997a)

---

towards a modernity in which there are both low numbers and proportions of workers on the land (labelled 'agriculturalists' in Figure 7.3). The stages in the model are shown in Table 7.3. Using data from 1960 onwards Hill (1997a; 1997b) compared the experiences of several Developing Countries from east Asia and Latin America, with respect to changes in numbers in agriculture and the proportions of economically active population in agriculture. For many countries this reveals a characteristic 'cross-over' pattern, as shown in Figure 7.4. In several cases (e.g. Philippines, Peru,

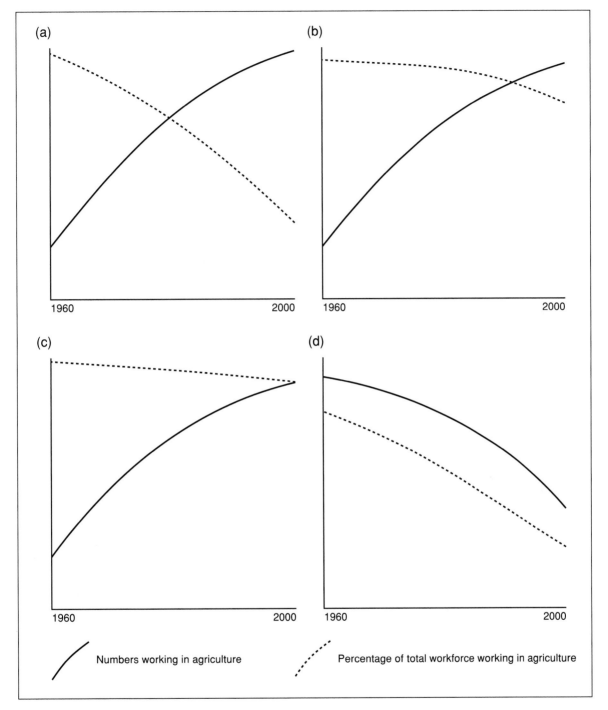

Figure 7.4 Evolution of numbers and proportion of agricultural workers in the workforce, from 1960: (a) cross-over in the 1970s; (b) later cross-over; (c) delayed cross-over; (d) early cross-over (based on Hill, 1997a)

Malaysia), the cross-over occurred in the 1970s (Figure 7.4a), and only slightly later in the cases of India and Vietnam (Figure 7.4b). Others (e.g. China, Laos) are yet to reach this stage (Figure 7.4c). However, in all of these countries mentioned as examples the size of the agricultural labour force has continued to rise. For more developed countries (e.g. South Korea and Japan), the agricultural labour force has been falling for some time (since the mid-1970s in South Korea), and hence they are at a different stage in the cycle (Figure 7.4d).

Hill argues that all the Asian and Latin American countries have passed through Stage One, and most through Stage Two. Most Asian countries and Pacific Rim Latin American countries (except Chile) are in Stage Three (a), with China, Malaysia and Ecuador about to move into the later phase of that stage.

Another key differentiating feature for the major Asian agricultural producers has been their experience with respect to growth of output. Between 1965 and 1996 half of these producers experienced negative productivity growth due to losses in technical efficiency and stagnation in technological progress (Suhariyanto and Thirtle, 2001). There was no evidence of a convergence in agricultural productivity for these countries, with the less productive falling further behind. Moreover, whilst sub-Saharan Africa is often portrayed as the economic 'sick man' of the world, and Asia overall had a greater agricultural total factor productivity than Africa in this period, three of the five African regions (east, central and south) grew faster than any of the Asian regions.

Rigg's (2001) analysis of the changes affecting agriculture in south-east Asia provides some detailed examples to accompany Hill's broad-brush picture. He argues that, even in areas that outwardly seem to have retained 'traditional' farming systems, there is an underlying current of change that often gives rise to strange paradoxes, such as agricultural labour shortages in a region with over 500 million people. These are reported even in densely settled localities in Thailand (Funahashi, 1996), the Philippines (Kelly, 1999) and Malaysia (Kato, 1994). The attractions of better-paid non-farm work have been strong, and have hastened a rural exodus as well as promoting agricultural

mechanisation and higher wages for farm workers (Rigg, 1998). However, farm work has also become much less attractive to the young, who often prefer no work at all to farm work. Hence the farm labour force is aging rapidly. In addition, traditional roles of men and women in farming are being reshaped by the changes, though not consistently across the region. In general, it is the poorer families who regularly and persistently cross culturally established gender boundaries (e.g. Elmhirst, 1997).

In some areas agriculture is being squeezed out by industrialisation of the countryside. This can be beneficial to rural communities because it is associated with the provision of more opportunities and can possibly be of benefit to those farms that survive by generating increased demand for agricultural output, promoting farm-based innovation and more capital for agricultural investment (Grabowski, 1995; Rigg, 2001, pp. 136–7). However, Hart (1996) sees the inter-linkages between farm and non-farm activities in the industrialising Asian countryside as less positive in terms of impacts upon farm households. This echoes Bryceson's (1996) findings for Africa in which 'de-agrarianisation' is occurring, typified by occupational adjustment, reorientation of livelihoods, social re-identification and spatial relocation. The evidence for this can be found in the growth of abandoned agricultural land, though Rigg (2001, p. 119) argues that this in part reflects the rise of land speculation in more diversified economies. There is no doubt, though, that the countryside of southeast Asia is changing rapidly, blurring divides between rural and urban, between different classes, and between the roles of men and women.

7.3.5 The new 'plantations': exporting high-value foods

In those Developing Countries with a history of exporting plantation crops, the traditional export commodities have often been complemented by exports of high-value foods, including fruit, vegetables, poultry and shellfish (see Grigg, 2001). In 1990 these accounted for 5 per cent of global commodity trade, one-third of which came from the Developing Countries where they exceeded the

Table 7.4 The impacts of large-scale agribusinesses and corporate retailers in Developing Countries

- They contribute to sharp inequalities in income, productivity and technology compared with the sector producing domestic staples
- Labour-saving innovations by agribusiness mean that numbers employed are reduced, with many becoming de-skilled, part-time or seasonal. Specialist export experts often brought in
- The benefits of intensive agriculture accrue disproportionately to foreign investors and, in the case of nationally owned estates, to urban-based people and local elite groups
- Large estates, especially those of agribusiness TNCs, often take the best land for export crops. This land absorbs the most inputs, investment and expenditure, even though it occupies only a small proportion of the total agricultural area. Unable to compete, localised agricultural production often begins to break down
- Many of the agricultural practices of agribusiness TNCs are unsustainable. Environmental degradation is a real problem and there are reported instances of groundwater depletion and agribusinesses moving into new areas when soils have become depleted.

(*Source:* Ilbery, 2001, pp. 260–1)

value of exports of coffee, tea, sugar, cotton, tobacco and cocoa (Ilbery, 2001, p. 260). Friedmann (1993) notes that five countries (Argentina, Brazil, China, Kenya and Mexico) accounted for 40 per cent of such exports. She termed them the 'newly agriculturalising countries' (NACs). Although this development reflects technical changes and the liberalisation of world agricultural trade, a vital catalyst has been dietary changes in the Developed World. Hence, the high-value foods are being produced and then exported to satisfy the tastes and lifestyles of certain groups of consumers in the Developed World. This has enabled pre-existing agribusiness TNCs to expand high-value food production in the NACs, utilising low-cost labour, government support (e.g. through structural adjustment programmes), globalised communications links and knowledge of the growing niche markets of the Developed World (Goodman and Watts, 1997). This development in the NACs is underpinned by intensive production techniques, with strong quality control from planting to storage, to transshipment and point of sale. This control has to satisfy corporate retailers who themselves need to satisfy the consumers' demand with respect to size, shape, colour and content. Marsden (1997, p. 177) notes that in establishing and implementing quality conditions a process of social and economic differentiation is reinforced in producing regions whereby smaller producers may be excluded from globalised food networks.

The corporate retailers can develop their own regulatory systems to control supply of key food products (Guy, 1994). This control can significantly affect regional economies and crop sectors in Developing Countries by linking them closely to global systems of food production and consumption. The multiplicity of impacts upon these countries can be summarised as shown in Table 7.4.

Concerns have grown over ethical aspects of production and trade, especially relating to renewable natural resources from Developing Countries, and to the treatment of workers and producers within farming systems there. 'Ethical trade' has included fair trade agreements, safe working conditions for disadvantaged producers and employees, and sustainable and environmentally safe natural resource management. In part the concerns have been both consumer and trade driven. For example, writing in 'consumer theory' refers to ethical consumerism as the 'fourth wave' of consumption, seeking to 'reaffirm the moral dimension of consumer choice' (Gabriel and Lang, 1995, p. 166). It is a consumer response that links the global and the local, but with global concerns as a key component (Bell and Valentine, 1997). There is much in common between this consumerist concern and the growing debate about the morals and ethics of international trade (Brown, 1993). For example, the creation of the WTO has highlighted the absence from the trade agenda of 'issues of sustainable resource management, the regulation of commodity markets, and poverty reduction strategies' (Watkins, 1995, p. 110).

The essence of ethical concern over production and trade is encapsulated in the following quote from Browne *et al.* (2000, p. 71):

> the improvement in trading relationships through ethical trading, enforced by organic concepts of production, contributes to the accumulation of both natural and social capital, through greater sustainability of natural resources and increased access by producer groups to networks of production and trade.

In developing these ethical arguments, Whatmore and Thorne (1997) have shown how traditional commercial networks for some 'plantation' products now exist alongside new networks associated with concerns for rural social justice. They focused upon trade in coffee. The long-established commercial network (Figure 7.5a) has been based on commercial imperatives in which an unequal power relationship exists between numerous small-scale producers of coffee and relatively few dealers. These dealers sell to processors who mass-produce coffee for sale as globally recognised 'brand' coffees. In this system profits heavily favour the end producer and the retailer. The alternative system that has emerged in the last two decades has a different arrangement or 'mode of ordering' (Figure 7.5b) based on partnership, alliance, responsibility and fairness. This gives greater power to small-scale growers working within locally based cooperatives. In this 'fair trade' coffee network, growers are paid a guaranteed minimum premium price whilst maintaining critical parameters of quality control and marketing deadlines.

Non-governmental organisations (NGOs) such as Oxfam have been crucial to the development of fair-trade networks, encouraging the growth of trading companies like Cafédirect to emerge as the key link between local producers in the Developing World and consumers in the Developed World. However, the changing nature of the market in the Developed World has also been significant, with increasing numbers of consumers prepared to pay a higher price for a fair-trade product, such as Cafédirect's ground and freeze-dried coffee as opposed to a conventionally marketed product.

Renard (1999) discusses the creation of the fair coffee networks which enable small producers to enter the global market under relatively favourable conditions. When purchasing coffee with the labels 'Fair Trade', 'TransFair' and 'Max Havelaar', the consumer is assured that the coffee has been purchased from democratically managed cooperatives of small producers, with the role of intermediaries reduced to a minimum, and a price that guarantees fair pay for the producers' work. Approximately 250 cooperatives in Latin America and Africa are included in the Max Havelaar–TransFair registry. However, the proportion of the European market taken by fair-trade coffee is still very small (< 5 per cent).

Coffee is also a commodity often associated with the global link to frontier development (Dean, 1995, pp. 178–90; Roseberry, 1995, pp. 3–10), for example with significant developments associated with this crop in Vietnam in the 1990s (Deng, 2000; Tan, 2000). The impact of the far-distant end-market for the green coffee beans in the Developed World can be seen upon the smallholder growers in terms of the demands upon their production processes and land management. Hence, although ultimate control of coffee production, marketing and processing may rest with corporate actors such as TNCs and the state, the global 'chain' starts with the small grower and their coffee growing to produce coffee of the requisite quality. Knowledge about that desired quality may be rudimentary at the new frontier of production, but it becomes enshrined within the surviving coffee-growing community by way of inter-personal exchanges in the first stages of the marketing of the crop. This usually involves a process of certification and sales through companies with export licences. So, whilst the corporate actors may not have exercised direct control over the colonisation process and smallholders' desire to grow coffee as a cash crop, there are a series of long-distance links that impact upon the farmers and tie them to the quality concerns of the global market.

This type of complex linkage between small producers in Developing Countries, at one end of the production chain, and the consumer in the Developed World, at the other end of the chain,

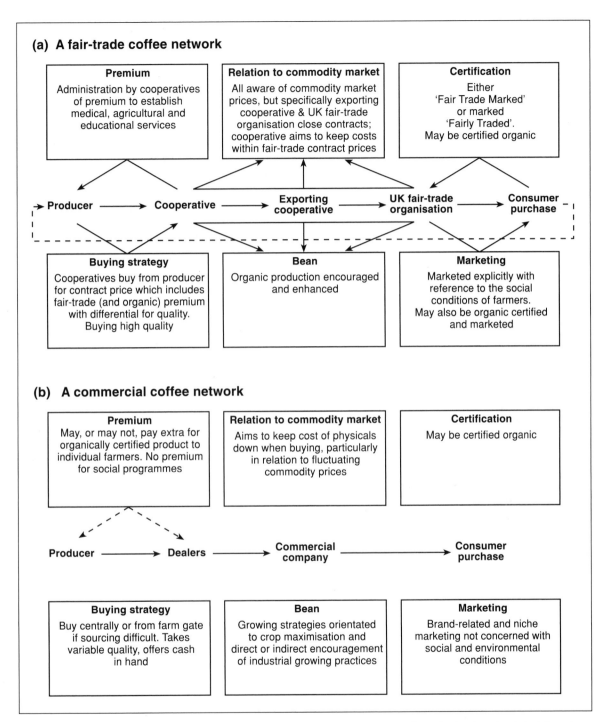

Figure 7.5 Network 'strengthening': fair-trade and commercial coffee networks exhibit distinctive 'modes of ordering' (based on Whatmore and Thorne, 1997, p. 300)

clearly involves many intermediaries, often with linkages regulated by national and international agreements that can affect the nature of production at farm level (see Eaton, 1998). The growing of 'ethical' coffee is one example of this in which relatively new considerations regarding producers' rights have entered the equation. However, a more common example is that provided by the 'traditional' plantation trade, in which the power of the growers in Developing Countries can be severely restricted (Maizels, 1997). In recent times the best example of this has been the growing of bananas in Developing Countries and their export to the markets of the Developed World.

7.4 Banana wars

7.4.1 The Lomé Convention and EU Council Regulation 404/93

On 1 July 1993 the EU implemented Council Regulation 404/93, a consequence of the Maastricht Treaty, which had mandated the creation of a Single Market for all goods and services among the (then) 12 member states. Regulation 404/93 attempted to deal with contradictions pertaining to common importation of bananas, thereby replacing the previous situation in which each member had its own banana importation policy. The contents of the regulations are summarised in Table 7.5.

The Regulation contributed to the EU granting of favoured nation status under international treaty to 71 African–Caribbean–Pacific (ACP) banana-

exporting countries. These are largely small producers who were granted a collective permit of a maximum of 857,000 tonnes of bananas per annum to be imported free of duty. This was intended to protect growers in countries such as the Windward Isles from being undercut by production from Central and South America where large TNCs, like Cincinatti-based Chiquita, have substantial mechanised plants to prepare bananas for export. As part of their protest over the distortions to free trade associated with this policy, the USA instigated sanctions against a range of EU products worth US$600 million, including Scotch whisky and French wine and cheese. Britain and Italy were targeted, with exports of US$100 million at risk initially.

Regulation 404/93 does not provide free trade in response to market conditions as it operates the equivalent of a quota system for three different categories of producer. For producers in the 'remote' EU regions (e.g. Madeira, Guadeloupe, Canary Islands) it specifies the amount of production eligible for duty-free entry to the CAP. It also defines the volume of 'traditional' ACP production, basing the volume on peak production years (e.g. Dominica has a market share of 71,000 tonnes). For 'third country' bananas (primarily from Latin America) it established a two million tonnes tariff quota, compared with 2.4 million tonnes from this source prior to the Regulation. When challenged before the GATT by Costa Rica, the EU increased the total tariff quota and implemented quotas by country for the Latin American producers, also contested by some producer countries, notably Ecuador.

Table 7.5 Key features of EU Council Regulation 404/93

1. Protection of EU consumers that are best served by a system that lowers or maintains price levels. However, the new policy does not accomplish this in several member countries, and it is largely negated by the following two points as bananas produced within the EU are heavily subsidised under the CAP, whilst imports from the ACP countries are more expensive than those from Latin America

2. Protection of EU banana production in 'remote' regions such as Crete, the Algarve, the Canary Islands, the Azores, Madeira, Guadeloupe and Martinique

3. The fulfilment of the EU's Lomé IV obligations to ACP producers

---

The Regulation established a new system for granting licences to import bananas from third countries, with 30 per cent of the two million tonnes tariff quota being reserved for European companies that traditionally operated in the ACP countries but not in Latin America (generally termed the 'dollar zone' in the EU). Therefore this provided new opportunities for European companies whilst increasing competition for those based in Latin America. However, if the WTO becomes a more effective enforcer of free-trade principles than the GATT, then this and the rest of the Regulation may be contested more strongly, and there is the prospect of a more open globalised trade in bananas. This is likely to be encouraged initially by the cessation of the series of Lomé Conventions when Lomé IV expired in 2000 (Wiley, 1998). The Lomé Convention, first signed in the 1960s and subsequently renewed four times, offered special access to the EU market for over 60 ACP countries.

7.4.2 Banana wars and Costa Rica

The EU Regulation had significant effects upon the banana industry in Costa Rica. In 1993 growers scaled down their operations following the restrictions on the EU market and the subsequent fall in price of bananas in the USA as this market received the surplus. However, there has been some recovery subsequently as Costa Rica has been able to fill the gap caused by the failure of EU and ACP producers to fill their allocations, albeit at a high tariff for the Costa Rican exporters. These exporters include a number of US-based TNCs, who have responded to falling prices in the USA and uncertainties in the US market by focusing on higher quality production and reductions in supply from 'independent' growers.

Costa Rica typifies the neo-colonial nature of banana production in Latin America and the Caribbean, based upon an initial plantation system run by the United Fruit Company (UFC) whose founder was granted 325,000 ha in the late 1890s in return for constructing the country's first railroad. The UFC had monopoly control of the industry for several decades, operating large monocultural units almost as a state within a state,

with few links to the rest of the regional or national economy. In part this separation or isolation gave rise to the term 'banana republic', but it also symbolised the political power wielded by UFC in the country and some of its near neighbours (e.g. Guatemala). The UFC took control of virtually all phases of the industry except for retailing; it controlled the labour supply to the plantations and used corruption of local and national officials to advance its interests. However, these interests were focused on generating profits for the business whose headquarters were based in the USA, so that little money accrued to the public exchequer in Costa Rica. In Costa Rica, banana production took place along the previously little developed and sparsely populated Caribbean coast. In contrast, banana-growing on several of the Caribbean islands was developed by the colonial powers in well-populated areas already utilised for agricultural purposes.

Under the UFC's dominant control, banana exports from Costa Rica became the second largest in the world, with a highly efficient system of production utilising a good railway network connecting the plantations to key ports at Limon and Golfito or with nearby Panamanian ports. The large plantations developed efficiencies in the purchase of pesticides, fertilisers and other inputs, such as cartons for export of the crop. Banana production has remained concentrated along the humid Caribbean coast, which now accounts for over 97 per cent of exports.

The UFC developed as a vertically integrated TNC ideally suited to deal with exporting a perishable product with a short shelf-life. Control over as many stages of production as possible was desirable and hence the development of strong forward and backward linkages in an efficient and unified structure. The UFC owned the plantations, the railroads transporting bananas to the port, the ships carrying the fruit to North America or Europe, and the ripening facilities and initial stages of marketing in the destination country. Additional profits were generated at all stages of the industry controlled by the company, a critical factor as most of the profit is realised after the crop is harvested. Only after the USA had passed anti-monopoly legislation in the 1950s, to promote greater competition between US-based companies in the region, was the UFC forced to relinquish its monopoly status in Costa Rican banana production. Del Monte and Standard Fruit, both US-based companies, set up similar vertically integrated enterprises. Independent growers also appeared at this time, purchasing land from the TNCs, but dependent on them for transportation and marketing of the bananas (Wiley, 1998, p. 71). This simply enabled the TNCs to spread the risks involved in the growing of bananas at a time when the UFC's plantations in Panama were experiencing disease problems.

As outlined by Wiley (1998), in 1974 a cartel called the Union of Banana Exporting Countries (UPEB) was established to help retain more of the income from banana production in the exporting countries. The cartel included the two leading exporters, Costa Rica and Ecuador, and created sufficient cooperation for several members to introduce a per-box export tax on bananas. This was strongly opposed by the TNCs. Its value has fluctuated, though Wiley (1998, p. 71) asserts that it represents 'the primary public sector financial benefit realised by the host countries from the industry'.

7.4.3 Banana wars and Dominica

Export production of bananas dominates the Caribbean island of Dominica and its Windward Islands neighbours. On Dominica, plantations using slave-labour were established by the British in the eighteenth century, growing coffee, cocoa, vanilla beans, oranges and grapefruit. When over-reliance on one of these failed, another major commodity was substituted. This was the case with bananas in the 1920s when lime trees had been damaged by hurricanes. In 1952 the UK's Colonial Office allowed Geest Ltd, a supplier of flowers and bulbs to the UK market, to be the preferred exporter of bananas from Dominica (and later from the Windwards) to the UK. This provided competition for the UFC who were the principal American-owned banana producers in the Caribbean, having taken over Fyffes Ltd who had operations in Jamaica. In 1966 Fyffes and Geest reached an agreement to supply the UK market on a 48:52 per cent basis, with the UK government supporting this via quotas and tariff-free entry of bananas, which helped offset the higher costs of Caribbean production compared with those of Latin America (Nurse and Sandiford, 1995, p. 45). However, although Geest was the sole exporter of Dominica's bananas, production was from independent growers whose numbers grew from the early 1970s via land reforms that broke up larger units and encouraged owner-occupier tenure by peasant farmers. This has helped accentuate the dominance of banana-growing on smallholdings: 5779 registered banana farmers (20 per cent of the labour force) cultivated 4676 ha of bananas in 1993, yielding 57,126 tonnes of bananas (Wiley, 1998, p. 70). Compared with most Latin American production, yields are lower and costs higher. However, the efficiency of Geest's operations and the regular output of bananas has ensured a regular cash flow for the growers. Until 1996 the growers sold their crops to Geest via the Dominica Banana Marketing Corporation (DBMC), though the price was largely determined by Geest based on its likely returns from sales in the UK. This meant an unfavourable pricing system for the farmers, to whom less than 20 per cent of the eventual retail price accrued (Nurse and Sandiford, 1995, p. 51). In 1996 Geest's operations in the Windwards and Dominica were purchased by a consortium consisting of the islands' governments and Fyffes. This can be seen as an attempt by the governments to maximise the outcome of the EU's banana import regime. It provides an opportunity for the industry to be overhauled, for diversification into other commodities to be introduced and for greater government control over key components of the industry, notably shipping.

Since Geest became involved in banana production in Dominica, the UK has imported over 80 per cent of the country's output per annum. With assistance from preferential access to the EU market through the Lomé Convention, the second largest overseas buyer has been Italy, which takes the majority of the remaining output. The Convention has supported the STABEX programme, which aims to smooth price fluctuations that have been such a problem for primary commodity producers. In effect the 12 banana industries supported under

Lomé have been enabled to obtain higher prices than under the operation of a free market, and hence references to its presence as representing 'banana aid' (Borrell, 1994, p. 33). The possibility of the removal of the preference system with the cessation of the Convention poses a major threat to Dominica and the other ACP producers (notably Jamaica, Belize, the Ivory Coast, Madagascar, Cameroun, Somalia, Cape Verde and the Windwards). Although ACP producers now have guaranteed shares of the EU market under Regulation 404/93, the greater competition under the new system has already reduced prices obtained by growers in Dominica, helping to drive some of them out of production. Increased competition from Latin American producers threatens to undermine banana production throughout the Windwards and highlights the precarious nature of the economics of countries reliant on exports of one staple crop. Wiley (1998) suggests three possibilities for the diversification needed by these countries:

- A move to products where competition is based on uniqueness and high quality rather than price and volume, e.g. Dominican coffee for specialist coffee retailers in Europe and North America.
- Use of temporal 'windows' for marketing selected crops during specific months to fill gaps left by larger suppliers, e.g. Mexico's exports of cucumbers to the USA during the window created by the seasonality of production in California and Florida.
- Development of relationships with individual retail chains in targeted countries. As yet none of these has offered the regular payments, transport arrangements or marketing network that bananas have provided and hence for Dominica and the other ACP banana producers the future is uncertain.

7.4.4 The aftermath of the banana wars

The USA has argued that the EU banana import regime, which allows preferential treatment for former European colonies such as the Windward Islands, is unfair to South American distributors, such as Chiquita (formerly United Fruit), most of whom are owned by US companies. This complaint has been taken to the WTO, which condemned an earlier EU preferential banana import system. EU spokespersons contend that changes to this system in 1998 now comply with WTO requirements whilst the Americans reject the changes as cosmetic and 'a mockery' of the global trading system and its rules. The USA may be prepared to support this view by doubling tariffs or establishing prohibitive duties of 100 per cent on selected EU goods, thereby undermining the authority of the WTO by acting prior to the WTO having the chance to make a new ruling on the dispute. Goods from Denmark and the Netherlands may be omitted from these measures as these two countries voted against the European Commission's revised banana imports regime in 1997. The political undertones of this potential dispute are clear, with the US government taking their claim to the WTO within 24 hours of Chiquita Brands making a US$500,000 donation to the governing Democratic Party, not to mention the opportunity presented to help reduce America's trade deficit which stood at a nine-year high at the end of 1998. American TNCs already control over three-quarters of the European banana market.

Compromise over trade in bananas was reached in 2001, prompting the USA to lift its US$191 million in sanctions against the EU, but this depended on special exemptions ('waivers') from WTO rules. Some key banana-exporting nations have objected to this, notably Guatemala, Honduras, Nicaragua and Panama, and so threaten an effective veto under WTO unanimity rules. The exemptions are required by the EU in order to maintain its preferential tariff quota for the ACP countries, which otherwise contravenes WTO norms. The EU requires the waivers for the lifespan of the Cotonou (ex-Lomé) Agreement between the EWU and ACP countries, which runs until the end of 2007.

Following the WTO agreement for the EU to increase its accessibility to non-ACP banana producers, the price of bananas worldwide has fallen. As a result producers in the ACP countries are being forced out of business: in 2000 only one-third of the Windward Islands banana producers recorded in 1993 were still growing bananas; half the members of the St. Lucia's Banana Growers' Association went out of business between 1992 and 1997, and one-third of the remainder have since ceased operating (Ryle, 2002). With the near removal of the EU quotas the banana exports from St. Vincent and The Grenadines have fallen from US$120 million in 1993 to US$50 million in 2001. These returns have been adversely affected by the increased market competition between the major supermarkets. This is most evident in the UK, where sales of bananas are worth £750 million per annum to the largest supermarkets, surpassed only by sales of petrol and lottery tickets. Competition between Asda, Tesco, Sainsbury's and Safeway has cut prices, reducing the producers' income. The effects have been especially great in the Windward Islands as all its bananas are sold to the UK, 40 per cent of which go to Sainsbury's, which has encouraged producers to take up EU irrigation and other programmes designed to make them more productive and competitive.

With the EU quota system finally being phased out in 2005 the ACP producers are having to consider alternative sales strategies to both maintain market share and income as retail prices fall. Strategies have included farm-based improvements, restructuring of growers' associations, reductions in administrative costs and, in the Windwards, the introduction of a certification scheme to guarantee quality and fair trade, and a campaign to persuade the British Caribbean community to 'Buy Windward'. The issues of quality and fair trade have become of increased significance since Agro-Fair, a Dutch import company, brought the first trademark 'Fair Trade' bananas into the UK in January 2000. Fair Trade bananas now account for 11 per cent of bananas sold by the Co-op in the UK, via an agreement with Agro-Fair, and 40 per cent of the packaged bananas. Sales of Fair Trade bananas are rising by 53 per cent per annum as shoppers show greater willingness to pay premium prices for 'fair trade' produce (Ryle, 2002).

8 Solving the world food problem?

8.1 Human hunger

If the world's food production was distributed so that every person had an equal share, nobody would be left starving. Yet there are more people starving in the world today than ever before in the history of humankind. Recent United Nations estimates refer to over 1,000 million people living on the very edge of existence, of which nearly three-quarters live in rural areas (World Bank, 2000). UNICEF and UNDP calculate that 800 million children globally are undernourished, and that two billion people (one-third of the world's population) exhibit effects of poor diet (UNICEF, 1998). These problems reflect both macro-nutrient (protein, carbohydrates and fats) and micro-nutrient deficiencies. This growing magnitude of human hunger is occurring despite continuing increases in world food production. Hence factors affecting both food distribution and production are central to the location and extent of hunger and starvation (Abraham, 1991; Evans, 1998; Smil, 2000; Young, 1996). This chapter briefly outlines the magnitude of the problem before considering the analysis of attempted 'solutions', drawing on recent work by agricultural and other geographers, in particular on so-called 'Green Revolution' packages tied closely to global interests of countries and businesses based in the Developed World, and various types of land reforms, usually representing one form of 'indigenous' solution to both low productivity and social inequalities.

Establishing the detailed spatial pattern and magnitude of hunger and starvation is difficult because of the lack of reliable data on food intake per capita in the world's poorest countries, though Grigg's analysis of UN statistics from 1930–90 shows how the scale of human hunger has grown (Table 8.1). However, estimates of intake, based on income levels or on food production, can be compared with subsistence levels of consumption. In the broadest terms, as shown in Map 8.1, such comparisons reveal that a significant problem exists in the tropics, especially in Africa south of the Sahara and parts of south and south-east Asia. The highest levels of under-nutrition are in west and central Africa, where daily calorie supply as a percentage of requirements is less than 90 per cent of the required intake. Over 40 per cent of the population in these countries are undernourished, leading to death, disease and malnutrition. The effects of limited food availability can be seen in the high levels of infant mortality (Map 8.2), with 15 million children in Africa and Asia dying each year from malnutrition coupled with diarrhoea and infectious diseases.

The falling levels of self-sufficiency in food in Developing Countries are generally explicable with reference to three factors (Ilbery and Bowler, 1996):

- Population growth has outpaced the rate of increase in food production in many Developing Countries. In conjunction with associated land degradation, this has necessitated imports to maintain per capita food consumption levels.
- In some Developing Countries agricultural exports have been increased at the expense of domestic food staples. Such exports have tended to grow in response to the need to

Table 8.1 Population estimated to be chronically undernourished, 1970–2010

Region	1970 mill	1970 %*	1975 mill	1975 %	1980 mill	1980 %	1990 mill	1990 %	2010 mill	2010 %
Middle America & Caribbean	21	23	21	20	16	15	20	13	15	10
South America	32	17	32	15	29	12	38	12	25	8
Near East & North Africa	32	23	26	17	15	7	12	5	29	6
Sub-Saharan Africa	94	35	112	37	128	36	175	37	296	32
South Asia	255	34	289	34	285	30	277	24	195	12
South-East Asia	101	35	101	32	78	22	74	17	20	5
China	406	46	395	40	290	28	189	16	57	5
Total	941	36	976	33	841	27	785	20	637	12

* %s refer to the percentage in each region chronically undernourished
(*Sources*: [for 1970–1990] Grigg, 1997, p. 198; [for 2010 estimate] Alexandratos, 1995)

earn foreign currency to repay debts incurred in funding development programmes.

- There has been a change in diets amongst the wealthier component of the population in favour of imported rice, wheat and livestock products (e.g. Robbins, 1999).

In combination, these factors have contributed to the Developing Countries becoming net food importers. However, the extent of their deficiencies in food and also of the degree of hunger and famine is strongly disputed. Neo-Malthusian views in the early 1990s predicted a 7 per cent fall in per capita food production through that decade (e.g. Ehrlich *et al.*, 1993; Myers, 1991). These estimates were based on falling global outputs of cereals per person since the mid-1980s, the 'failure' of the Green Revolution of the 1960s and 1970s in terms of its lack of continued development and spatially uneven impacts, falling rates of adoption of high-yielding varieties of grain, and predictions of recurring drought akin to that associated with the 1972–4 'world food crisis'. More optimistic scenarios pointed to estimates of a fall in the numbers of malnourished people in the 1980s: by 108 million, whilst the world's population rose by 841 million in the decade (Dyson, 1996, p. 75). However, there was general agreement that the frequency and demographic impact of famines was being confined increasingly to parts of sub-Saharan Africa where 'traditional' approaches to economic development had yielded disappointing

outcomes (see Simon *et al.*, 1995; Simon and Narman, 1999).

In the 1990s it appears that the proportion of chronically undernourished people was reduced, as were the absolute numbers, mainly through the great advances made in China since the 1970s (Evans, L. J., 1998, p. 186). The proportion of the world's population suffering from malnutrition had fallen to less than 20 per cent, though the proportion is higher for children, especially in south Asia and parts of sub-Saharan Africa. Specific dietary deficiencies also remain very high in some areas, e.g. iron deficiency amongst 60 per cent of south Asian women, and vitamin A deficiency in 40 countries, blinding one million pre-school children each year.

8.2 Approaches to the study of hunger and famine

In general terms, three different types of approach to the study of hunger and famine can be recognised:

- Entitlement and capability. This approach is typified by the work of Amartya Sen (1981) who has stressed the role of differential command over food. This emphasises the importance of ownership of the means of production, with ownership equating to endowment or exchange entitlement (Sen, 1990). Endowment can have legal, economic

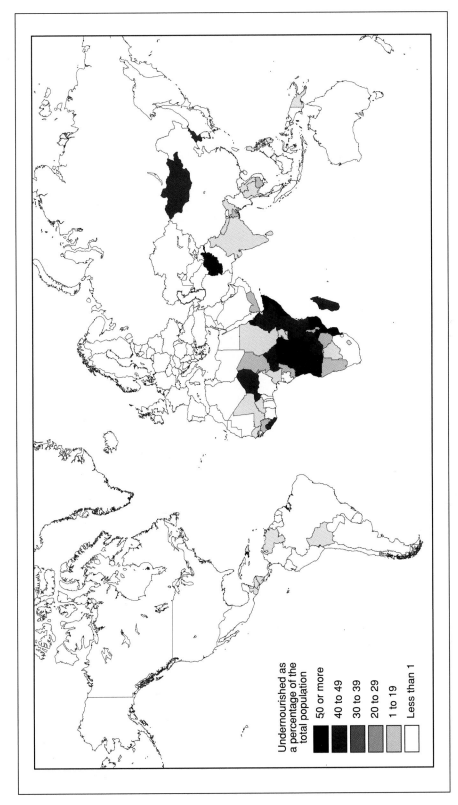

Map 8.1 Distribution of the world's 'malnourished' population (data source: www.unstats.un.org/unsd)

Undernourished as
a percentage of the
total population

- 50 or more
- 40 to 49
- 30 to 39
- 20 to 29
- 1 to 19
- Less than 1

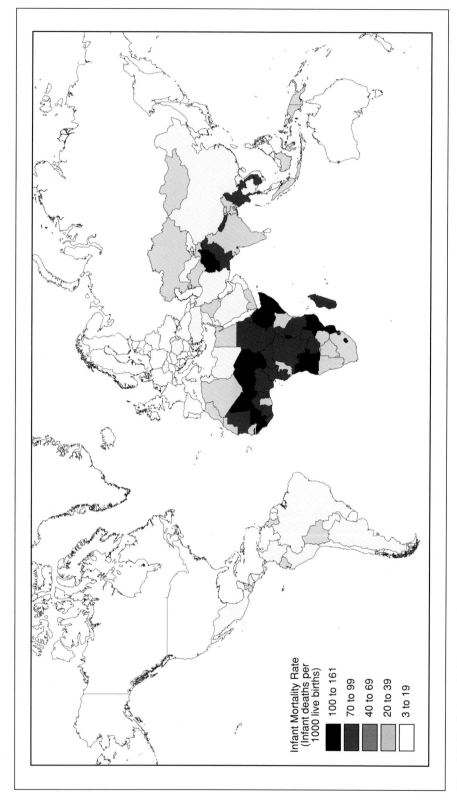

Map 8.2 Distribution of infant mortality (data source: www.un.org/depts/unsd/social/health)

Infant Mortality Rate
(Infant deaths per
1000 live births)

100 to 161
70 to 99
40 to 69
20 to 39
3 to 19

and political manifestations. Its significance can perhaps best be seen with reference to severe food crises associated with radical shifts in entitlement as opposed to declines in total availability of food (e.g. Dreze and Sen, 1989). Literature drawing upon entitlement theory has tended to focus either upon food security and coping strategy models on the one hand (e.g. Alamgir and Arora, 1991; Mortimore, 1991) or social security theories on the other (Hirtz, 1994). Hence food security can be considered in terms of food availability at different social and spatial scales, with entitlements varying according to scale (e.g. individual, household, sub-national, national and global). There are socially specific ways in which entitlements are mobilised during critical periods of food crisis (Corbett, 1988). In terms of social welfare, the relative weakness and incapacity of the state places greater emphasis upon local forms of assistance and welfare, perhaps embracing reliance on kinship groups (e.g. Warriner and Moul, 1992). Complex forms of familial, social structural and community institutions are important: sometimes termed a 'moral economy' (Scott, 1976). There is both a complementarity and tensions between state and non-state social security laws and forms of distribution.

- Empowerment and enfranchisement. This rests upon notions of power by linking famine and hunger analytically with powerlessness. Hence malnutrition is linked fundamentally to powerlessness, which helps explain why women and children are more likely to suffer from malnutrition as they occupy subordinate positions within the patriarchal family unit. This argument was extended by Collins and Lappé (1986) to link hunger to problems of democracy and a lack of political rights among the poor and marginalised. Political rights determine who can make claims on public resources as the basis for food security and the maintenance and defence of entitlements. Enhancement of entitlements should be considered as part of the way in which political power is exercised.

This entails considerations of power in the domestic sphere, at work and in the public–civil sphere. The domestic level tends to involve patriarchal exercise of power over women and children whilst the work and public–civil spheres can involve power relations affecting the property rights of the rural peasantry. An additional 'global' sphere can be recognised in which individual countries might be rendered powerless to gain access to food through the 'international food order' (Friedmann, 1990) or through social burdens imposed by the structural adjustment stabilisation programmes of the World Bank and the IMF.

Appadurai (1984, p. 481) employed the concept of 'enfranchisement' to refer to 'the degree to which an individual or a group can legitimately participate in the decisions of a given society about entitlement'. Hence he saw changes in south Asia in the 1970s as bringing a move from a relatively secure entitlement to a minimum level of food (a 'subsistence ethic') towards a partial enfranchisement in which political exclusion of certain groups increased food scarcity. Indeed, others have argued that protection of entitlements can only be delivered via a politicisation of civil society in which food issues are addressed openly in the public arena (Wisner, 1988). In developing these arguments, Watts and Bohle (1993, p. 50) asserted that 'empowerment is invariably grounded in specific property rights or specific bundles of entitlememts'.

- Class and crisis. This views farming, poverty and hunger as class phenomena (Alamgir, 1982), and draws upon the Marxist tradition to use class to refer to the appropriation and distribution of surplus from the producers (farmers), rather than focusing upon the link between class and property, and class and power. Hence, the concerns of this approach are with how the distribution of assets in peasant societies is related to the appropriation of surplus labour via taxes, interest, unequal exchange and patriarchal relations (De Janvry, 1980). According to

Watts and Bohle (1993, p. 51) these class-based analyses of famine include three basic propositions: the social relations of production are historically specific; the historical character of the manner in which surpluses are appropriated and distributed provides a basis to distinguish the broad character of the political economy(-ies); the political economies have their own 'crisis tendencies' which can be seen as market failures or other difficulties. According to Harvey (1982) these tendencies arise under capitalism through class conflict, and other conflicts relating to the conditions of production and accumulation. These propositions represent structural preconditions that shape the famine process.

In effect, arguments based on class draw upon historical and structural forces to account for how and why particular patterns of entitlement and empowerment are produced and reproduced in society. Hence there is a concern with how the market develops in Developing Countries as part of the processes of modernisation and commercialisation (Harrison, 1982). This type of concern is also expressed in the seminal political ecologies by Blaikie (1985; 1995; Blaikie and Brookfield, 1987), in which soil erosion was linked to three key political economic concepts: the incorporation of local systems into broader regional, national and global ones; the creation of a proletariat; and the marginalisation of various groups within society. The 'extraction of surplus' can be traced from the extraction of surpluses from cultivators who, in turn, extract surpluses from the environment (and hence produce soil degradation). However, this and similar class-based analysis, can fall into the trap of overlooking how individuals and groups develop different strategies to cope with their changing situation.

Generally, explanations of the wide extent of human hunger and the continuing recurrence of famines have tended to emphasise the destabling effects of the calamitous contact between the colonial powers and indigenous systems of production. However, more probing research, such as the classic study of famine in northern Nigeria by Watts (1983), has revealed the complexity underlying both recurrent famines and the various problems affecting agriculture in Developing Countries. Research from several different disciplinary perspectives has revealed a range of factors impacting adversely upon food production. Amongst those highlighted in Watts' book were the following (Giordano and Matzke, 2001):

- the tendency for the colonial powers to undermine existing 'indigenous' mechanisms to combat food scarcity without replacing them with new institutions;
- the long-term shift towards production of export crops and away from crops for local consumption;
- the impact of external markets, introducing price fluctuations that small producers are ill-equipped to handle;
- the monetisation of taxes and the economy in general, including the imposition of tax collecting systems that have encouraged corruption;
- the imposition of colonial boundaries that disrupted trade routes;
- the break-down of risk-spreading labour systems;
- increased levels of indebtedness amongst small farmers.

Hence it was demonstrated that famine is not 'natural' and that, far from reducing the scale of famine already present in pre-colonial times, colonialism changed the nature of famine by altering its causes, often in ways that increased the incidence of food scarcity. However, in addressing the extent of famine today, the search for 'solutions' is often complicated by the fact that the scenarios referred to above cannot be readily translated into simple statements of cause and effect (Macmillan, 1991).

To take one of the factors as an example, it is often argued that once farmers have access to credit (and hence raise their levels of indebtedness) they can engage in productive investment, promoting the type of efficient capital utilisation associated with much of the 'progressive' agricultural development of the western world. In the Developed World this has contributed to specialisation towards production of tradable commodities and has been a characteristic of farming development for several centuries, with considerable generation of wealth. Perhaps the key to understanding why these processes of development have not had the same impacts around the world is to recognise that the presence of different institutional arrangements outside North America and Western Europe have significantly altered the ways in which the various factors and processes operate. Hence utilisation of credit and the growth of rural indebtedness in some parts of the Developing World has simply helped to fuel an exodus from the land and hastened rural-urban migration and the growth of land-lessness or it has promoted greater differentiation amongst farmers (e.g. Sturm and Smith, 1993).

In particular, Watts (1983) argued that the colonial system changed the nature of famine from an issue of absolute to relative scarcity, and contributed to a growth in the incidence of food scarcity. What Watts' work did not address was the incidence of 'mega' famines, as in the Soviet Union in the 1920s and 1930s, China in the 1950s and Cambodia in the 1970s, each of which resulted in the deaths of over one million people, and which were linked to totalitarian 'socialist' regimes attempting to pursue state-sponsored economic transformations (Giordano and Metzke, 2001, p. 624). In addressing this subsequently, Watts (2001, p. 627) argues that 'it was not that capitalism or colonialism *grosso modo* caused famine – as some of the 1970s' Marxist anthropology seemed to argue – but rather that particular sorts of markets were constructed and specific forms of commoditisation rendered certain segments of the peasantry vulnerable to climatic perturbations' (Watts, 2001, p. 627).

One well-known analysis that highlights problems associated with colonialism is that by Blaikie (1985), who argued that most soil conservation programmes in Developing Countries do not succeed. He identified that most of the origins and ideological assumptions behind these policies date back to the colonial period and, as such, are unpopular with bureaucrats at all levels, as well as farmers and pastoralists. His analysis highlighted the complexity of most environmental 'problems' in Developing Countries whilst emphasising the strengths of a political economy perspective in

disentangling both the factors underlying the problems and the 'solutions' adopted. At the heart of the complexity are three sources of uncertainty within the debate about world-wide environmental deterioration:

- lack of accurate and widespread measurement over a sufficiently long period to show trends;
- difficulty in distinguishing human influence upon soil erosion from other effects such as climatic change and 'natural' erosion processes;
- the wide variety of views in which environmental deterioration may be regarded. The various views relate to widely differing political judgements and hence soil erosion must be regarded as a political-economic issue and not just an environmental process.

8.3 The vulnerable groups

Within the population experiencing starvation and malnutrition, it is possible to distinguish particular groups more at risk. These are generally termed the 'vulnerable groups' and principally include infants, young children, pregnant women and the elderly. However, it is important to disaggregate these vulnerable groups if a clearer understanding of who is most at risk is to be gained (Curtis *et al.*, 1988; Kabeer, 1991). Indeed, Watts and Bohle (1993) argue that recognising and then reducing vulnerability is as fundamental an objective as reducing poverty. They champion the need for better definitions of vulnerability, involving an account investigating the realms of causality and necessity, which frame the study of famine, hunger and human deprivation.

Vulnerable individuals, groups, classes and regions are those most exposed to perturbations (for example in food supply), who possess the most limited coping capability, who suffer the most from crisis impact and who have the most limited capacity for recovery (O'Hare, 2002). There are spatial dimensions to vulnerability (e.g. Robinson, 1985b), but it has also been studied in terms of ecology (Liverman, 1990), in relation to political economy and class structure (Susman *et al.*, 1984), and as a reflection of social relations (Harriss, 1990). This range of work highlights the multi-layered and multi-dimensional nature of vulnerability.

Watts and Bohle's (1993) analysis of vulnerability focused on a tripartite causal structure operating through the intersection of three causal powers:

- command over food (entitlement);
- state–civil society relations in political and institutional terms (enfranchisement and empowerment);
- structural–historical forms of class relations within a specific political economy (surplus appropriation and crisis proneness).

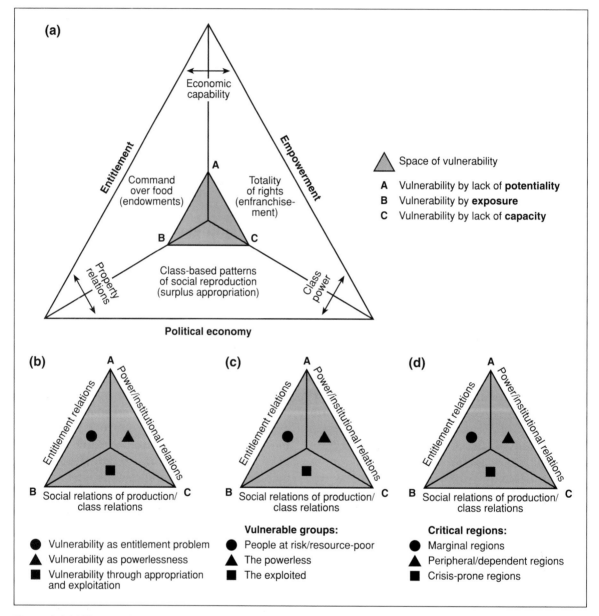

Figure 8.1 (a) The causal structure of vulnerability; (b) the social space of vulnerability: mapping the space of vulnerability through social relations; (c) the social space of vulnerability: mapping vulnerable groups in the space of vulnerability; (d) the social space of vulnerability: mapping critical regions in the space of vulnerability (based on Watts and Bohle, 1993)

This situation is illustrated in Figure 8.1a, with the intersection of the three causal powers producing three parallel analytical concepts: economic capability, property relations and class power. 'Famine results from violent short-term changes in these same mechanisms' (Watts and Bohle, 1993, p. 54). Vulnerable groups and regions can be located with respect to the three co-ordinates as shown in Figure 8.1c and d, though it must be acknowledged that all three 'spaces' of vulnerability coexist simultaneously so that all may operate to different extents in any given situation. When translated into spatial contexts, the portrayal shown in Figure 8.1d can be produced. Again, the exact situation for any region is dependent upon locally and historically specific configurations of class, entitlement and political processes.

Two examples from South Asia are shown in Figure 8.2. Figure 8.2a refers to the transition from pre-colonial to post-colonial times, with changing entitlement relations playing the key role in rendering dependent groups more vulnerable. Figure 8.2b summarises the situation for the Green Revolution in southern India through which there has been a shift from partial entitlements to disentitlement. Tenant farmers in particular have often been disadvantaged as landlords have sought to take more land in hand to benefit from applications of the new technology. This emphasises the key role of changing entitlement relations. This situation may be compared with that for the regular seasonal fluctuations and irregular drought-induced crises in western India. Here vulnerability reflects patterns of entitlement and enfranchisement. A range of different groups can be affected by drought crises, with neither traditional nor formal social security systems proving very effective for peasants, labourers, traders, artisans and people involved in service provision, though the last three groups may be able to develop risk-avoiding coping strategies (Chen, M., 1991).

There may also be significant temporal dimensions to vulnerability, for example, referring to hunger in South Asia, Appadurai (1984) distinguished between pre-colonial, colonial and post-colonial vulnerability. This pattern may be repeated in other parts of the world, with replacement of a dominant subsistence economy by growing commercialisation and market integration during colonial times. This could produce a substantial class of sharecroppers and wage labourers who could be vulnerable to the vicissitudes of the market. Vulnerability depends on restrictions to individuals' integration into formal social security systems, as demonstrated in the Great Bengal Famine in 1943–4 (Sen, 1981, pp. 52–85). In postcolonial times the gradual and incomplete shift from a caste to a class society conferred new types of enfranchisement in India (Gough, 1981), but it systematised dependency of some groups to market volatility and appropriations by landlords and merchants. Further work on this key theme of vulnerability and notions of entitlement appears in Watts (2000), Fine (1994) and Hartmann and Boyce (1983).

One of the 'solutions' to hunger and starvation has been that of food aid, whereby transfers of food have been made from areas of food surplus to areas of deficit. This has taken many forms and with various deferrals of payment employed, including supply gratis. However, what might seem on the surface as a simple and sensible measure, in terms of supplying food to the needy, is in fact a complicated and controversial one.

8.4 Food aid

In the second half of the twentieth century, agricultural output in Canada and the USA grew by 2 per cent per annum, making these two countries the world's highest volume agricultural producers and with the widest range of farm products. The substantial excesses of production of food and fibre over domestic demand has enabled this region to maintain its long-term role as a major trader in world markets, especially in cereals, whilst also donating heavily to international food and aid programmes (Hammond, 1994; Tarrant, 1980b). Between 1960 and 1995 approximately US$40 billion (in constant 1988$) were provided to Developing Countries in the form of aid of various kinds. However, of this only approximately 4 per cent constituted food aid, and this proportion has fallen steadily in the last decade as priorities have been reassessed. Atkins

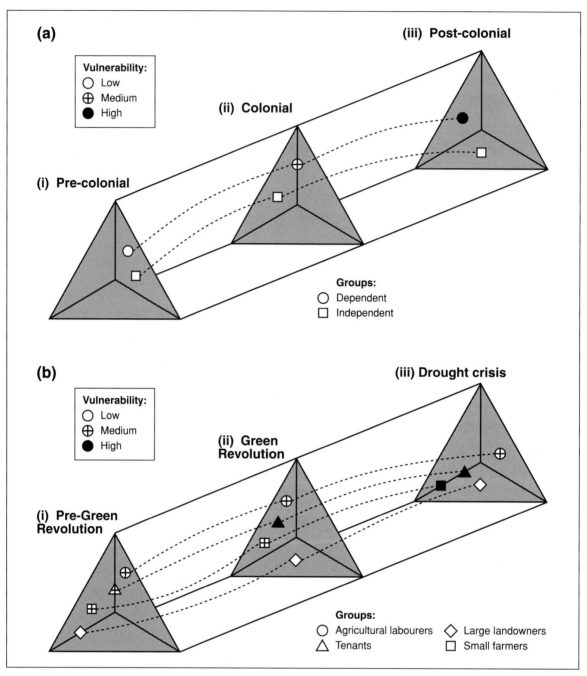

Figure 8.2 (a) Demise of the moral economy in South Asia (derived from Appadurai, 1984); (b) Green Revolution in south India (derived from Bohle, 1990)

and Bowler (2001) recognise three types of food aid:

- Emergency aid. This is disaster relief provided by individual governments, NGOs and international relief organisations such as the World Food Programme (WFP), a UN agency established in 1961. In some cases such aid has been delivered on a long-term basis to refugees fleeing military conflicts. In general, such aid cannot provide more than a temporary palliative and does not represent part of the necessary structural political and economic measures required to alleviate hunger. Nevertheless, despite frequently adverse media reports, food aid can be a vital life-saving measure in 'disaster' situations, e.g. the Feed The World (Band Aid) programme to the Sahelian zone of Africa led by pop star Bob Geldof in 1985 (Blundy and Vallely, 1985).
- 'Programme' food aid. This is targeted at the macro-economic level in recipient countries, intending to address shortfalls in basic foodstuffs in difficult years. A typical example is the structural adjustment measures of World Bank/IMF programmes (e.g. Zekeri, 1992). This does not necessarily imply 'free' food to either the recipient government or the population. Food aid programmes operated by the USA have typically provided food to governments at special rates, for sale in urban markets so that some of the proceeds could then be repaid to the USA or used for general development purposes (Tarrant, 1980b; 1980c).
- 'Project' food aid. This is targeted at specific groups in society, through such programmes as Food For Work (FFW) or Mother and Child Health (MCH) centres. Hence this is part of development policy aimed specifically at reducing food insecurity, and is often applied in rural areas where it is linked to projects such as road construction and dam building. Its weakness is that it may be associated with payment of very low wages or diversion of labour from important agricultural tasks. Moreover, beneficiaries

may not be the labourers themselves but rather the local elite. It has impacted differentially upon women in sub-Saharan Africa, as they are frequently the labour in these projects. More favourable results have been reported in South Asia and with monetisation, in which the labour force is paid wholly or partly in cash and the project represents an integral part of a well-defined rural public works programme (Tarrant, 1985a). Weaknesses are associated with MCH projects in which MCH meals come to replace meals provided in the family, and the poorest families may be by-passed.

Although much criticised, food aid in the 1950s and 1960s did successfully transfer large quantities of resources to the poorest nations and was of benefit to people facing hunger and starvation. The combination of food 'surpluses' in some of the world's wealthiest nations and a growing awareness of the need for humanitarian assistance helped to provide the impetus for initiation of food aid distribution channels. These channels were established not only for 'emergency' situations associated with particular 'disasters' but for more widespread dissemination of food aid, encouraged by a strong coalition of political constituencies that included farm commodity groups, shipping organisations and relief agencies. However, after rising to 18 million tonnes per annum at the end of the 1960s, the annual average subsequently has been around 10 million tonnes, falling in the early and mid-1990s so that in 1997 it was at 6.3 million tonnes. Its subsequent rise to 11.2 million tonnes in 1998 (Figure 8.3) parly reflects the use of food aid for political purposes by the USA, with 3.3 million tonnes of this aid (29 per cent) going to the Russian Federation and North Korea (see also Tarrant, 1985b).

In terms of the content (by weight) of food aid since 1971, wheat formed the majority (63 per cent) followed by coarse grains (16 per cent) and rice (11 per cent). Non-cereals represented only 4 per cent of the content. The major suppliers have been the United States and the EU, though the proportion of cereal aid as a percentage of

Figure 8.3 World food aid donations, 1970–2000 (data sources: FAOSTAT and Atkins and Bowler, 2001, p. 168)

national production has fallen below 1 per cent for both suppliers. In the late 1960s as much as 5 per cent of US cereals were shipped as food aid (Tarrant, 1980c).

Prolonged criticism of food aid policies, combined with improved food security in key recipient countries such as Bangladesh and India, contributed to a sharp fall in food aid donations in the 1990s: by two-thirds in volume since 1993 and four-fifths in value between the high point in the mid-1980s and 1997 (Figure 8.3). A higher proportion of food aid has been focused on emergencies in recent years, averaging over 80 per cent of aid under the UN's World Food Programme since 1996 (Figure 8.4).

Food aid remains the primary instrument for remedial action in 'disaster' situations under emergency programmes, but its role in alleviating general hunger and poverty has been replaced with financial aid, credits and bi-lateral aid programmes.

Food aid's share in overseas development assistance has declined from 25 per cent in the 1950s to less than 4 per cent in recent years. The overall value of food aid (in constant dollars) has declined to about half of its value in the 1950s.

In explaining this dramatic fall in the amount of food aid, Reutlinger (1999) highlights a series of fundamental changes occurring in the environment in which food aid operates:

- The main donor countries no longer have a need to dispose of food surpluses. Policies promoting production have been replaced by ones aimed at other outcomes.
- The aim of alleviating hunger and poverty, formerly closely identified with food aid, has become a central objective of all foreign aid.
- Most aid for both short- and long-term alleviation of hunger is supplied in the form of cash rather than food.

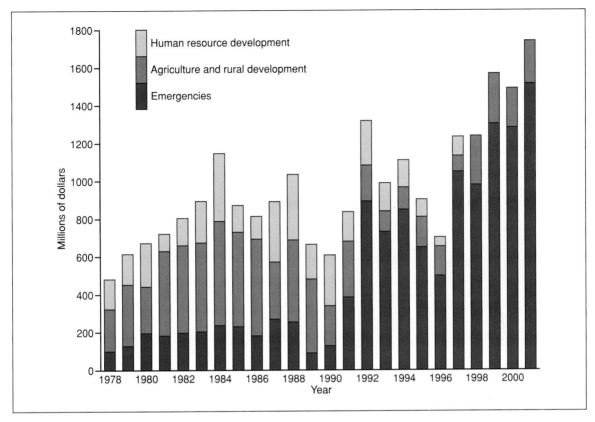

Figure 8.4 World Food Programme commitments, 1978–2001 (US$ millions) (data sources: www.un.org/wfp, and Atkins and Bowler, 2001, p. 168)

- The additional political benefits to the donors from food aid have dwindled since the primary motivation has shifted from surplus disposal to humanitarianism, with an increasing share supplied by countries other than the United States. However, some political additionality remains for certain donor countries.

Table 8.2 reveals that there has been a poor correlation between the leading recipients and those most in need, at least at the aggregate country level. Leading countries in terms of receipt of total grain food aid per capita in 2000 were Eritrea, the Gaza Strip, North Korea and Jordan. The next highest were Ethiopia, Angola, Tajikistan and Georgia. So in both groups there was a mixture of countries experiencing food crises and those where a polit-

ical significance would seem to be the key vehicle for food aid.

Geopolitically insignificant African countries south of the Sahara seem to fare worst in terms of the link between food aid and overall wealth. The distribution of food aid seems tied inexorably to geopolitical considerations and especially the foreign policy aims of the donor countries. A good example of this was the move of the USA in the early 1990s to divert a portion of its food aid to Eastern Europe as a means of supporting the newly democratic regimes as they emerged from their communist stranglehold. As a result food aid of any description is often regarded as tainted and possibly harmful to its recipients. The complex framework within which the USA's food aid policies have been developed is illustrated in Figure 8.5. Essentially, the selective pattern of US

Table 8.2 Differences between receipt of food aid and GDP per capita, 2000

Country ranking*	Country	GDP per capita p.a. (US$)	Food aid – cereals only (million tonnes)	Food aid – cereals only (mt per capita)
229	Sierra Leone	510	38,659	7.12
228	Somalia	600	23,520	3.14
226	Ethiopia	600	1,360,648	21.63
225	Congo Democratic Republic	600	50,602	0.94
224	Tanzania	710	101,567	2.89
223	Eritrea	710	237,563	64.93
221	Burundi	720	29,978	4.82
220	Madagascar	800	38,478	2.41
219	Afghanistan	800	239,713	11.01
218	Yemen	820	196,596	10.71
217	Mali	850	11,969	1.09
215	Guinea-Bissau	850	10,931	8.31
214	Zambia	880	24,146	2.47
213	Rwanda	900	65,381	8.94
212	Malawi	900	36,166	3.43
211	Nigeria	950	0	0
209	Sudan	1000	195,818	6.30
208	Niger	1000	43,498	4.20
207	Mozambique	1000	147,491	8.06
206	North Korea	1000	1,542,433	69.27
205	Gaza Strip	1000	120,455	92.66
204	Chad	1000	32,414	3.72
203	Burkina Faso	1000	16,989	1.38
202	Angola	1000	204,416	15.56
201	Benin	1030	13,092	1.99
200	Uganda	1100	66,121	2.76
193	Tajikistan	1140	177,274	29.12
185	Kenya	1500	367,272	11.98
183	Bangladesh	1570	653,944	4.76
172	Haiti	1800	105,007	12.90
171	Ghana	1900	96,863	5.02
170	Vietnam	1950	113,139	1.45
161	India	2200	201,976	0.20
160	Yugoslavia	2300	174,701	16.56
146	Indonesia	2900	256,079	1.21
138	Sri Lanka	3250	97,511	5.15
135	Jordan	3500	193,836	39.45
136	Morocco	3500	277,521	9.29
130	Philippines	3800	108,908	1.44
120	Peru	4550	149,019	5.81
119	Georgia	4600	106,450	20.23
84	Russian Federation	7700	351,776	2.42

* Countries listed in descending order from lowest GDP per capita, 2000. Comoros, Kiribati, Mayotte and Tokelau omitted for lack of data. (*Sources*: FAOSTAT, www.worldfactsandfigures.com)

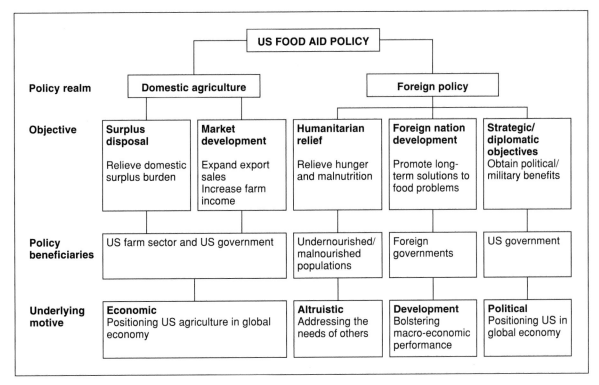

Figure 8.5 The context of US food aid policy (based on Kodras, 1993)

food aid is largely a function of the locations of its foreign priorities at any one time.

There are two broad arguments presented against food aid:

- Food aid, and especially grains, may dominate local markets, thereby acting as a disincentive to local producers who may be unable to compete with respect to quality and price. This may be a deterrent to local production, the very opposite effect of what is desirable in the long-term. Moreover, local populations may develop a preference for the imported cereals, reducing the market for locally produced counterparts. The various deterrents need not apply if the most vulnerable, food deficient groups are carefully targeted or if counterpart funds can be used to support local producer prices. Alternatively, donor countries can purchase food produced in the recipient country and distribute it as supplementary food aid, though in 1997 such local purchases represented only 9.5 per cent of cereal food aid and 4 per cent of non-cereal aid arranged by the WFP (Atkins and Bowler, 2001, p. 167).

- Food aid perpetuates the power of the North (Developed) over the South (Developing) both in terms of the images of the South as portrayed by the media in the Developed World, and in terms of food aid's lack of overall contribution to solutions to the world food problem (Gibb, 2000).

Whilst food aid is essentially an intentional or bilateral response to food shortages in Developing Countries, there are also internal programmes of food distribution and subsidies, the latter having been promoted during the 1990s by the World Bank (1997) as part of the targeting of vulnerable groups for additional support in times of economic adjustment. For example, India has had some form of food distribution system (FDS) since the Bengal Famine of 1943, though it has expanded substantially, in area covered, in quantity of food grains handled and in costs involved (Bhatia, 1991).

From the mid-1960s its FDS has been applied to emergency situations, such as droughts; it has been used to distribute food at fair prices to vulnerable people; and it has aimed to guarantee remunerative prices to farmers (Mooij, 1998).

The impact of food aid on socio-economic development in Developing Countries must be set alongside the effects of other measures applied in attempts to transform the economies and societies of these countries. Two sets of measures in particular have been applied directly to agriculture, namely land reforms and the Green Revolution. These 'solutions' are now examined in turn.

8.5 Land reforms

Particularly in Latin America, pre-1945 guerilla movements, peasant uprisings and spontaneous land occupations were the precursors to land reforms that offered one of the few hopes of overall economic reform. However, throughout Latin America the land reforms had to deal with some of the world's most extreme property relations in which Portuguese and Spanish colonists had left a legacy of extensive large estates (*latifundia* or *haciendas*) with their exploitative sharecrop tenantry system, and a large number of very small holdings (*minifundia*) on a minority of the land area (Burger, 1994, p. 263; Sobhan, 1993).

In some parts of Latin America various types of land reforms have a history approaching one hundred years, but scarcity of land and rural poverty remain rife, with alienation of land from the rural poor now associated with power in the hands of small local elites rather than colonial rulers. Indeed, the wave of Latin American land reforms in the 1960s, involving political intervention to support family farms, did not generally lead to substantial redistributions of wealth (Morner and Svensson, 1991, pp. 20–42). Elsewhere land reforms may have an even longer history, as in the case of the enclosure movements in Western Europe, or they may be a more recent and ongoing phenomenon as in parts of Africa. In many recent examples they are intended to increase

production and further social justice by eliminating large unproductive estates and absentee landownership through land redistribution.

Several different systems of land reforms have been applied throughout the tropics, with distinctions usually drawn between schemes establishing peasant proprietorship, creating numerous owner-occupier small farms (land-to-the-tiller), as in the long-running scheme in Malaysia operated by the Federal Land Development Authority (FELDA) (Sutton and Buang, 1995), and cooperative or collective farming systems, associated with revolutionary changes to a country's economic and political system. The latter have frequently been based on the Russian or Chinese systems of collectivisation, in which communal or state ownership has replaced both large private estates and numerous small, uneconomic privately owned or tenanted holdings.

In some countries there have been combinations of these two basic types of reform. In El Salvador, for example, land reforms were introduced in the 1970s in three phases, initially aimed at confiscating land holdings of over 500 ha (Phase I) and establishing co-operatives to be run directly by the beneficiaries (Phase II). These were largely estates producing export crops and it was felt that it was important to maintain the estates intact. Opposition by large landholders and outbreak of civil war disrupted these reforms (Pelupessy, 1997; Thiesenhusen, 1995), as shown in Table 8.3. A more recent reform, Decree 207 (Phase III), provided land-to-the-tiller, with beneficiaries taking on a mortgage, and landowners reimbursed. However, the holdings created were often very small and on land of poor quality. Again, the reality was that only one-tenth of the intended beneficiaries received land. Overall, instead of reforming all of the agricultural land in the country, just 18 per cent were affected (McReynolds, 1998). Nevertheless, the extent of the reforms was more wide-ranging than for any other Latin American country, with the exception of Nicaragua (Brown, 1994), and the new farm operators under Decree 207 more than doubled the rate of hired labour utilised. Nevertheless, as in most other cases, it is difficult to determine the extent to which the reforms have

Table 8.3 Extent of agricultural reforms in El Salvador, 1970s and 1980s

	Beneficiary households				Agricultural land			
	Expected		Actual		Expected		Actual	
	Households	% of total	Households	% of total	ha	% of total	ha	% of total
Phase I*	50,000	11.9	30,268	7.1	223,000	15.4	207,372	13.7
Phase II	50,000	11.9	0	0.0	343,000	23.6	0	0.0
Phase III	117,000	27.9	47,001	11.1	198,000	14.0	71,600	18.2
Total reform	764,000	52.6	77,269	18.2	764,000	100.0	1,508,206	18.2
Total agriculture	420,000	100.0	425,000	100.0	1,452,000	100.0	1,508,206	100.0

* All land and beneficiaries were organised in 322 production cooperatives.
(*Source*: McReynolds, 1998, p. 462)

succeeeded in terms of their effect on production, wealth creation and the distribution of wealth.

It is impossible to do justice to the range of impacts accruing from land reforms in Developing Countries, but two brief examples give some flavour of the problems and on-going issues associated with land reforms.

8.5.1 Land reforms in Mexico

The longest history of land reform in Latin America occurs in Mexico where the first reforms occurred in the second decade of the twentieth century when 92 per cent of the rural population were landless and one per cent of the population owned 97 per cent of the agricultural land. The Mexican revolution under Emiliano Zapata was supported by many poor rural workers seeking to break up the hacienda system of large estates. The revolutionary Congress of 1917 adopted a constitution that included commitment to reform of land ownership and control of land. New laws permitted expropriation of large estates (above 100 ha of irrigated land, 200 ha arable, 150 ha irrigated plantation land, 300 ha non-irrigated plantation and 5000 ha poor-quality land in mountainous regions). Provision was made for establishment of communal villages (*ejidos*), strongly influenced by the desires of Indian groups who had fought with Zapata. All land within 7 km of a new ejido and larger than the specified limits on holding size could be confiscated

and transferred into the ejido to be worked collectively or individually by an ejido family.

Laws encouraging development of ejidos were passed in 1920, 1922 and 1925, but more significant were agrarian reform laws in 1934 and 1942. These determined the character of the ejidos for the next 50 years by setting out their structure in the law and fostering the large-scale land appropriations that followed. These reforms were designed to serve the political purpose of stopping peasant rebellions, especially by indigenous, Indian communities. Initially, collectivisation was the principal form of operation, but over time this has given way to individual family farms operating within the ejido 'umbrella'. Today approximately 50 per cent of Mexican agricultural land is within the ejido sector, and it is dominant in the unirrigated southern part of the country. However, the output from these southern ejidos tends to be very low and incomes for the inhabitants (*ejidatarios*) are correspondingly poor. Hence rural poverty has not been eliminated by the reforms (Morgan, 1990).

In many cases the reforms were only partial, producing an *ejido pejugal* in which small plots of land were created essentially for day labourers (*jornaleros*) on the large estates. This was common in the area around Mexico City. Elsewhere, larger amounts of land were redistributed, e.g. in eastern Morelos, one of the heartlands of Zapatismo. In the 1940s more communal redistributions were effected (under the Cárdenas regime) only to be challenged by neo-latifundia measures in the 1950s and 1960s. Social unrest by the peasantry in the 1970s brought a new wave of reforms (under Luis Echeverria), legalising some land invasions, e.g. on foreign-owned estates in Sonora. A new credit system was instated for small farmers and Green Revolution measures were intensified, building upon schemes first initiated in the 1940s and 1950s with US funds. However, there remained a pronounced division between the small ejidos and the large commercially operated estates. So there has continued to be a strong demand by the peasantry for more land.

New laws were passed in 1992, reversing several key provisions of the 1917 Constitution. There was an end to the government's formal obligation to distribute land to the peasantry and to maintain production guarantees on basic grants for land recipients. *Ejidatarios* were permitted to privatise and sell land and to enter into a variety of partnerships with the private sector as a means of improving their lands. The deabte over these reforms tended to break down into a discussion on the simple dichotomy between privatisation and state ownership (De Janvry *et al.*, 2001). Meanwhile the rural inhabitants faced a bewildering array of local interpretations and adaptations to the reforms. So it is not surprising that there were many notable cases of strong resistance to the Constitutional amendment, ranging from the violent protests of the Ejercito Zapatista de Liberación Nacional (EZLN) in Chiapas (Harvey, 2000, p. 73; Rovira, 2000) to the Barzonista movement, which is a militant organisation started by private-sector farmers in western Mexico who found themselves unable to repay their loans to the banks (Benton, 1999). Overall, change was much slower than the government may have anticipated, particularly because some *ejidatarios* had already treated their lands as semi-private assets for a long time. So, bypassing the highly inefficient bureaucratic hurdles, they had routinely bought, traded and rented ejido lands (Cornelius and Myhre, 1998). However, ultimately the reforms are likely to help accelerate the impacts on the countryside of urban development, the restructuring of financial markets and increased production of export crops.

8.5.2 Land reforms in Kenya and Zimbabwe

In many parts of Africa land reforms have followed independence post-1945, with the aim of tackling disparities in ownership between European settlers and the indigenous population, though with the exception of Senegal there have been few reforms in Francophone West Africa. In many cases the progress of reforms has been controversial and strongly contested, often attempting to reverse removals of land from the indigenous population that were effected during colonial times (e.g. Kalabamu, 2000) or, in the case of South Africa, under minority rule (May, 1992). In Kenya, for example, prior to independence in 1962, the Mau-Mau revolt of the 1950s targeted the large holdings of European settlers on which the local Kikuyu people were sharecroppers and the land area they could use for subsistence was limited to a minimum amount by law.

One solution offered by the British colonial power during the Kenyan dispute was a settlement project, repealing the ban on land purchase by Kikuyu, transferring title to sharecroppers for the land they worked, and consolidating fragmented sharecropper land. However, it was much more wide-ranging reforms that brought substantive change post-independence. This provided for distribution among black farmers of about 400,000 ha of the 3.1 million ha occupied by white settlers. The actual outcome was the transfer of 570,000 ha occupied by white settlers to 36,000 black farmers within five years of independence. This was known as the Million Acres Scheme. The new settlers paid a compensatory rental for their land. However, many had little experience of farming and did not make sufficient money from their first efforts on the land. Hence there were numerous settlers who could not maintain the rental payments and so were forced to quit. As a result some of the former European-owned holdings were reconstituted by Europeans, and so a significant proportion of this land remains in European hands. There have been ongoing problems in developing an accurate land register, in overcoming corruption, in recognising women's rights in the reform process and in dealing with conflicts between graziers and crop producers. Nevertheless, the Kenyan reforms are the most comprehensive market-oriented land reforms within Africa.

When Zimbabwe gained independence in 1980 an initial target was to create 162,000 farms occupied by Zimbabweans of African descent by 1985. A problem, though, was that the Lancaster House Agreement, which brought about independence, contained a clause stating 'willing seller, willing buyer', and hence it was mainly the less attractive and cheaper of the commercial farms in more marginal areas that the government was able to acquire.

In the first five years following independence in 1980, 52,000 African-run farms were created, largely on land formerly occupied by farmers of European descent. These were primarily leasehold properties, but following the creation of a new constitution in 1990, the tenants were able to purchase their holdings if they wished. The stated policy intention has been to create a settled peasantry, eliminating previous customs whereby many men would leave the land to take seasonal or full-time employment in distant towns. Hence, the new farm tenants were not allowed to farm whilst holding urban-based employment. Nevertheless, migration from communal lands has continued, with some long-distance commuting reflecting the high cost of urban housing.

In establishing new farms and new rural communities in Zimbabwe, various settlement models have been followed, including both a co-operative system and the creation of individual holdings. The latter have generally comprised 5 ha of arable land, use of common grazing, and a farmstead within a nucleated settlement comprising 30 to 35 farms per village. Farmers have been granted a permit for residence, depasturing for stock-keeping, and cultivation rights. However, there have often been problems related to the fact that individual communes have been unable to supply all the needs of their resident farms with regard to fuelwood, tobacco flumes and materials for brick-making (Gibb and Lemon, 2001). Nevertheless Harts-Broekhuis and Huisman (2001) in a study in Insiza District, near Bulawayo in the south-west of the country, note that the overall impact of resettlement on agricultural production and rural poverty has been positive. Resettled households have mainly been able to obtain higher output and income levels than households in the communal areas. However, there has been a considerable amount of squatting in areas designated for resettlement, placing great pressure on natural resources.

The reforms have not eliminated substantial areas of land owned by farmers of European descent – nearly 4500 farmers of European descent occupy 11.2 million ha or 35 per cent of the country's commercial farmland. They also control large areas of game parks, forestry and agro-industries. Although the European-descent population is only 1 per cent of the population (130,000) they employ nearly two-thirds of total workforce. One million African Zimbabweans live on 16.3 million ha of communal land, but only 9.4 per cent of this land is suitable for specialised or intensive crop farming (compared with 30 per cent for the land owned by farmers of European descent) (Gibb and Lemon, 2001, p. 25). This has caused resentment and political agitation for this land to be swiftly transferred to the indigenous population.

Between 1980 and 1990 a total of 71,000 families were resettled. To speed up land transfers to black Zimbabweans the 1992 Land Acquisitions Act allowed the government to confiscate farms from white Zimbabweans and to decide on the appropriate level of compensation. This damaged the country's image overseas, especially as some of the land acquired went to government ministers, provincial governors and prominent government supporters. As a consequence the UK withdrew financial support for the land reforms. In 1997 the Zimbabwean government published a list of 1471 white-owned farms covering 4.5 million ha that would be purchased compulsorily with some compensation provided to the former owners for 'improvements' made, though not the full market value. Following the establishment of a new Constitution in 2000 compulsory acquisition of land without compensation was introduced. This Constitution was rejected in February 2000, but a critical Clause (no. 57) was subsequently passed in parliament, ushering in some violent direct seizures of land, especially within a 150 km radius of the capital, Harare, with the country's President, Robert Mugabe, lending his support to such actions. Under these conditions little attention is being paid to either the social justice implications of the outcome or the impact upon agricultural production. At present this is severely affecting Zimbabwe's chief export crop, tobacco, which is largely grown by farmers of European descent. Deaths of both white farmers and their black labourers have occurred. Agricultural output has plummeted and famine conditions are taking hold in some more remote rural districts, especially in the south-west where the people belong to a different tribal group to that of the ruling power. This instability is severely limiting the impact of the imaginative Communal Areas Management Programme For Indigenous Resources (CAMP-FIRE) focusing on autonomous development of local resources (Logan and Moseley, 2002).

Even from these very brief examples it can be concluded that, at best, land reform schemes can be said to have experienced mixed results. Often the initial upheaval associated with reform and reorganisation of farm holdings has actually reduced agricultural production. The long-term results of reforms may also fail to increase food production. For example, in Tanzania the reforms of the 1960s created new villages throughout the country, in a process known in Swahili as *ujamaa* ('villageisation'). This certainly assisted the increased provision of rural schools, hospitals and clinics, but the associated establishment of collective farms and the uprooting of the formerly dispersed rural population decreased agricultural output. Today many of the collectives have been returned to peasant proprietorship in which each farmer farms his or her own land (Forster and Maghimbi, 1999). It is difficult to provide any general conclusions about a process that has taken so many different forms around the world, but it is pertinent to note that land reform processes, despite very mixed results, are continuing on all the inhabited continents as part of the ongoing struggle over who controls land resources and who determines their management (Young, 1998).

8.6 'Green Revolution' solutions

Accompanying the commercialisation of agriculture in the tropics there have been attempts to transform production by bringing the inputs to agriculture closer to those of farming in Western Europe and North America. This has comprised the promotion of a move from labour-intensive farming to capital-intensive farming, involving a direct transfer of technology from rich countries to poorer ones in a package generally termed the Green Revolution. The package has usually consisted of high-yielding crop varieties (HYVs), fertilisers, pesticides, machinery and irrigation, promoted largely by funding from the wealthy nations. The staple cereals, maize, wheat and rice, have been targeted through the establishment of special American-funded plant breeding centres in Mexico (for maize and wheat) and the International Rice Research Institute (IRRI) in the Philippines, from which the improved varieties have been distributed (Robinson, 1993c).

So-called 'miracle rice' (strain IR8), bred in the Philippines in the 1960s, has been part of the package spread throughout south and south-east Asia. IR8 is a cross between two strains of rice, Petan – a tall, vigorous Indonesian variety, and Dee-geo-woo-gen, a short, stiff-strawed Chinese rice. It produces a vigorous dwarf or semi-dwarf variety capable of giving increased yields per unit area. By 1990 the IRRI had bred 150 dwarf and semi-dwarf varieties of rice, and there were over 500 such varieties in total in the world. As shown in Table 8.4, the spread of HYVs of wheat continued for at least a quarter of a century, indicating that the Green Revolution was not just one single event. Rather it was a series of adoptions of particular innovations, characterised by a recognisable number of different phases. Atkins and Bowler (2001, pp. 222–5) refer to four of these:

- Phase One: Initiation and early dissemination. The spread of HYVs was
  rapid in areas of suitable climate. In Mexico, South Asia and the Philippines not only did increases in output exceed rates of population growth, but harvest fluctuations were also reduced. The latter encouraged storage of grain against future problems. Moreover, more employment was generated as the HYVs required high labour inputs for

Table 8.4 Percentage of wheat area planted to modern varieties in the Developing World

Region	1970	1977	1983	1990	1994
Sub-Saharan Africa	5	22	32	52	59
Middle East	5	18	31	42	57
China	42	69	79	88	91
Rest of Asia	n/a	n/a	n/a	70	70
Latin America	11	24	68	82	92
All LDCs	20	41	59	70	78

n/a not available
(*Sources:* Atkins and Bowler, 2001, p. 222; Pingali and Rajaram, 1998)

weeding, fertiliser application and moisture control. With more grain being produced, retail prices of grain fell, benefiting consumers. As farmers could produce the same amount of grain as in previous years, but now from less land, they could put more of their land into other cash crops.

- Phase Two: Emergence of significant problems. After the high hopes of the late 1960s, significant difficulties and limitations emerged that revealed inherent weaknesses in Green Revolution strategies. In particular, it became clear that outside the most favoured areas climatically there could be considerable problems. For example, the brown leaf hopper, which carries the grassy stunt virus, caused widespread damage to IR8 crops in Indonesia and other Asian countries. This could only be combated through use of large and expensive inputs of pesticides. It also became apparent that the HYVs were quite location specific. For example, the dwarf rice varieties were unsuited to areas subjected to annual water inundation of significant depth, as in the Ganges delta. Some varieties did not mill or cook as satisfactorily as their traditional counterparts and so were not as acceptable to the consumer. In other areas they were not well adapted to local cultivation practices. Moreover, the Green Revolution had little effect outside those areas where maize and rice were the main staples. The Revolution was revealed as overly reliant on cash and technical assistance from the rich countries, often simply turning potential recipients into targets for sale of costly equipment. Where the package was adopted there were claims that it accentuated differences between the poorest farmers, unable to afford the new technology and improved crop varieties, and those more commercially orientated farmers more capable of affording them.

- Phase Three. A second generation of HYVs was developed from the mid-1970s, applying to crops other than rice and wheat, such as sorghum in Maharashtra (India) and maize in Malawi and Zimbabwe. Small farmers gradually adopted the new varieties, though often without taking the full Green Revolution 'package'. More research was done on linking the HYVs to the overall conditions under which they might be grown. For example, in 1971 a consortium known as the Consultative Group on International Agricultural Research (CIGAR), from the World Bank, various regional banks, several UN agencies, charitable foundations and national governments, established a network of countries, largely funded through the public and charitable sectors, which developed research on farming systems. This extended knowledge into new areas, especially giving emphasis to understanding pest and disease resistance and reactions of plants and animals to different environmental conditions within the tropics. Some of these results were translated into developments targeted specifically at small farmers, thereby endeavouring to reach parts of the farming community by-passed in the earlier phases of the Green Revolution.

- Phase Four. From the early 1980s it became clearer that the traditional plant breeding methods that had underpinned the initial stages of the Green Revolution were incapable of generating further sustained increases in yields. Significantly, in pioneering areas such as the Indian Punjab, yields of HYVs were being restricted by pests resistant to chemical treatments and by environmental problems (Leaf, 1987). New technologies were developed in attempts to extend gains in output, e.g. tissue culture and embryo culture to modify livestock breeding. However, from c.1990 the technological solution that gained widespread media attention was genetic modification (GM), and hence the new term, the Gene Revolution. This is beginning to have an impact upon the growing of certain food crops in Developing Countries, though as yet on only a limited scale (see Chapter 10). Hence it is the HYVs from the various phases of the Green Revolution that remain crucial to the well-being of crop production in many Developing Countries. Some of the limitations of the early days of the Revolution remain in place, but lessons have been learned that have led to increased production and levels of output that crucially have increased food availability.

There are many illustrations of the way in which the spread of HYVs has been spatially restricted. Nowhere, though, is this shown more starkly than in Africa. For example, the first HYVs of maize were released in Zimbabwe in 1960, five years before the Green Revolution got underway in India. Hybrid maize seed was available in Kenya in the late 1960s. In both Kenya and Zimbabwe uptake of the hybrid seed by large commercial farmers was substantial and rapid. Adoption of the improved seed spread to Zambia, but there were numerous reasons why the great majority of farmers failed to be drawn into this 'revolution'. Byerlee and Eicher (1997) refer to a combination of political factors, lack of credit and poor delivery systems for required inputs, non-availability of suitable hybrids for specific agro-ecological zones and resource availabilities, weak incentives, and insufficient government funds and commitment to sustain the required breeding and agronomic research programme. Perhaps in Africa, even more than other parts of the Developing World, scientists and agricultural advisors failed to recognise the vital limiting factor to the adoption of new seeds and Green Revolution technology that relates to smallholders' tendency to be risk-averse and both labour- and cash-constrained (Watts, 1989).

8.6.1 The Green Revolution in Sri Lanka and the Indian sub-continent

In Sri Lanka irrigation was at the heart of attempts to facilitate economic development in the country's northern 'Dry Zone', which had lagged behind the 'Wet Zone' in the south-west quadrant of Sri Lanka during colonial times (Farmer, 1977). Irrigation linked to settlement schemes pre-dated independence in 1948 but gained renewed importance as part of projects to intensify agricultural production and increase utilisation of the country's largest river, the Mahaweli Ganga. The various projects embracing irrigation and agricultural production in the Dry Zone have included several different objectives: relief of poverty resulting from landlessness; increasing food production and thereby enhancing national self-sufficiency; protection of peasant farmers; employment creation; increasing GNP; and the establishment of growth centres. Within these projects there have been conflicts between the broader social objectives of settlement and production goals. However, there has been some success in terms of the first four objectives distinguished, but little for the last two (Harriss, 1981, p. 317).

Low levels of productivity and high costs characterised the schemes developed in the 1970s, related to various factors but often to problems of water management, which in turn affected other aspects of the projects. Key problems have been poor irrigation design, inadequate structures and lack of suitable measuring devices, conflicts between the interests of individual farmers, high water use and social interests, better regulated use, cultivation deals, wet rice growing on inappropriate permeable soils, lack of effective management and political interference. In addition, there have been problems related to the organisation of settlement schemes in the Dry Zone that have affected social cohesion and restricted productivity on some farms. In part this reflects the implementation of policy prescriptions that have ignored inherent inequalities and differentiation within Sinhalese village society, reminiscent of similar failings in the ujamaa villages of Tanzania (Rakes, 1982) or the idea of the 'village republic' in the community development approach to rural development in India in the 1950s. The outcome has been to reinforce the power of those already powerful individuals in villages and to restrict the collective action required to produce efficient operation of irrigation schemes. Indeed, collective modes of action have tended to conflict with the small-scale property holding and small-scale production of Sri Lanka's settlement schemes.

There are some parallels with the implementation of Green Revolution schemes in India, where the advent of irrigation has often been a crucial factor. For example, in parts of India's Deccan plateau in the absence of irrigation, rice cultivation would be restricted to valley bottoms that are flooded by the south-west monsoon rains between July and October. Hence there are irrigation works dating from the colonial period in the early years of the twentieth century. Irrigation provided a transition from a pre-industrial agrarian society dependent largely on dry-land crops grown for subsistence, to a society where farmers became increasingly reliant upon technology, capitalism and wet-land crops of sugar cane and rice as their main source of livelihood. However, society was sharply stratified into those farmers with sufficient land and wealth to benefit from irrigation and the rest who had little alternative to working for their larger neighbours. This differentiation and bias in favour of the large farmers, having been accentuated by irrigation, was furthered by the Green Revolution from the 1960s, with new rice varieties developed by the IRRI first becoming available in 1966.

At the same time there was large-scale funding of the Indian Agricultural Research Institute (IARI), which had its main headquarters in Delhi, and nine regional research centres. Initially this developed dwarf wheat based on Mexican strains, producing 64 new varieties (Slatford and Fishpool, 2001). The IARI also developed training in farming practices and helped to establish co-operative farms. It proved very difficult to develop monsoon-dependent varieties of wheat, with variations in micro-climate proving crucial, especially changes in the water regime through irrigation and water management. These new HYVs were most successful when grown in conjunction with irrigated water management. Hence their adoption tended to be accompanied by extensions of the canal network. For example, Bayliss-Smith (1984, p. 161) reported a 60 per cent increase in 'wetland' in the village of Wangala in Mandya district, Sri Lanka, between 1955 and 1970. However, only the more prosperous peasant farmers actually improved their position substantially. HYVs tended to attract more pests, required more fertilisers and therefore could increase pollution. In places their slow adoption reflected the innate conservatism of the subsistence producers.

The Green Revolution raised yields considerably in irrigated areas, enabling both HYV rice and sugar cane to become cash crops. However, inputs also rose, as wet-land crops are more labour intensive and far more capital intensive. Hence, in real terms net incomes rose little, and greater differentiation between the wealthier peasant farmers and the rest of society increased. For the wealthy, farming ceased to be a way of life and became more of a commercial business. Meanwhile, the income of the poorer castes declined. Though labour on the lands of the wealthy farmers remained plentiful, there was also the growth of an exodus of rural workers to the cities. This partly reflected the tendency for the more innovative farmers in southern India to move towards a western model of change rather than a more labour-intensive alternative. This choice is distinguished as the Green Revolution model in Figure 8.6, and tends towards the industrialisation tendency associated with many Green Revolution schemes, e.g. commercial rice production in Surinam, where labour has been substituted on a large-scale by fossil-fuel-based technology. This contrasts with the tendency towards 'involution'

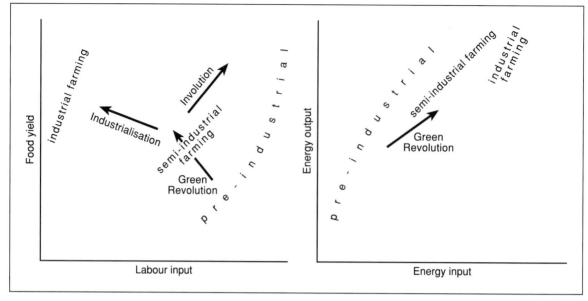

Figure 8.6 The relationship between yield, energy and labour inputs in rice-growing systems (derived from Bayliss-Smith, 1984, p. 169)

associated with rice-growing in Java in the 1950s, as identified by Geertz (1963). This process reflects the capacity of wet rice to permit greater labour intensification in the face of population growth. In some circumstances this can sustain yields substantially higher than those from the Green Revolution scenario.

There are conflicting findings regarding the impact of Green Revolution-related technological changes upon employment. Jayasuriya and Shand (1986) claimed that modern technology in Asian agriculture had enabled more labour to be utilised on the land, but that, over time, various chemical and mechanical innovations would result in net reductions of agricultural labour use. The latter was observed in the Philippines by Estudillo and Otsuka (1999), though others have observed that HYVs can mean more labour inputs are needed (Hazell and Ramasamy, 1991). In Bangladesh, Hossain (1989) concluded that modern techno-logy had increased the size of the labour market, with increased demand for hired labour. Further-more, there was a greater use of high-wage casual labour as opposed to low-wage permanent labour. However, Rahman (2000) notes that, in Bangla-desh, Green Revolution technology has displaced

rural women in the post-harvest processing sector, especially with the use of mechanised husking of rice. The impacts on the supply of female labour in crop production have been highly variable, though incidence of female labour hire has remained low, as shown in Table 8.5. Hired female labour is only of significance in the cultiva-tion of cotton, and only in the growing of vegeta-bles does the overall application of female labour come close to equalling that of men.

8.6.2 Whither the Green Revolution?

Many different interpretations have been placed upon the outcomes associated with the Green Revolution (e.g. Conway and Barbier, 1990; Shiva, 1993), partly depending on the time scale employed in the evaluation. For example, Cross (1987, p. 45) dismisses it as a failed technical fix of the 1960s whilst Hayami (1984, p. 394) is more posit-ive and refers to resultant social inequalities as being the result of 'insufficient progress in the tech-nology'. Lipton (1989, p. 1) claims that 'history records no increase in food production that was remotely comparable in scale, speed, spread and duration'. The contrasting views partly reflect the

Table 8.5 Labour input in crop production in Bangladesh

Crops	Family labour		Hired labour		Total labour		
	Men	Women	Men	Women	Men	Women	no./ha
Local rice	41.0	10.8	47.3	0.9	88.3	11.7	156
Modern rice	37.5	9.8	51.5	1.2	89.0	11.0	191
Local wheat	46.2	16.1	35.6	2.1	81.8	18.2	143
Modern wheat	44.1	13.2	41.2	1.5	85.3	14.7	136
Jute	38.3	5.3	55.5	0.9	93.8	6.2	227
Potato	52.4	12.5	33.8	1.3	86.2	13.8	311
Pulses	41.3	25.7	31.2	1.8	72.5	27.5	109
Oilseeds	38.5	18.3	40.4	2.8	78.9	21.1	109
Spices	50.4	18.8	27.8	1.0	78.2	21.8	276
Vegetables	34.9	46.5	17.4	1.2	52.3	47.7	186
Cotton	50.1	15.3	23.1	11.5	73.2	26.8	295

(*Source*: Rahman, 2000, p. 499)

nature of the process in which the Green Revolution was more of an evolution, further developing improvements in cereal varieties, e.g. 'improved' varieties of rice were introduced to China from Vietnam in AD1000, and advanced fertilisation techniques were widespread in the Medieval period. Nevertheless the Green Revolution represented a dramatic acceleration in the pace of change (e.g. Rigg, 1989; 1990).

In the short-term there have been some significant successes, with growth in food output in many parts of Asia and Latin America exceeding growth in population. For example, in India since the mid-1960s food production has risen at the rate of 2.4 per cent per annum whilst population numbers have grown at 1.58 per cent per annum (Slatford and Fishpool, 2001, p. 5). India now exports HYVs to Algeria, Bangladesh and Yemen. Chapman (2002, p. 157) refers to average rice yields in South and south-east Asia in 1991–3 as being 83 per cent higher than in pre-Green Revolution times in the mid-1960s. However, he notes the social divisiveness of the new technology (termed the 'talents effect'), initially favouring larger farmers. This has not been apparent everywhere and there has possibly been some 'catching up' with more opportunities for poor farmers to adapt as there

have been improvements in the supply of seeds, fertilisers, pesticides, credit extension schemes and agents, with evidence of a 'filter down' adoption cycle. This pattern may now have altered with associated technology proving more scale neutral, especially that associated with irrigation, though adoption of irrigation is often a key differentiating factor amongst small farmers (e.g. Sivertsen and Lundberg, 1996). However, there remain significant environmental problems, including loss of indigenous crop varieties, especially of wheat (Shand, 1997), over-reliance on artificial fertilisers (with increasing dosages having to be applied), build-ups of pesticides which are especially harmful in padi-rice systems, reduced straw output (impacting on production of animal fodder) and salinisation. These environmental impacts have disproportional impacts upon the rural poor.

The Green Revolution has rarely extended its focus to key subsistence crops such as cassava in Africa and millet in India. Indeeed, large swathes of farmers in Africa have been by-passed (Binswanger and Townsend, 2000; Watts, 1989). Similarly, areas amenable to irrigation practices have fared much better than those reliant on rain-fed agriculture. There appear to be two broad future scenarios associated with reactions to the 'technological fix': one is a search for a better integration between traditional and new methods, sometimes termed Low External Input Agriculture, incorporating integrated pest management (discussed further in Chapter 10), more disease-resistant crop varieties and an empahasis on the greater use of community resources, e.g. more equitable access to and use of water. A related 'solution' is a greater emphasis upon agri-forestry in a farming system that is more sensitive to the needs of the local environment and community. However, some community-based interventions have been criticised for ignoring intra-community differences, especially regarding access to land and capital (Leach *et al.*, 1997). This type of local complexity can be difficult to build into policy formation and hence generally is ignored by policy-makers (Edwards, 1994). In recognising this, Batterbury and Bebbington (1999) suggest five options for input from Western researchers to be relevant to local needs:

- use of a local political ecology based on an agenda set by local people, e.g. the work of Lane (1998) on Ma'asai land rights in East Africa;
- partnership research in conjunction with development organisations who can effect change, though this restricts the independence of the research as it may have to incorporate the organisation's own agenda (Batterbury, 1997);
- selective application of methods recognised by policy-makers in order to overcome established views, e.g. the use of scientific analysis to demonstrate causal processes of land degradation (Sillitoe, 1998);
- use of participatory studies in which the agenda set by local people supersedes that of government or a development agency (e.g. Kelly, 2001; Scoones, 2001);
- dissemination of research findings beyond the confines of researchers and government to involve individual farmers and community leaders in conjunction with politicians and civil servants (Chambers, 1997).

Aspects of these options have become increasingly familiar in the 1990s within the broad field of international development assistance, often under the guise of seeking to implement a 'sustainable initiative'. This notion of sustainability is discussed further in chapter 10.

The second 'solution' to the ever rising demand for more food, as population numbers in many parts of the Developing World grow (e.g. around 800 million more in China and India alone by 2050) is the expansion of biotechnology and the development of genetically modified (GM) organisms (discussed more fully in Chapter 10). This means that the vanguard of the new technology lies with the TNCs based in the Developed World, and there are major fears that these corporations will overlook both the needs of the poor in Developing Countries and will ignore major environmental concerns associated with GM organisms (Chapman, 2002). The application of GM 'solutions' is regarded by many as the antithesis of a sustainable option for further agricultural development. It is also seen as the antithesis of participatory development in which local people take greater responsibility for their own future (Mohan and Stokke, 2000; Pretty, 1995), changing the balance of power in the development process, and placing more emphasis upon local knowledge systems. As Mohan (2002) notes, however, it is difficult to provide a satisfactory definition of participatory development, but keywords applied in literature on the topic include involving the beneficiaries, self-help and individualism, less reliance on the state, co-determination of internal and external agendas and participation.

An example cited by Mohan (2002) is that of Village Aid, a small UK-based non-governmental organisation (NGO) working in West Africa. It seeks to develop a situation where 'village communities set the agenda and outside agencies become responsive' so that the requirements of the village are met (Village Aid, 1996, p. 8). This approach extends beyond a narrow definition of participatory development, but is quite typical of many schemes, especially in sub-Saharan Africa (Kelly, 2001; Wiggins, 2000). There are limitations, though, including: tokenism in which empowerment of locals is limited (Goebel, 1998); the lack of attention to the presence of social heterogeneity within rural communities, and especially the significance of gender differences; the development of competition between local organisations; and the inadvisability of attempting to solve all problems purely through local actions – some change needs to occur at national government level, e.g. the rules of trade (Chambers, 1997). Moreover, Parnwell (2002, p. 116) concludes 'it is unrealistic to suggest that "alternative development" is set to become the mainstream paradigm'. Furthermore, many governments have a limited view of participation and are unwilling to encourage it because of its implications for the distribution of power and resources (Desai, 2002; Gilbert and Ward, 1984).

9 Land use competition

9.1 Losses and gains of agricultural land worldwide

Between 1995 and 2025 a conservative estimate is that the world's agricultural systems will have to adjust to feed a further 2.5×10^9 people (WRI, 1996). The majority of these people will be in Developing Countries where farming systems are already struggling to provide sufficient food (Mannion, 1997). Mannion (1998, pp. 24–5) presents two options for the future of world agriculture to meet this challenge: intensification of existing systems and/or expansion into natural and semi-natural habitats (extensification) (Figure 9.1). The former is likely to require substantial increase of fossil-fuel inputs, but both would have profound implications for the environment and for associated socio-economic systems.

Based on current FAO estimates, approximately 11 per cent of the earth's land area is under cultivation whilst a further 24 per cent is under permanent pasture. The potential for extending the cultivated area is disputed, though the extent of this potential is generally estimated at between doubling (e.g. Tolba, 1992) and trebling (e.g. CRM, 1969) of the current area. However, much more conservative views have also been advanced (e.g. Pimentel *et al.*, 1987). The ability of humankind to effect any expansion rests upon the capacity to overcome a range of environmental constraints, but most notably those of leaching, laterisation and desertification.

Laterisation relates to limitations on cultivation imposed under tropical climates with distinct wet and dry seasons. When forest is cleared in areas with these climatic conditions and cultivation begins, unless soil nutrients are carefully managed, leaching can rapidly occur. This can then be followed by a drying process that promotes the formation of a red or yellow hard layer of clay, known as a pan or crust. Both leaching of nutrients and pan formation can seriously reduce crop yields, the former being even more severe if soils are not carefully managed once forest cover has been removed in areas with rainfall throughout the year. These are limitations that can greatly restrict the outputs resulting from cultivation of land formerly under rain forest and dense tropical vegetation cover.

The risk of desertification and salinisation is high in areas with an excess of evapotranspiration over precipitation, that is, in arid and semi-arid regions. In order to extend agricultural land in such areas provision of water for plants has to be managed very carefully. This can entail installation of expensive irrigation equipment. Similarly, the maintenance of soil fertility in areas with strong leaching can involve purchase of expensive artificial fertilisers, notably involving input of phosphorus. Only one-third of the land area that is potential arable land if irrigated can realistically be irrigated under existing technological conditions. To irrigate even this one-third would involve major costs.

Data from the FAO suggest that the world's arable area doubled between 1870 and 1970, with major expansions of arable land before 1930 through advances of the agricultural frontier in mid-latitude temperate grasslands in the Soviet Union,

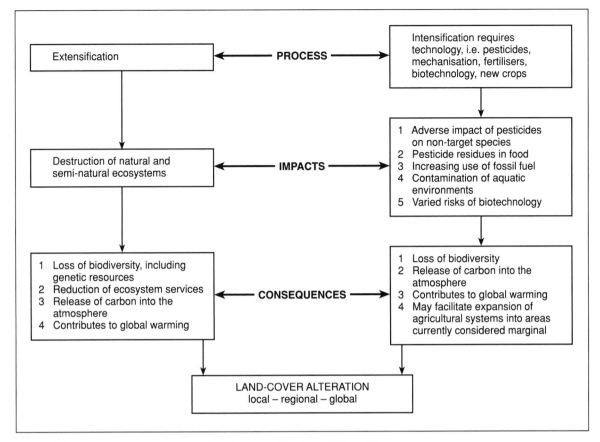

Figure 9.1 Future options for world agriculture (based on Mannion, 1998, p. 25)

North America, Australia, Argentina and Uruguay. This expansion slowed in the 1940s and 1950s as the 'pioneer phase' of agricultural settlement in these areas came to a halt, and with reduced need to exploit these new lands. Instead of increasing the amount of land under cultivation in order to raise output, the required output was generated from increasing other inputs to the agricultural system, notably fertilisers, machinery and improved farm structure/management. These changes meant that the same amount of land could produce more output, and hence less land, rather than more, was required in the agricultural system to feed the same number of people. This helps to account for the post-1945 trend for the amount of arable land in the Developed World to decline (Baldock and Conder, 1987), especially through losses to urban sprawl. Even so, in the 1950s and 1960s there

were three areas in the Developed World where new land was brought under cultivation: in the Soviet Union, through the northwards expansion of grain cultivation, especially during the Kruschev era; in the Canadian Prairies, again through northwards expansion associated with the development of fast-maturing grain varieties and influenced by the availability of overseas markets; and in Australia, through the development of drought-resistant grain varieties, again produced for sales on the world market.

Meanwhile, in the 1950s and 1960s many parts of the Developing World expanded their arable land to meet the growing local demand for food. Without the Developed World's access to other inputs, this expansion of land under cultivation represented one avenue for increasing output. Notable increases occurred in Latin America, especially

Table 9.1 Land-use change, 1975–1999 ('000 ha)

	Arable and permanent crops			Agricultural land		
	1975	1999	+/–75–99 %	1975	1999	+/–75–99 %
Developed World	671,906	647,928	–3.57	1,883,982	1,856,504	–1.45
Eestern Europe	49,389	46,231	–6.39	69,367	65,849	–5.07
Western Europe	92,525	86,956	–6.02	160,155	147,955	–7.62
Former USSR	232,400	217,514	–6.41	550,240	573,945	+4.31
North America	232,318	224,700	–3.28	497,846	493,288	–0.94
Developing World	734,926	853,524	+16.14	2,786,218	3,104,785	+11.43
Africa	171,719	201,784	+17.51	1,065,270	1,093,605	+2.03
Asia	454,763	511,727	+12.53	1,133,874	1,649,340	+45.46
China	100,637	135,361	+34.50	401,637	535,362	+33.29
ASEAN[1]	74,641	88,905	+19.11	90,548	105,807	+16.85
South Asia[2]	201,320	205,051	+1.85	222,269	224,199	+0.09
Far East[3]	381,590	434,618	+13.90	859,534	988,070	+14.95
Near East[4]	84,273	94,153	+11.72	394,622	504,617	+27.87
Central America + Caribbean	36,127	43,428	+20.21	127,473	142,222	+11.57
South America	90,570	116,131	+28.22	559,607	619,255	+10.66
Oceania	47,018	52,978	+12.68	505,568	472,377	–6.57

1 Brunei, Cambodia, Indonesia, Laos, Malaysia, Myanmar, Philippines, Singapore, Thailand, Vietnam
2 Bangladesh, Bhutan, Indonesia, Maldives, Nepal, Pakistan, Sri Lanka
3 ASEAN and South Asia plus China, East Timor, Korea (North and South), Mongolia
4 Afghanistan, Bahrain, Cyprus, Gaza, Iran, Iraq, Jordan, Kuwait, Lebanon, Oman, Qatar, Saudi Arabia, Syria, Turkey, United Arab Emirates, West Bank, Yemen
(*Source:* FAOSTAT)

through advancing the agricultural frontier into forested areas in Amazonia.

Despite the presence of inhibitors to extensions of agricultural land as mentioned above, since 1980 the extent of arable land in the Developing World has increased by over 15 per cent and the area under pasture by 6 per cent (Table 9.1). In contrast, there has been a decline in the amount of arable land in the Developed World (by over 3 per cent) and a less than one per cent increase in permanent pasture (Mannion, 1998, p. 12). The most rapid advance of the agricultural frontier has come in areas of tropical rainforest through a combination of both planned and unplanned settlement. Examples of the former include the opening up of Rondonia in the Brazilian Amazon (Walker *et al.*, 2000) and the Repelita programme in Indonesia (IPF, 2001). In both cases road construction for mining and logging operations has facilitated the spread of settlers into pristine forest. Despite the

substantial impact of large-scale clearance of forest for cattle ranching in Brazil, Fearnside (1993) estimates that one-third of deforestation in Amazonia is caused by small farmers who each clear less than 100 ha. Commercial arable production has been a significant factor in other Developing Countries, as virgin forest is converted by growing 'plantation' crops or other export crops (e.g. Symon *et al.*, 1997).

The decline in the cultivated area has also extended to some parts of Asia (Bowler, 2001), notably China where government data shows that the area of cultivated land fell by 4.73 million ha between 1978 and 1996 (Yang and Li, 2000, p. 73). Losses were most pronounced in coastal and central provinces where land is relatively fertile and incidence of multiple cropping is high. In observing the losses extending over several decades, Lester Brown (1995) predicted that by 2030 China would be unable to feed itself. Several government schemes were introduced in the mid-1990s to halt the losses and to promote grain production. These policies have continued the government's long-term aim of maintaining a very high level of self-sufficiency and avoidance of any reliance on the international market (Yang and Zhang, 1998). Unfortunately, both the official data on land use and the policies relating to grain production are based on inadequate and inaccurate official estimates, as revealed by remote-sensing techniques from the late 1980s (see below, Section 9.1.1). These show an under-estimation of the cultivated land by up to 40 per cent (Ash and Edmonds, 1998; Smil, 1999), and FAO data actually suggest substantial increases in the amount of both arable land and pasture in the last quarter of the twentieth century (Table 9.1). However, the magnitude of these increases (over 30 per cent) probably also reflects earlier under-estimation (e.g. in 1975). Moreover, there are several well-documented losses of cultivated land in China in recent years and, when set alongside the high pressure of demand for production from the land in cultivation, this raises serious issues regarding future output. In particular, a similar concern arises to that voiced throughout the Developed World, namely whether increased output from existing cultivated land or new land brought into production can offset any further withdrawals of land from cultivation (Hill, 1997b).

Yang and Li's (2000) analysis of losses of cropland in China showed that, whilst some losses were to non-agricultural uses such as construction, others involved land transfers to different agricultural uses such as orchards. This has had a negative impact upon grain production in some of the more fertile areas. Significant losses in western and northern China were of land in areas more marginal for grain production, but in such areas there had also been land reclamation schemes that had added little to overall production and had also been associated with negative environmental effects.

In another, small-scale, example, Khan (1997) reports a loss of 30 sq km of agriculturally productive farmland between 1951 and 1991 around the city of Aligarh in Uttar Pradesh, India. This rate and scale of urban expansion is having major impacts in Developing Countries. This is illustrated in New Bombay, Maharashtra, India. Urban expansion here involved the purchase of land and houses in 33 villages and just the land of a further 62 villages. Apart from the displacement of those individuals losing both their home and their land, there was substantial disruption to the lives of those remaining. The availability of new opportunities (e.g. rickshaw-driving, quarries, small shops/stalls, industry) increased the social and economic differentiation among the villagers, advantaging village leaders and larger landowners. Employment in farming, fishing and salt-making was virtually eliminated. The latter was particularly important for women, who therefore became marginalised in the labour market. However, only one-third of households were able to obtain urban-related employment. Hence, unemployment rose from 48 per cent to over 60 per cent, and the proportion of households with a monthly income less than 80 rupees (£1) rose from 9 to 29 per cent. However, those households receiving over 280 rupees (£3.50) a month also increased, from 9 to 29 per cent. This enabled these households to extend the level of material possessions. Compensation money from the developer enabled some families to build houses of bricks and cement, but other promised compensation in the form of better amenities did not appear (Bowler, 2001, pp. 135–6; Parasuraman, 1995).

The pattern of changes in Table 9.1 confirms the contrast between the Developed and Developing Worlds in the last quarter of the twentieth century, with the latter increasing the area under arable and permanent crops, especially in China and Latin America, but only marginally in South Asia. Substantial increases of grazing land were recorded in parts of Asia (notably in the Near East) and also in the former USSR, largely at the expense of grain production. There was also a sharp decline in rangeland pasture in Australia (Holmes, 2002). However, as mentioned with respect to China, detailed interpretation of this table requires closer examination of data quality and of both local and regional processes of land-use competition. To appreciate the impacts of the land-use changes upon food production and consumption also requires more detailed analysis of the composition of the food supply in the different regions. For example, the USA uses 2.9 billion bushels of maize in feeding 39.6 million calves and 22 million cattle (Lappé and Bailey, 1999, p. 87) whereas virtually all the grain produced in China is used for direct human consumption. Moreover, the data in Table 9.1 make no mention of the yields obtained from agricultural land. For example, UN estimates refer to 300 million ha of farmland worldwide severely degraded and 1.2 million ha showing moderate fertility loss (Earth Summit + 5, 1997).

9.1.1 Using remote sensing to estimate the extent of agricultural land

In his book on agricultural geography, published in the mid-1970s, Tarrant (1974) discussed the potential offered by satellite remote sensing in the study of land-use change. Over the last two decades this has been realised in various studies as the potential has grown considerably, advancing significantly since the initiation of the earth resources technology satellite (ERTS) programme in 1964. The subsequent development of various programmes utilising different types of imagery, has added greatly to data from aerial photography, which had been used by geographers since the First World War (e.g. Fuller et al., 1994).

Vertical air photograph interpretation was widely applied during the Second World War, with the Royal Air Force providing complete coverage of the UK in 1946–7 at a scale of 1:10,000 in black-and-white photography. This type of aerial photography was also developed extensively in surveys of tropical regions in the 1950s, though the classes of land use distinguishable were relatively crude and required substantial reinforcement from ground-based surveys in any mapping exercise. From the 1950s colour photography has enabled different classes of land use to be identified more accurately. Identification has been aided by using the reflectance of different types of surface in narrow bands within the visible spectrum (e.g. Board, 1965).

The new satellite technology presented significant advances to investigation beyond that of traditional analysis of aerial photographs. Over time the evolution of higher spatial and time resolution of remote sensing, and at reduced cost, has meant that there is a genuinely useful and affordable tool with which to monitor changes in land use, especially in areas where certain sources of information, such as agricultural censuses, may be absent (Clark, 1992; Shueb and Atkins, 1991). Use of satellite imagery relies upon classification of both farmers' fields and surrounding vegetation, ideally in situations in which there is a limited number of land cover classes, i.e. where each pixel of the satellite image can be assigned to a specific class. Hence there has been much research on classification methods for allocating the pixels in Landsat and SPOT images to land cover classes in intermingled cultivated areas in Europe and North America where each pixel is likely to belong to one of a limited number of classes (Rasmussen, 1992). However, the task tends to be somewhat harder in tropical areas where individual fields are small compared with the spatial resolution of the images. Also, the plant/crop cover of the fields is often heterogeneous so that it is difficult to separate the fields from surrounding 'bush'.

The basic characteristic of satellite remotely sensed images is the ability to distinguish between different types of surfaces. The nature of this differentiation varies with that part of the electromagnetic spectrum that is being analysed. Any given surface absorbs radiant energy from the sun and re-radiates some of it back as radiant energy, the extent of this re-radiation being termed emissivity. Some radiation is also reflected directly from the earth's surface back into space. The relative absorption of the surface determines the character of the reflected radiation, which can be detected by the remote sensor. Detection has focused on four areas of the electromagnetic spectrum:

- visible light;
- reflected infrared, sometimes termed false-colour photography;
- thermal or emissive infrared. This measures energy from the sun that is re-emitted in the form of radiation of longer wavelength. Different radiation levels reflect differences in the surface temperature of objects and the differences in their degree of emissivity.

Analysis of this type of radiation has been employed in terrain studies for over three decades (see an early summary by Board, 1968);

- microwaves, as in the principle of radar whereby microwave radiation is generated and bounced off objects at the earth's surface. Variations in the surface affect the level of reflected radiation from the surface.

The Second World War provided a stimulus to the use of false-colour photography, primarily to detect camouflaged vehicles. Infra-red imagery has proved extremely useful for the remote sensing of vegetation, as plants reflect a high proportion of received infra-red radiation. Different intensities give rise to clear distinctions between certain plants, notably broad-leaved and needle-leaved trees. Hence reflected infra-red photography can be used as an effective means of distinguishing between different types of land use (Cooke and Harris, 1970).

As different parts of the reflected electro-magnetic spectrum can be used in remote sensing, much of this sensing uses multi-band sensors operating in both the visible spectrum and in the reflected infra-red. This multi-waveband photography can utilise the fact that different plants are detectable using different wavelengths. They have their own signature or pattern of reflectance over a particular range of wavelengths. However, the signatures vary according to moisture content, seasonality, presence of direct sunlight and shade, so that interpretation can be complicated. This means that accurate classification is heavily dependent upon good ground-truthing.

In the 1980s classifications were combined with software improvements to transform this type of analysis, especially through the use of NASA Landsat multi-spectral scanning (MSS) imagery, and subsequently an improved MSS called the Thematic mapper (TM). Examples include work on estimating land cover in both lowland and up-land environments in Britain (Fuller and Parsell, 1990; Harding, 1988), the Large Area Crop Inventory Experiment in the United States, yielding unbiased and precise estimates of the area under wheat in the Great Plains (MacDonald and Hall,

1980), monitoring of rangeland in Australia (Kelly and Hill, 1987; Graetz *et al.*, 1988), land use inventory in France (Hill and Meiger, 1988), and integration of agricultural census data with remotely sensed images in analysing agricultural development in Belize, Central America (Robinson *et al.*, 1989).

Each Landsat satellite passed over the same area on the earth's surface during daylight hours every 18 days, providing TM data in seven spectral bands (blue, green, red, near-infrared, mid-infrared, thermal infrared and mid-infrared) using a 30 m × 30 m ground resolution cell. Computer-assisted analysis of Landsat data provided an appropriate synergy between remote sensing and Geographical Information Systems (GIS) in analysing the imagery, as demonstrated in Xu's work on land-use change in south-east Scotland (Xu, 1992; Xu and Young, 1989; 1990). This showed the tremendous scope of remote sensing allied to GIS for mapping and monitoring land-use change. Subsequently this potential for linking remote sensing and GIS has been widely realised, building on the growing variety of types of imagery available, both for semi-natural vegetation and agricultural land (e.g. Cook and Norman, 1996; Gooding *et al.*, 1993; Grainger, 1996; Green *et al.*, 1996). Indeed, the use of remotely sensed data and GIS has become commonplace in many types of research on changes in land use, both in areas where ground-truthing is problematic, e.g. tropical rain forests (Boyd *et al.*, 1996), and in densely settled areas such as around major cities (Michalak, 1993). Even so, problems of allocating remotely sensed data to distinct land-use categories restrict its utility.

9.2 Land-use competition in the rural–urban fringe

9.2.1 The stimulus of the von Thunen model

In the 1960s, when agricultural geography was dominated by simple conceptualisations of the economics of agriculture, one of the most commonly utilised models was that of Johann Heinrich von Thunen, a Prussian landowner, who, in the early nineteenth century, had formulated a model depicting the competition for agricultural land in the vicinity of a major market for agricultural produce. This model emphasised the notion of economic rent to account for the spatial pattern of agricultural land-use zones around the market (Hall, 1968).

Von Thunen's model assumed soil fertility to be constant and that the only variable cost was the transport of agricultural produce from farm to market. Therefore, it followed that the type of land use and the intensity of cultivation were functions of distance from the market. This yielded a series of concentric zones of land use around the market centre. Despite its excessive simplification and reliance on one over-riding variable, transport costs, Chisholm (1962) argued that this model possessed two key attributes:

- it represented an attempt to obtain a theoretical explanation of land-use patterns, with a distance-cost relationship that is largely valid and which can be modified to include other factors;
- it could offer an explanation for contemporary land-use patterns around villages and farms throughout the world.

Critrics argued that these two attributes are extremely limited in terms of the model's applicability to an understanding of actual land-use patterns. Hence attempts were made to incorporate additional variables, such as freight rates (Haggett, 1975, p. 399), government policies, land tenure, soil fertility and behavioural factors (Blunden, 1977; Wolpert, 1964). Indeed, there were several notable examples of research that examined deviations of actual agricultural land-use patterns from those 'predicted' by the von Thunen model (e.g. Atkins, 1987; Blaikie, 1971; Griffin, 1973; Hill and Smith, 1994, Peet, 1969).

The focus on the significance of farmers' decision-making processes in the late-1970s and the subsequent emergence of political economy approaches have largely consigned the von Thunen model to the historical backwaters of agricultural geography. However, the potential for land-use competition inherent in the model's notion of economic rent offers a simple introduction to the topic of competition for agricultural land and between agricultural land uses (Clark, 1999). As shown in Figure 9.2a, the rate of fall of production expenses per ha per km from the city (market) for crop x is greater than for crop y. If all the farmers select

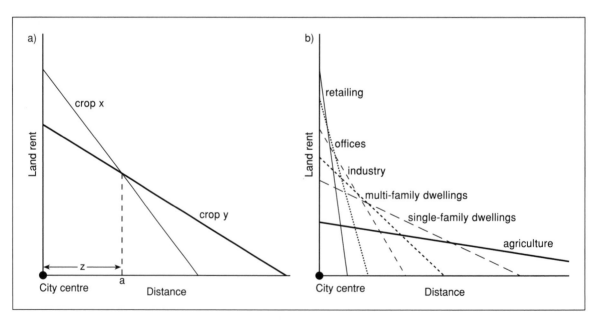

Figure 9.2 The von Thunen model: (a) land rent versus agricultural land use; (b) land rent versus a range of land uses

land use to gain the best land rent then crop x yields the greater return as far as z kms from the market. Further away from the market the rent is greater for crop y and hence this will become the preferred crop. However, around point a, the reality is that factors other than economic rent are likely to blur the sharp boundary between the two crops so that they do not maintain such a regular distribution. These factors can include the imperfect economic knowledge possessed by farmers, physical variations that will differentially favour particular crops, variable intensity of cultivation, economic constraints operating within an individual farm, and variable downstream and upstream relations developed by particular producers (Bryant, 1992, pp. 291–4).

This conceptualisation can be extended to help explain the arrangement of land use on an individual farm (Blunden, 1977) or to consider competition between agricultural and non-agricultural land uses (Bryant and Johnston, 1992, pp. 79–86). The latter is illustrated in Figure 9.2b, showing different bid-rents for land uses with increasing distance from a city centre. The implication is that agriculture can only produce a competitive economic rent beyond some considerable distance from the city centre. The reality, though, is that on either side of the junction between agriculture and land taken by housing there will be an intermingling of both agricultural and non-agricultural land uses under the influence of factors other than simple economic rent.

9.2.2 The rural–urban fringe in the tropics

The intermingling of land uses referred to in the previous section can be seen in many cities in the Developing World. However, in many parts of the tropics the very rapidly growing urban areas are also being sustained by 'urban' agriculture in the same manner that agriculture was integral to urban areas in nineteenth-century Europe. Drakakis-Smith (1994) made a simple distinction between home gardens and the cultivation of the urban periphery. However, others have asserted that the structure of tropical urban agriculture is more complex (Binns and Lynch, 1998; Ellis and Sumberg, 1998; Obosu-Mensah, 1999). There has

been increasing promotion of urban agriculture as a solution to the rising concerns regarding urban food security (Smith, 1998). As summarised by Lynch et al. (2001), the main reasons commonly given for promoting urban and urban-fringe cultivation are:

- provision of vital or useful food supplements in close proximity to the main area of demand;
- environmental benefits;
- employment creation for the jobless;
- providing a survival strategy for low-income urban residents;
- making use of urban wastes in urban agriculture.

Ellis and Sumberg (1998) refer to a number of studies concerned with empirical research on tropical urban agriculture. These reveal various problems in promoting cultivation in urban areas:

- water shortages, particularly in arid or semi-arid urban environments;
- health concerns, particularly from the use of contaminated wastes;
- conflicting urban land uses;
- a focus on urban cultivation activities rather than its position in relation to broader urban management issues;
- inability of urban agriculture to contribute substantially to urban food needs;
- a lack of clarity over whether it is best to produce in the city or country.

There appear to be a number of interacting variables that distort the classical von Thunen model of high value and perishable crops being produced close to the urban market in the tropics. In particular, urban agriculture is often closely associated with issues of contested access to the land. 'It is not uncommon for many farmers to cultivate land in and around African cities over which they have no formal rights' (Lynch et al., 2001, p. 162). Hence Orchard et al. (1997, p. 17) argue that a key factor in the development of agriculture around Addis Ababa (Ethiopia), Enugu (Nigeria) and Kano (Nigeria) has been secure access to land for farmers. In many African cities there is great ambiguity about legal rights and a problematic tenure system.

This may provide some opportunities for farmers in the short-term but there may also be difficulties related to problems in obtaining credit (which is often tied to possession of legal ownership of land) (Freidberg, 2001). The land market can be manipulated both by government and development agencies, sometimes to the advantage of farmers, but often 'spaces' are created that cannot be used by farmers (e.g. Maxwell, 1994). Lynch *et al.*'s (2001) survey of Kano (Nigeria) revealed large amounts of fruit and vegetables grown within the built-up area, but with the issue of insecurity of land tenure a major problem for many cultivators. So most agricultural production is performed by squatters who are threatened by the sale of these undeveloped lands to urban land developers by government agencies (Binns, 1996; Binns and Fereday, 1996; Olofin *et al.*, 1997).

The presence of the urban area can also have a dramatic effect upon the labour market for the agricultural workforce in Developing Countries, notably by raising the opportunity cost of working in agriculture (Bowler, 2001). The highest rates of movement of farm workers to full-time urban–industrial occupations tend to occur in rural–urban fringe locations. Moreover, as suggested above in the von Thunen model, the close presence of the urban market can have a major impact upon agricultural activity. For example, it has long been recognised that there is a gradient of rising agricultural intensification with decreasing distance to urban markets. This is related to reductions in transport costs in both purchasing farm inputs and marketing farm products, and access to urban markets for perishable crops and livestock products such as fruit, vegetables and fresh milk. In the rural–urban fringe of tropical cities there often tend to be a preponderance of a large array of products and services that exploit proximity to urban consumers, including farm stalls, shops and other direct marketing schemes.

9.2.3 Fringe belts and the city's countryside

Competition between agricultural and non-agricultural land uses is at its most intense at the interface between the urban and the rural, frequently resulting in a succession of land uses

at any one location over time. However, most analyses of such successions emphasise not only the fluctuating strengths of urban and rural economies (which alter the bid rents) but also other factors affecting land use, notably the changing directions of agricultural policy and countryside planning policies (e.g. Blair, 1987). The relationship between the urban economy, the agricultural economy, agricultural policy and countryside planning is illustrated in Figure 9.3, which focuses on the identity of the actors involved in the land market, their relative power, and the intervention of constraints on the development process, including planning controls (e.g. Pacione, 1989).

This argument was elaborated by Worthington and Gant (1983), contending that an understanding of the politics of land development and conversion processes is crucial to any interpretation of land-use change, but especially in the rural–urban fringe where conflicts of local interest can be very strong. They contend that to fully understand such conflicts (and resulting land-use patterns) it is necessary to determine the extent to which the course and outcome of a conflict can be explained by the role of agencies (individuals, organisations, companies) and how far by structural processes (wider impersonal forces that shape society). In areas where farmland is under threat from development interests of various sorts, there are frequently contests between different parties for control over planning regulation in an increasingly uncertain planning system in which, throughout the Developed World, more powers are being vested to local communities.

The range of factors impinging upon the decision-making affecting rural–urban fringe areas was conceptualised by Veldman (1984) in terms of the interaction between the physical-spatial structure and the socio-spatial system in rural–urban fringes (Figure 9.4). This stresses interconnections between components of the physical environment, economic activities and the resident population in an open system emphasising the three components of space users, used space and space use.

Conzen (1960, p. 125) defined these 'fringe-belts' as 'a belt-like zone originating from the temporarily stationary or very slowly advancing edge of a town and comprised of a characteristic

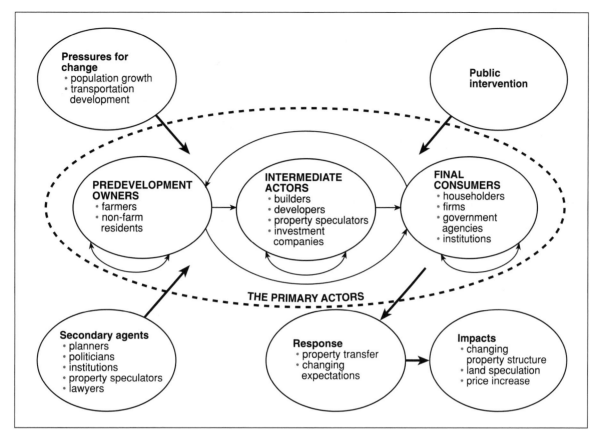

Figure 9.3 Actors and the land market (based on Pacione, 1989)

mixture of land use units initially seeking peripheral location'. The type of land uses produced in these belts is illustrated in the von Thunen-style model shown in Figure 9.5. Whitehand (1967, pp. 229–32; 1972) recognised belt formation and change as part of a two-stage framework of formation and modification (or occlusion), linked closely to the broader context of nationwide housebuilding cycles, which alter the value of land in the belt. These values, and developers' ability to exploit them, are controlled by planning laws, which take different forms throughout the world, contributing to distinctive patterns of 'mixed' land use in the rural–urban fringe. The example of the UK's Green Belt legislation in affecting the fringe is discussed below in Section 9.3.1.

Studies of land use in rural–urban fringes have typically revealed a large variety of uses, often with a high degree of inter-mixing that reflects the haphazard growth of the city at different rates in different directions. Irregular urban growth has produced an incoherent land-use pattern including agriculture, rural settlements, modern residential estates, industry, out-of-town retail and service centres, derelict land, cemeteries and sewerage works (Gant and Talbot, 2000).

The processes of urbanisation have also created what Bryant *et al.* (1982, p. 3) termed 'regional cities', covering a built-up core to the urban area and a more dispersed urban form which, in the case of the largest cities, may have a radius of over 100 km. This dispersed form has developed rapidly since 1945 as increased private ownership of cars has permitted the extended separation of workplace and residence. Hence there has been substantial growth of suburbs attached to the urban core as well as the growth of both existing and new settlements in the area beyond these

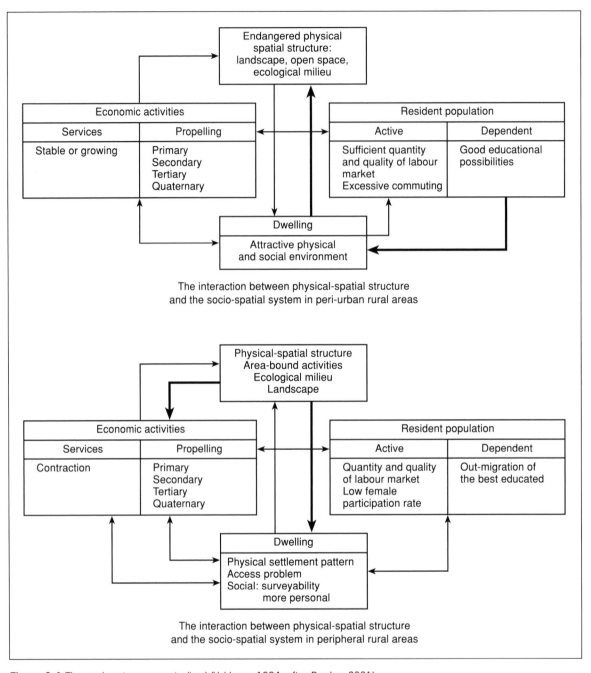

Figure 9.4 The rural system conceptualised (Veldman, 1984, after Bowler, 2001)

suburbs. Within this simplistic division of core, suburbs, and outlying settlements, Russwurm (1975; 1977), amongst others, recognised a more complex mixture of different land uses character-

ising the area beyond the suburbs. This mixture includes ribbon development, uses requiring large amounts of space (e.g. stockyards), and various types of development of residential property. Other

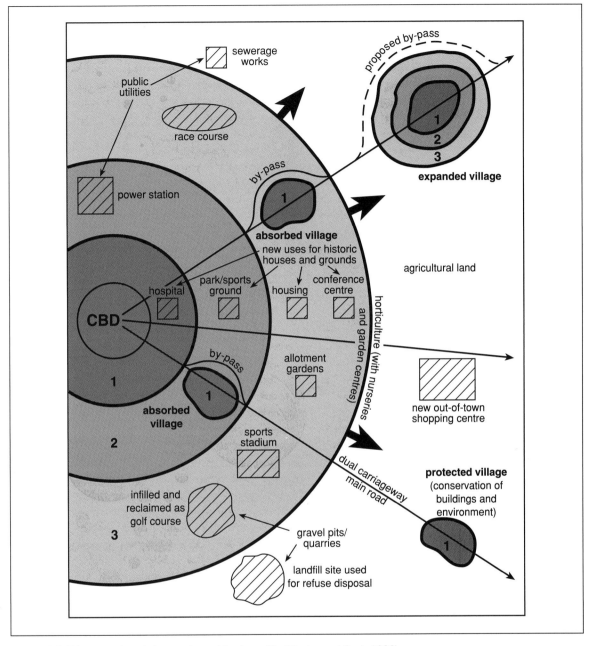

Figure 9.5 Urban growth and changes in rural land use (Worthington and Gant, 1983)

more subtle forms of urban influence on land use occur by way of increased non-farm ownership of land and modifications to agricultural land use. Second-home ownership and development of tourist accommodation have also become features of this area beyond the suburbs, perhaps more aptly referred to as the 'city's countryside'.

One characteristic of the emergence of this extended urban impact has been the growth in population of non-metropolitan areas. From the

1960s in several parts of North America non-metropolitan areas have grown at nearly double the rate of their metropolitan counterparts (Berry, 1976; 1980) as decentralisation and migration from larger to smaller centres has formed the basis of a shift in the focus of population growth (Coffee and Polese, 1988). Decentralisation reflects a complex web of decision-making relating especially to the growth of the service sector and diseconomies of location in large cities. These have produced a strong desire in the middle-classes to seek homes beyond the urban core and suburbs. Changes in transport, communications, and creation of a more diffuse urban form, have placed increased pressure on agricultural land whilst also changing the composition of communities in areas affected (Smit and Conklin, 1981).

Forces contributing to this urbanisation of the countryside include increased population mobility, economic factors associated with the costs of transportation, land, housing and local taxes, and the lifestyles and desires of individuals. In particular, changing demographic and environmental contexts, housing availability, employment opportunities, and public policy have been key determinants of dispersal (Herington, 1984).

The various processes that have shaped the more diffuse urban form have been the subject of numerous geographical studies, key issues being land-use competition, especially the loss of prime agricultural land; the changing nature of communities, including a focus on the groups in society that are involved in migrations associated with the dispersal of population; and the planning of the changes occurring in the city's countryside, especially the regulation of land-use competition through legislation affecting the land and housing markets.

As the city has taken on this ever more dispersed form, new patterns of social and economic life have affected communities in receipt of the influx of newcomers from the city. The in-migrants have frequently located in settlements previously dominated by farming households, which were at the centre of life in small, rural communities. Hence farmers and ex-urbanites increasingly live in the same locale. This proximity of urban development to farming in the Developed World has had four major influences upon agriculture:

- loss of farmland for urban development;
- reduction in the amount of farmland produces smaller farming units and the fragmentation of holdings;
- speculation in anticipation of development can lead to a deterioration in farming standards. Similar deterioration may be associated with increased trespass and vandalism;
- spread of what Carter (1981) referred to as 'rural retreaters' or hobby farmers who farm on a part-time basis.

The significance of these influences can be seen in a variety of ways, but perhaps most notably in attempts to preserve farmland from urban sprawl and the growth of part-time farming.

9.3 Protecting farmland from urban development

In the Developed World there has been considerable debate regarding the impacts of the loss of prime farmland to urban development, acknowledging that this represents an irreversible reduction in the physical resource base for food production. However, national-scale enquiries have generally concluded that there is a land surplus in terms of there being sufficient land to maintain a desired level of food output (e.g. NALS, 1980). The critical assumption made by such studies is that there will continue to be gains in agricultural productivity from the surviving farmland, associated with technological progress. In the Netherlands the conclusion that there is a surplus of farmland was accompanied by predictions of a reduced role for food production and related employment, but enhanced roles for urban land, environmental conservation and recreation (NSCGP, 1992). Nevertheless, farmland in the 'green heart' of the country has been strongly protected by planning constraints upon urban sprawl. Similar measures have been pursued in the post-war period in the UK whereas the protection of farmland in North America has become an emotive issue leading to mobilisation of citizens' groups.

9.3.1 Green Belts in the UK

Green Belts are areas of open and low-density land use surrounding existing urban settlements where further continuous urban sprawl is deemed to be in need of strict control. Although there have been formal schemes incorporating this type of measure in many parts of the Developed World for over a century, in the UK current incorporation of the concept into planning controls dates to a ministerial circular in 1955. Since then the Green Belt has been extended to cover over 1.65 million ha in England (or around 13 per cent of the land surface) (Map 9.1) (DETR, 2000). This popularity stems largely from the high degree of success in halting losses of agricultural land (Munton, 1981).

Planning guidance notes in 1995 stated explicit objectives for use of land in the UK's Green Belts:

- to provide opportunities for access to the open countryside for urban populations;
- to provide opportunities for outdoor sport and outdoor recreation near urban areas;
- to retain attractive landscapes, and enhance landscapes, near to where people live;
- to improve damaged and derelict land around towns;
- to secure nature conservation interests;
- to retain land in agriculture, forestry and related uses.

The pressure on the Green Belt increases land values on both sides of the Belt, thereby tending to encourage several types of developmental change. Within the urban area there is greater pressure for site redevelopment in the suburbs as suburban spread is restricted by the legislation. Beyond the Green Belt there may be increased pressure for building development on farmland, in effect a translocation of pressures unsatisfied within the Green Belt that produces this 'leap-frogging' effect (Fothergill, 1986). In recent years the change in attitudes towards the importance attached to preserving agricultural land for productive farming purposes has led to various suggestions for relaxing planning controls within the Green Belt. In south-east England pressure has been strong for relaxations to accommodate the demand for house-building, with several major clashes over proposals to build new housing estates on Green Belt land (notably in Hertfordshire and West Sussex). These disputes have frequently pitched the national interest, as represented by concerns of central government, against the wishes of local residents who have often embodied NIMBY (not in my backyard) and conservationist attitudes.

In a detailed study of development pressures in south-east England in the early 1990s, Murdoch and Marsden (1994) noted the various economic and social changes affecting the settlements in the extensive Metropolitan Green Belt and adjoining countryside. They observed the 'leapfrogging effect' produced by the presence of the Green Belt, noting that the success of urban containment policies was quite varied in spatial terms. Moreover, even in areas where the appearance of the countryside looked relatively unchanged, e.g. Aylesbury Vale in Buckinghamshire, the pace of change was still rapid. This was having the effect of maintaining pressure on local authorities to convert some areas of farmland to housing, industry and recreational use. The result was a piecemeal erosion of controls and further losses of agricultural land to development. The significance of the role of large private landowners as bulwarks against development was identified. However, the desire to retain a 'green' countryside has not necessarily extended to a concern for maintaining a viable agricultural industry. Instead local planning has tended to permit land use changes from agriculture to other activities provided that these meet specified aesthetic and environmental criteria (p. 128). One new use that has been permitted has been conversion of farmland to golf courses, provided they were on land of lower agricultural, landscape and ecological value (p. 149). Between 1987 and 1992 Aylesbury Vale District Council alone received 24 applications for golfing facilities, of which 17 were approved. However, this golf course 'boom' gave way to 'bust' in the late 1990s, partly because of insufficient take-up of the new golf course memberships and tighter controls introduced by planners regarding such developments. Nevertheless, a range of other leisure uses has appeared, often encouraged by farmers themselves as part of farm diversification. The most popular developments have been equestrian-related activities and nature trails.

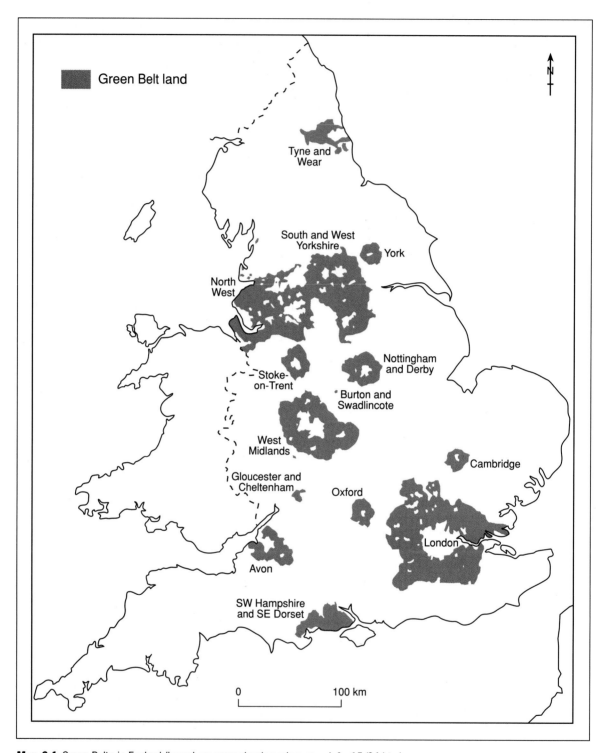

Map 9.1 Green Belts in England (based on: www.planning.odpm.gov.uk/luc15/04.htm)

9.3.2 Farmland preservation in North America

The amount of cropland in the USA post-1945 has declined at a rate of about 1 million acres (405,000 ha) per annum, though primarily in more marginal farming areas with soils of low inherent fertility and topography unsuited to modern farm machinery. The amount of farmland in the USA peaked at 469.85 million ha in the late 1940s and by the late 1990s 2.18 million ha had been lost, though there have been arguments over the rate of loss and how it can be measured (Theobald, 2001). Of the overall land on farms lost, about two-thirds was woodland sold by farmers to non-farming organisations (Hart, 2001, p. 530). Moreover, if the actual amount of cropland harvested is considered, the area in 1998 was similar to that of the mid-1960s, though some 15 per cent below that of the mid-1940s (Figure 9.6).

Change has been smallest (often under 5 per cent per county) in the Midwest, the US agricultural heartland. Indeed, in one in every seven counties the area of cropland has actually risen since the late 1940s, especially through extensions of irrigation works (e.g. on the Great Plains, along the lower Mississippi and in the swamps of southern Florida). Concentrations of cropland lost occur in the Southern states and in conjunction with urbanising centres of the eastern seaboard, the Great Lakes and coastal California.

There is a strong but not systematic urban influence in this pattern of loss (Furuseth and Lapping, 1999), but this factor sits alongside others in the South, notably the impact of stagnant demand and loss of market for cotton, tobacco, peanuts and traditional crops (plus increases in yields reducing the area of cropland necessary to produce the same amounts of these crops) (Hart, 2001). The most populous counties have lost much greater

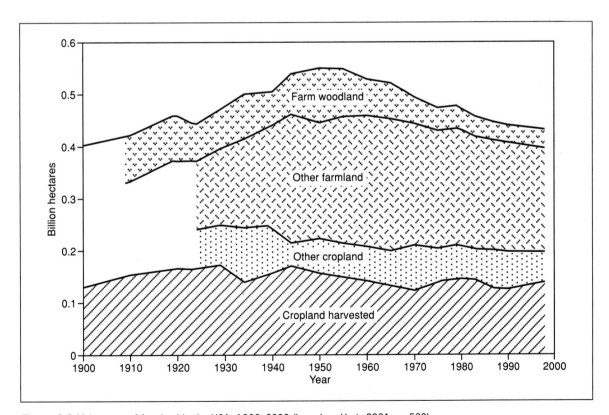

Figure 9.6 Major uses of farmland in the USA, 1900–2000 (based on Hart, 2001, p. 530)

percentages of their cropland (Hart, 2001, p. 538) as the 'bow wave' of metropolitan expansion consumes farmland irrespective of its quality (Hart, 1976; 1991; 1998).

The concern over the destruction of high-quality farmland by urban sprawl has been expressed most vocally in California, but the Census of Agriculture reveals that this state has more land under orchards, vineyards and vegetables today than in the late 1940s (+271,145 ha in the case of vegetables). Hence Hart (2001) concludes that concern over loss of farmland to urban expansion, even in California, is generally misplaced, and is based more on aesthetic concerns than on an absolute need to retain high-quality farmland for reasons of maintaining agricultural output – 'over-production' remains a more significant concern than lack of production. Others have taken different views (see Katz and Liu, 2000; NALS, 1981; Platt, 1985), reflecting the fact that metropolitan counties still account for one-quarter of agricultural output in the USA (Daniels, 1999). The range of opinions is affected by the lack of adequate knowledge of processes in the rural–urban fringe, especially as it is difficult to map/measure land use changes there. Coarse analytical units, such as counties, do not capture the fine-grained pattern of land use. Moreover, problems exist regarding clear definitions of hobby or 'non-farm' farms (Hart, 1992), for example what production value is required of a ranch or farm before it is no longer considered 'true agriculture', and at what point should it be regarded as low-density residential land use? The growth of low-density ex-urban residential development has exacerbated the problem (Furuseth and Lapping, 1999).

Recently, attempts have been made to use remote sensing to monitor land-use changes in the USA. In 1982, 1987, 1993 and 1997 the National Resources Inventory (NRI) was conducted by the Natural Resources Conservation Service (NRCS) of the USDA. This surveyed non-federal land resources using aerial photographs and field surveys of over 800,000 points, using 11 classification types (Nusser and Goebel, 1997; USDA, 2000), but results are only reported at an aggregate level (state or county). The US Geological Survey uses photographs and Advanced Very High Resolution

Radiometer (AVHRR) satellite imagery at 1 km resolution. This yields land cover characteristics and is combined with 30 m resolution Landsat TM imagery to give 21 categories of land cover (Vogelmann et al., 1998). A remaining problem is the uncertain relationship between land cover and land use, as remarked by Theobald (2001, p. 548): 'land-cover databases derived from classified satellite imagery substantially underestimate lower-density settlement'.

Bunce (1998) recognises four ideologically based lay discourses with respect to the continuity of state controls for farmland preservation and the farming lifestyles they maintain in urban–rural fringes in North America:

- farmland as a finite productive resource – productivism;
- food production in the national interest – progressive agrarianism;
- farmland preservation as environmental protection – ecological environmentalism;
- farming as a connection between nature and society – romantic agrarianism.

With reference to the last two, he concludes that 'the language of farmland preservation articulates ideals for which farmland itself acts as a physical symbol and thus elevates the meaning and significance of agricultural life and landscape above that of a basic productive resource' (p. 235).

This indicates that there are significant differences of attitude towards farmland preservation in North America and Western Europe. In the latter in recent years there has been a growing general view that agriculture itself is responsible for environmental disbenefits, and hence farming should not retain a special place in general countryside planning. In contrast, in North America there continues to be a contentious debate about the desirability for and efficacy of farmland preservation policies. This difference reflects basic geographical, historical and political differences in the nature and use of rural land that have produced divergent meanings of countryside. However, it also reflects the comparatively recent emergence of the conversion of agricultural land as a planning issue in North America (Bunce, 1998).

It was not until the 1970s that the loss of prime farmland to urban sprawl became a major public issue in the USA, though Maryland had adopted limited measures to reduce urban pressure on state farmlands in the 1960s (Lehman, 1995). Awareness was raised by several studies that revealed rapidly increasing rates of conversion (e.g. Reilly, 1973; Simpson-Lewis et al., 1979). This brought a response via several more government-sponsored studies in both Canada and the USA, and initiation of farmland protection policies at the local, state and provincial level (Furuseth and Pierce, 1982). These tended to use farmland preservation as a management tool to restrict urban growth. Bunce (1991) notes that this urban-centred perspective on farmland preservation has been a consistent theme in local land use planning, though much of the public concern was related to the impact of sprawl on agricultural land itself, especially on the productive capacity of the resource base. This was essentially a productivist argument voicing concerns over the threat to food production, despite little supporting evidence (e.g. Cocklin et al., 1983). However, in regions with scarce and specialised agricultural land, as in parts of New England, the Okanangan in British Columbia, and the Niagara Peninsula, the case for strong farmland preservation strategies was supported by the clearly evident attendant negative impacts of urbanisation upon farming. Elsewhere, arguments focusing on the loss of productive land were accompanied by other factors, including control of urban sprawl, preservation of countryside amenity, protection of the rural environment, maintenance of rural communities and the farming way of life, and guarding national identity. Indeed Lehman (1992; 1995) claims that the farmland preservation issue has really been an ideological struggle related to the socially constructed primacy of farmland. In this context, farmland preservation acts as a political symbol for wider arguments about the value of rural life that extend beyond conceptions of farming as an economic activity.

Bunce (1998) notes how much of the concern over the loss of farmland in the 1960s and 1970s was linked to neo-Malthusianism within the burgeoning environmental movement. Hence the 'threat' of loss of farmland forever to urban sprawl quickly attracted the attention of a public already used to the language of 'threat' related to global famine and general resource scarcity. Urbanisation was generally presented as a permanent threat to productive capacity whereas land degradation through increased intensification of farming methods was identified as redeemable through improved management practices. The popular media and academia presented the image of the USA 'running out of productive farmland', with surveys such as that by the National Agricultural Lands Study (NALS) (1980) reinforcing this perception. In Canada, the Ontario Institute of Agrologists (1985) warned of domestic food shortages. Similarly, the Lands Directorate of Environment Canada, established in 1971, monitored Canada's land resources for the next 15 years with a strong preservationist ethic that affected both public awareness of and policy towards agricultural land. Yet, there was little evidence brought forward to relate the actual loss of farmland to declining output.

Losses of farmland to other land uses were undoubtedly substantial and occurred over a long time period in some parts of North America. For example, by 1970 eastern Canada had lost 40 per cent of its 1940 farmland extent and around 150,000 farms. Nearly two-thirds of the farmland had been lost in Nova Scotia and New Brunswick. Overall, the farmland area in the country halved between 1941 and 1991, including nearly 40 per cent of the improved farmland base (Troughton, 1997, p. 285). The largest proportional losses occurred in the Maritime provinces, but the largest absolute losses were in Quebec and Ontario: 6.5 million ha of farmland (51 per cent) and over 200,000 farms (72 per cent). Some of these losses were due to urban sprawl, especially on high agricultural capability lands in southern Ontario (Robinson, 1991d), but much land has also been converted to forest or simply reverted to an unproductive use. However, this is not a nationwide trend, as in western Canada the farmland area has risen by over 7 million ha (14 per cent) since 1941. This has been part of a more than doubling of the size of holdings, as mechanisation combined with farm amalgamations and new biotechnology has produced larger holdings and an ability to bring more land into cultivation. This has encouraged movement of the agricultural frontier northwards in the Peace River district (in Alberta and British Columbia) whilst also bringing new land into production in some southern parts of the Prairies.

The concern over loss of prime Canadian farmland was at its height in the early 1980s when, for example, there was a rapid extension of the spatial extent of Toronto's commuter belt (Robinson, 1991d). This prompted action in the form of land-use monitoring activity, notably the Canada Land Use Monitoring Program (CLUMP), which focused on 'urban-centred regions' (UCRs) as one of four different types of area selected for special attention, with detailed analysis of the pattern, type and rate of change of land use, analysis of loss of high-capability agricultural land and other resource lands to urbanisation, and the correlation of changing land use with growth of population. This monitoring revealed a loss of just over 300,000 ha of rural land to urban uses in the UCRs between 1966 and 1986. The largest losses were associated with expansion of the major cities (population over 400,000) where a one per cent growth in population produced a 1.5 per cent increase in land under urban use (Yeates, 1985, pp. 6–7).

Other rationales for farmland preservation reflected different concerns. In Canada there was a view that urbanisation was taking the best farmland and therefore pushing agriculture onto more marginal land (McQuaig and Manning, 1982). This view can be seen as part of broader environmental considerations encompassing both loss of farmland and soil degradation. These were presented as parallel threats that detracted from the primacy of the land ethic, which presented an argument for ecologically sustainable agriculture (Berry, 1981; Leopold, 1949). Support for this came from mainstream environmental organisations, such as the Sierra Club in the USA, who were instrumental in forming the American Farmland Trust in 1981.

Pierce and Seguin (1993) have observed that, in addition to the environmental and adequacy of resources arguments, lobbies also developed focusing on local amenity protection and conservation of communities. There were notable examples of

this in New York and New Jersey in the 1960s and early 1970s (Esseks, 1978), but subsequently more widespread local amenity activism came to dominate farmland preservation campaigns through initiation of preservation policies and programmes at local government level (Peters, 1990). Wright (1993, p. 269) argues that the growth of farmland preservation movements has been a consequence of the 'failure of governmental land-use planning programs to protect cherished places from urbanization'. This has embedded farmland preservation within grass-roots initiatives to protect rural environment and character. This can be seen in the 'Save America's Countryside Campaign' operated by the National Trust for Historic Preservation (Stokes, 1989). This perspective of 'rural heritage' is embraced by the land trust movement that has been so influential in rural America in the 1990s. There are around one thousand of these trusts, frequently including farmland as a key part of a broad conservation package intended to act as a bulwark against urban development. Critics of this movement point to the control of the trusts and related organisations being largely in the hands of non-farm people for whom 'the protection of open space and of rural character is inextricably bound up in lifestyle and property values' (Bunce, 1998, p. 240).

Closely related to concerns over rural amenity are agrarian ideals that remain entrenched in American political and social culture (Dalecki and Coughenor, 1992). These ideals can be traced to Jeffersonian views that the true wealth of the nation is drawn from the land, and the conception of the yeoman farmer as the source of moral and civic responsibility that lay at the heart of a harmonious yet bountiful agrarian republic. These and related ideals of agrarianism have been appropriated by various groups championing farmland preservation, often assimilating it into what Dorst (1989) refers to as 'traditionalisation', through the extension of heritage preservation to a living contemporary agrarian society. This is a romanticism that has ties to the anti-urbanism and anti-industrialism of the back-to-the-land movements of the early twentieth century in the USA (Danbom, 1991; Naples, 1994). In academic work, Goldschmidt's (1978) thesis that the demise of the family farm equates to the decline of the rural community lends support to this traditionalisation, i.e. if the traditional family farm culture is re-established then the traditional rural community can also be recreated. This idealises the family farm as a 'national icon' and a 'sacred object' (Molnar and Wu, 1989). Bunce (1998, pp. 241–2) argues that this is best illustrated in the high levels of interest in and popularity of Amish and Mennonite culture areas and images because they are equated with continuing adherence to traditional ways of agrarian life. Farmland preservation in this context is directly related to maintenance of productivity, character and quality of life, as in the case of the Lancaster Farmland Trust in Lancaster County, Pennsylvamia, a centre of 'Amish country'. This link between farming culture and farmland preservation has been explicit in the politics of the American Farmland Trust (AFT) throughout the 1990s (e.g. AFT, 1990). It casts farmers in the role of land stewards, operating in the local interest to protect landscape and heritage, but also satisfying demand for local produce through farmers' markets, farm-gate sales and pick-your-own operations.

Although the preceding discussion demonstrates that there are various conflicting, overlapping and merging ideologies underlying the arguments for farmland preservation, it is also clear that farmers are central to these arguments. Their involvement in preservation campaigns has been both voluntary and coerced. The latter has occurred through land use regulations restricting farmers' rights of land disposal, thereby intending to keep farmers on their land to serve various farmland preservation objectives. Farmers have also been co-opted to the farmland preservation agenda through incentives and voluntary agreements. This has been most apparent in the operation of negotiated agreements between conservation and land trusts and farmers. There are contradictions, though, between the independence of the family farm and the interests of society, and hence individual cases of farmland preservation vary in the means whereby farmers' support has been obtained. In general, farmers have been strongly supportive of farmland preservation when neighbouring non-farm uses impinge on their activities (Pfeffer and Lapping,

1994), but they have been resistant if significant restrictions on their development rights are threatened. In noting this, Bunce (1998, pp. 244–5) recognises that the main control over the agenda of farmland preservation in North America has been largely in the hands of non-farm people in the rural–urban fringe and their social construction of the primacy of farmland. However, this agenda has failed to address other issues associated with farming, notably the maintenance of farmers' incomes, and it has failed to integrate farmland preservation with broader rural land-use policy.

9.4 Hobby farming in the rural–urban fringe

As part of the increase in farm diversification and pluriuactivity, as well as through pressures specific to the rural–urban fringe, in much of the Developed World post-1945 non-commercial farming has expanded, through mainly small and low-income farms whose operators have their principal occupation outside farming. There are various types of operator involved. Some are long-term rural residents now employed outside agriculture or retired, but who continue to farm on a part-time basis. Others have other sources of income, and profits from farming are not a prime consideration. In many cases, often in areas beyond the rural–urban fringe, the farms function as retirement homes, second homes or vacation homes and the farms are 'hobby farms' (Daniels, 1986). For long-term ruralites the combination of part-time farming and a local non-farm job enables them to continue a preferred lifestyle whilst avoiding migration to a city.

It is difficult to determine the extent of 'hobby' farming because of the limitations of agricultural censuses (Lund, 1991). For example (from the early 1990s), the census definition of a farm in the USA is very broad: 'a holding of any size where $1000 or more of agricultural products were produced and sold in the preceding 12 months, together with all farms with actual sales of less than $1000 but having the production potential for sales of $1000 or more' (USDC, 1993). This means that a property marketing a few steers, a hectare of corn or breeding a couple of horses can be classified as a farm (Hart, 1992, p. 166). Nevertheless, data relating to those 'farmers' claiming that farming is not their main occupation have risen. For example, in 1992, of the 1,925,300 farms in the USA, 872,150 (45.3%) were operated by persons whose principal occupation was not farming (Archer and Lonsdale, 1997, p. 407). Nearly half (906,517) provided under $10,000 worth of agricultural sales per annum (78% of all farms in West Virginia) and one-third provided less than $5000 per annum in sales.

The growth in the extent of part-time farming was first noted by several researchers in the 1960s and 1970s, and interpreting this as more than just a phenomenon of the rural–urban fringe (Gasson, 1966; 1967; Lewis and Maund, 1976; Mignon, 1971). In categorising part-time farmers, a historical chronology was distinguished (Robinson, 1994a, p. 114):

- In pre-industrial times farmers supplemented their income from farming with that derived from craft industry and rural trades. There has been some persistence of this, e.g. crofting in the Scottish Highlands where farming is combined with fishing.
- From the early nineteenth century there was the emergence of a worker-peasantry, in which factory work provided an additional income to farming. This combination of farm-work and factory-work has persisted in some parts of Europe, though on a small scale (Franklin, 1969).
- Farming as a hobby, which has been primarily a post-1945 development, with concentrations in some rural–urban fringes.
- A modern phenomenon, in which the lack of an adequate income from farming alone has prompted farmers to diversify or become pluriactive by seeking other gainful activities both on and off the farm.

It should be recognised, though, that hobby farming is not synonymous with part-time farming. For example, in the UK there are at least five distinct groups of part-timers recognisable today, extending earlier categorisations (Aitchison and Aubrey, 1982; Bryden et al., 1992; Gasson, 1984; 1986; 1988a):

- crofters (in the Highlands and Islands of Scotland), a statutorily recognised group who combine farming with fishing, craft industry or seasonal unemployment;
- farmers able to maintain a living from working less than official designations for full-time farming, usually by combining multiple income streams;
- institutional control, which may involve farming that violates definitions of full-time farming;
- farmers who have reacted to pressure from changing economic circumstances to diversify their farms, but by developing secondary activities that take their labour away from strictly agricultural work, for example horse-breeding (see Robinson, 1996, for an example of this in Australia), farm shops or farm holiday schemes;
- situations in which those engaged in agricultural activity derive income from employment unrelated to their farm.

Those falling into either of the last two groups have been categorised into other groups in various studies, based largely upon the degree to which their income is derived from off-farm activities. However, if the presence of any form of off-farm income is considered then this can include a substantial proportion of farmers. For example, in 1978 44 per cent of American farmers worked off their farm for 100 days or more, and 92 per cent of American farm families held some form of non-farm related income in 1979 (Albrecht and Murdock, 1984). In England and Wales in the late 1980s farm households in which farming was only a secondary source of income accounted for around one-fifth of the entire farming population (Gasson, 1988a, p. 156).

The expansion of part-time farming has been especially prominent at the lower end of the farm-size spectrum. So, whilst the total number of farms has been falling, the number of small, part-time enterprises has been growing. In some cases there have been structural factors favouring this development. For example, in the USA major tax incentives have made purchase of small farms attractive, through deduction of mortgage interest payments and local property taxes from federal taxable income, farm use-value property taxation, and incorporation of farm operations for investment tax credits and tax sheltering (Daniels, 1986). Other factors include the growing preference for living in small towns or rural settings, relative cheapness of rural land and houses, improvement of transport and commercial networks that have enabled greater separation of work place and residence (as well as ease of access to urban amenities for rural inhabitants), and the ruralisation of industry has increased the number and range of job opportunities in non-metropolitan areas thereby helping to create higher disposable incomes which can be used to support hobby farming.

The growth of hobby farming in recent years also reflects the growth of interest in small-scale food production as part of a broader set of lifestyle choices emphasising rural living, food quality, working with animals and the land, and ideas of health and fitness associated with farm-work and country life (Holloway, 2000a). This growth reflects what Thirsk (1997) refers to as the emergence of a phase of 'alternative' agriculture or, in more common parlance, post-productivism that incorporates a re-emergence of favourable attitudes to farming as an enterprise not solely on economic grounds but because of the satisfaction and benefits it provides. This may have echoes with earlier 'back to the land' ideals present in so-called agrarian communes, cottage farming and farm colonies (Marsh, 1982) or the Organicist movement of the 1930s (Matless, 1998). There may also be some continuation of ideals associated with self-sufficiency and the morality of self-sufficiency as typified in books by John Seymour (1991; 1996), which 'presents a domestic farming image centred very much on the home and its small piece of land, as a site of family-scale food production, processing and storage, and as highly independent of "modern" urban society' (Holloway, 2000b, p. 6). This is typified in the UK television programme 'The Good Life', first aired in the 1970s, but being reshown in the early 2000s. This portrayed 'suburban farming', but the reality is that areas at some remove from the suburbs or the rural–urban fringe have more frequently been the scene of 'good life' experiments, and that hobby farming is by no means the preserve of urban fringe locations.

10 Twenty-first century agriculture: towards sustainability?

10.1 'Industrial' farming and environmental well-being

There has been a steady growth in the academic study of linkages between farming and environmental issues, especially with a focus upon resource management problems and notions of sustainable development (e.g. Buttel, 1992; Cocklin, 1995; Drummond and Marsden, 1999; Hassenein and Kloppenburg, 1995; Liepins, 1995; McHenry, 1996). In part this growth reflects awareness, evolving over several decades, of environmental impacts arising from productivist agricultural policies (Table 10.1). These environmental impacts have varied according to the particular ecological characteristics of individual farms, but in general terms and especially in many parts of the Developed World, under productivist policies there has been a sharp decline in the area of semi-natural habitat and in the numbers and range of farmland plants and animals characteristic of both the semi-natural and of more traditionally managed 'artificial' habitats. Indeed, agriculture has become a central factor in the loss and decline of biodiversity. Where distinctive semi-natural areas survive it tends to be despite agricultural practices rather than as a result of them (as it once was).

One example of this loss of biodiversity through intensified farming activity occurs throughout the temperate zone, where some bird species have been especially badly affected by the use of agricultural chemicals and changes in habitat caused by new farming practices (Gregory and Baillie, 1998). For

---

Table 10.1 Environmental impacts of productivist policies

- Loss and fragmentation of semi-natural 'infield' habitats through agricultural improvement or 'arablisation'
- Abandonment or undermanagement of extant semi-natural 'infield' habitats (mainly in the uplands)
- Overgrazing of semi-natural habitats (mainly in the uplands)
- Loss or mismanagement of 'interstitial' habitats (e.g. hedgerows, field margins, ditches)
- Drainage or drying out of wet-land habitats due to water over-abstraction
- Pollution and eutrophication of surface and ground-waters leading to loss or degradation of aquatic ecosystems
- Loss of crop rotations and arable-pasture mosaics leading to severe reduction in characteristic farmland species
- Shift from spring-sown to autumn-sown cereals leading to loss of winter stubbles and to loss of arable weed species, invertebrates and thereby food sources for other wildlife groups
- Universal application of artificial fertiliser leading *inter alia* to the loss or degradation of characteristic hedgerow and field margin vegetation

(*Source:* Tilzey, 2000, pp. 280–1)

---

example, in the UK the number of skylarks has declined by three-quarters in the last quarter century, largely because of the change from spring-sown to autumn-sown cereals (Gregory *et al.*, 1999) (Figure 10.1; Table 10.2). Spring-sown cereals allow the stubble of the previous crop to be left unploughed over the winter, providing food and cover for the birds that is denied in the move to autumn sowing. This type of 'bird-friendly' farming is also advocated as part of the greater adoption of organic farming methods (Sears, 1990).

Another example is the removal of field boundaries. In the UK this has been well documented, especially for hedgerows (Rackham, 1990). Shoard (1980) quotes a figure of about 192,000 km (120,000 miles) in all, or about 7200 km (4500 miles) a year of hedgerows removed in England and Wales between 1946 and 1974. A loss of around 6400 km (4000 miles) per annum in England was reported in surveys in the mid-1980s (Barr *et al.*, 1986). However, this has not been a simple one-way process and measures designed to promote hedgerow replanting have had some success. One estimate reports that 25,600 km (16,000 miles) of hedgerow were replanted in England between 1984 and 1990, though around 85,000 km (53,000 miles) were grubbed-up during the same period (Bryson, 1993) and there was a further net loss of more than 10,000 miles (16,000 km) in the first four years of the 1990s. Measures such as the

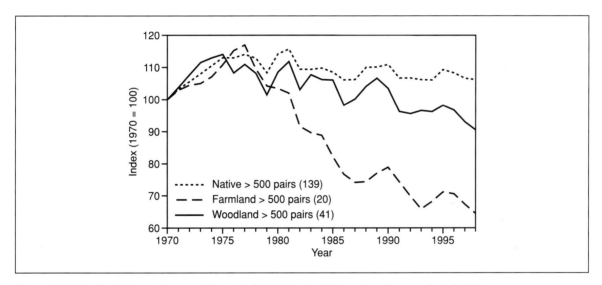

Figure 10.1 Headline indicators for populations of wild birds in the UK (based on Gregory *et al.*, 1999)

Table 10.2 The decline in bird species in the UK, 1972–96

Species	% UK population decline 1972–96	Species	% UK population decline 1972–96
Grey partridge	78	Linnet	40
Turtle dove	85	Reed bunting	64
Skylark	75	Corn bunting	74
Song thrush	66	Yellowhammer	60
Tree sparrow	76	Lapwing	46

(*Source:* Gregory *et al.*, 1999)

Farm and Conservation Grant Scheme (1989–93) and the Agriculture Improvement Scheme (1985–9) operated throughout the 1970s and 1980s to encourage replanting and other desirable conservation and landscape improvements (Ghaffar and Robinson, 1997; Miller, 1991; Robinson and Ghaffar, 1997). These have now been succeeded by the Hedgerow Incentive Scheme, launched in 1992, and intended to restore c.900 km (c.560 miles) of hedgerow per annum. This is part of the Countryside Stewardship Scheme, offering ten-year agreements to restore damaged or neglected hedges (see Section 10.4.3).

Numerous other examples of environmental disbenefits linked to agricultural intensification and the adoption of industrial methods by farmers can be found throughout the Developed World. Since the presence of these disbenefits was first brought to wider attention in the 1960s following the publication of Rachel Carson's book, *Silent Spring*, in 1962, there has been a huge surge in media and public awareness of the 'ills' of modern farming (Fox, 1996). This has undoubtedly led to a major shift in the public perception of farmers, but this negative view has been reinforced periodically by health 'scares' directly related to the downstream consequences of farming activity. In particular, concern over such issues as the presence of salmonella in chicken and certain types of cheese and the spread of bovine spongiform encephalopathy (BSE or mad cow disease) in the late 1980s provoked calls for a different type of evaluation of agriculture, beyond the narrow economic viewpoint that generally underscored government policy (Blakemore, 1989; Dwivedi, 1986; North and Gorman, 1990). The continued occurrence of such problems partly illustrates continuing deficiencies in the nature of environmental regulation, but there have been some regulatory changes, in many cases associated with non-governmental agents, such as supermarkets and consumer associations.

From the myriad of examples that could be used to illustrate ongoing problems with respect to agriculture and health, two are outlined here, one directly related to human health problems and one related to the spread of disease amongst farm animals, and both drawing upon geographical research in the UK.

10.1.1 Bovine spongiform encephalopathy in the UK

It appears that BSE first occurred in cattle in the UK in the mid-1980s when an infective agent in sheep was able to jump the species barrier to infect cattle. This was most likely made possible following the use of high protein feed for cattle based on the ground remains of sheep. Although the use of such feed had been common for several years, the introduction of anti-pollution measures in the early 1980s led to protein feed being manufactured using a lower heat treatment. Hence, the infective agent survived, and was subsequently identified as scrapie, a condition first recognised in sheep in the early nineteenth century. Cattle consuming this infected feed were thus at heavy risk of contracting BSE. The highest incidence of infection with BSE in the UK has been amongst dairy calves, representing 81 per cent of all cases (MAFF, 1999).

The government response to the disease was strongly criticised for its inadequacy, especially in terms of the delay in introducing wide-ranging slaughter and culling schemes. A specified Bovine Offal Ban was introduced in 1989, preventing bonemeal feed from being sold to farmers, with prevention also of brain and spinal column material from entering the human food chain. However, this did not stop some farmers with stockpiles of feed from continuing to feed it to their cattle. Meanwhile, some abattoirs failed to exclude potentially infected material from processed carcasses (Evans, 2000). Despite fears that the infective agent could jump another species barrier and infect people eating beef, an extensive programme of tracing and slaughtering infected cattle was not introduced initially. A mixture of lack of unequivocal scientific evidence regarding the risk to people from eating infected beef, desire to maintain consumer confidence in British beef and the fear of damaging farmers' livelihoods restricted government action. However, in March 1996 the government's Spongiform Encephalopathy Advisory Committee (SEAC) announced that a link had been established between infected beef products and the occurrence of the human brain disease, new variant Creutzfeld Jacob Disease (nvCJD). Immediately the price of beef plummeted and there was a total worldwide ban on export of British beef and derived products. Various emergency measures were introduced to help support beef producers, contributing to a substantial reduction in production. Measures included programmes of slaughter to prevent cattle products from entering the human food chain, compensation payments, slaughter of young cattle for which a market no longer existed, payments to abattoirs and cutting plants to dispose of unsaleable stocks and for additional slaughter, support for maintenance of the rendering industry, and a Beef Assurance Scheme for disease-free herds (Gaskell and Winter, 1996).

In the UK, consumption levels of home-produced beef fell by 40 per cent after the SEAC announcement in March 1996, but eight months later they had recovered to a level just 10 to 15 per cent below those at the beginning of the year (Evans, 2000, p. 98). By summer 1999 consumption had returned to a level comparable with summer 1995, though prices for producers were still up to 20 per cent lower. The ban on exports to the rest of the EU was rescinded in 1999, but the actual amount of exports re-established by autumn 1999 was very limited, and at this time France and Germany were still refusing to take any British beef. The French government may be fined by the EU for its continued refusal to permit imports of British beef.

In assessing the broader impacts of the BSE crisis and other 'scares' or 'food crises', Lang *et al.* (2001) refer to the potential for a new ecological health model to be implemented. This new model refers directly to concerns regarding food production, and addresses the interface between social policy and environmental policy (Raven *et al.*, 1995). It acknowledges that there have been responses at both national and supra-national level, e.g. creation of national food agencies, such as the UK Food Standards Agency (FSA) from April 2000, the EU's White Paper on Food Safety (2000) proposing a new European Food Authority (EFA) and a new EU health and consumer protection Directorate General. Regional food and nutrition action plans have also been developed by the World Health Organisation. Hence it is possible to recognise the advance of 'multi-level food governance', though it is debateable whether this really rep-resents the move from an 'old' to a 'new' model of food policy as recognised by Lang *et al.*'s model depicted in Table 10.3. Nevertheless, with the BSE crisis costing the UK taxpayers £4 billion in direct costs (Lang, 1998), it is clear that there are ongoing pressures from the public regarding the nature of the food supply chain, and this may encourage moves towards the 'new' model. An additional factor in fostering such a development is the even more recent 'scare' in the form of the UK's foot-and-mouth epidemic.

10.1.2 The foot-and-mouth outbreak in the UK, 2001

The last reported outbreak of foot-and-mouth disease (FMD) in the recent epidemic amongst cloven-hoofed animals in the UK, on 30 September 2001, was case number 2030. During the outbreak nearly four million animals were slaughtered (600,000 cattle, 3.2 million sheep, 147,000 pigs, 2000 goats and 1000 deer). Through the Livestock Welfare Disposal Scheme farmers were paid over £126 million compensation, and over £70 million were allocated to a Business Recovery Plan to assist rural communities and businesses affected by FMD.

Perhaps the most striking characteristic of the outbreak was its dissimilarity to the previous large-scale FMD epidemic in the UK in 1967/8. In the earlier outbreak more farms were affected (2364), but less than half a million livestock were slaughtered, reflecting the smaller herds and flocks at the time and the prevalence of smaller farms. Moreover, the previous outbreak was confined almost entirely to one area (Cheshire, Lancashire, Shropshire and north-east Wales). The contrasts also reflect the significant changes to the agri-food system in the intervening 33 years. Most notably, the recent outbreak highlighted the extent to which the pattern of livestock marketing and slaughter has changed, moving right away from the local system of markets and slaughterhouses that had been characteristic for centuries (Broadway, 2002).

First identified on 20 February 2001 at an abattoir in Essex, the likely source of the disease within the UK was traced to diseased pigswill on a farm in north-east England, over 550 km away. However, before authorities were aware of its

Table 10.3 Some features of the old productionist and new ecological health models of food policy

Key policy feature, by area	'Old' productionist model	'New' ecological health model
Economic policy	Increase production and supply by application of science and capital. Consumers have right to choose	Reducing inequality by state action provides health safety net. Citizenship requires both skills and protection
Health policy	Health stems from prosperity. Availability and some equity of distribution; rising prosperity makes health services affordable	Population approach; ill health stems from entire supply chain; degenerative diseases suggest how food is grown and delivered is important
Environment policy	Should not dislocate market forces; long supply chain; global reach for affluent consumers	Has to be built into food practices; short supply chains where possible; bio-regionalism for all?
Social policy	Family responsibility; plus welfare safety net	Population approach; the state applies corrections to imbalances between individual and social forces
Morality	Individuals should be responsible for food within market rules	Societal responsibilities should be based on citizenship
Price policy	Cheapness of food may externalise costs	It is false accounting if costs are externalised to other budget headings; costs should be internalised where possible
Policy coordination	Primacy of economics; fragmented specialist decision-making	Social goals as significant as other policy goals; new mechanisms for integration.

(*Source*: Lang *et al.*, 2001, p. 541)

occurrence, infected livestock had been sent across the Pennines to a market in Carlisle, from where the purchasing dealer took them to Devon in the south-west and shipped some to the East Midlands. Hence from the outset, the pattern of spread reflected the breakdown of localised systems and replacement by trading in livestock that involved long-distance transport. Ironically, greater distances in livestock transportation had been a consequence of the closure of many small abattoirs following new EU health and hygiene regulations introduced in response to the BSE crisis. Broadway (2002) refers to the increasing concentration and specialisation among fewer slaughter companies as an element of the third food regime. But he notes that the slaughter industry has not become more geographically concentrated in livestock producing areas, except for pig slaughtering. Cattle in particular are generally not slaughtered in the areas where they are raised and fattened, largely because

the location of new slaughterhouses has been greatly determined by the availability of government grants.

In terms of the 2001 FMD outbreak the extent of contagious diffusion of the disease was relatively restricted (as in 1967), but there was the operation of a spatially extensive hierarchical process of diffusion to markets and abattoirs handling infected animals. However, the principle of contagion was uppermost in the government's response to the disease, involving the slaughter of healthy animals on farms adjoining those where diseased animals had been detected. Yet, possibly as few as 10 per cent of farms where FMD was contracted were infected by diseased livestock on a neighbouring farm. The principal factor in the initial basic pattern of disease occurrence was the mass transit of animals over long distances (Ilbery, 2002).

The spatial and temporal occurrences of the outbreak are shown in Map 10.1 and Figure 10.2.

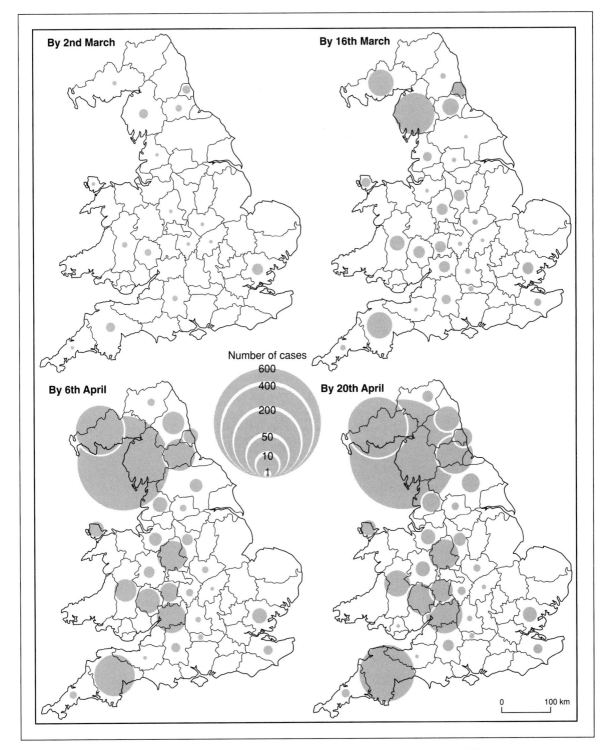

By 2nd March

By 16th March

By 6th April

Number of cases

600
400
200
50
10
1

By 20th April

0   100 km

Map 10.1 Spatial occurrence of the foot-and-mouth outbreak in the UK, 2001 (based on Ilbery, 2002)

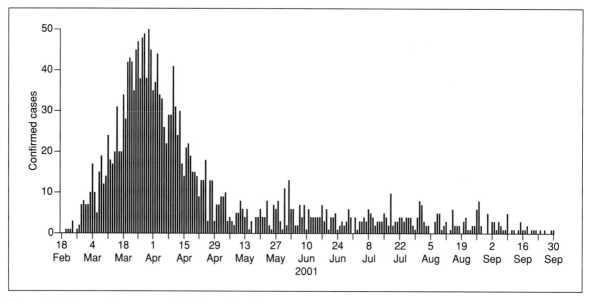

Figure 10.2 Temporal occurrence of the foot-and-mouth outbreak in the UK, 2001 (data source: www.defra.gov.uk/animalh/diseases/find/cases/histogram.htm)

The peak number of new outbreaks occurred only six weeks after the first report of the epidemic. However, the gradual decline in the number of newly infected farms shows a number of smaller peaks, e.g. in early April, early June and mid-July 2001, which tended to reflect spread of the disease around new 'hot spots'. This meant that after initial concentrations of the disease had formed in Cumbria, Dumfries (Scotland) and Devon, secondary smaller concentrations developed in the Welsh Borders, Crickhowell (Powys), Settle and Thirsk (North Yorkshire), Clitheroe (Lancashire) and Allendale/Hexham (Northumberland).

Benton (2001) views both the BSE outbreak and the foot-and-mouth epidemic as strong illustrations that welfare-inspired concerns of the public have not been addressed by government. He maintains that this is the case in the UK despite the creation of the new international food standards agency and the establishment of a new rural affairs ministry (the Department for Environment and Rural Affairs – DEFRA) in place of the MAFF. He views these two developments as very partial attempts to deal with the environmental impacts of food production and contrasts them with New Labour's support for new biological technologies

(notably GM foods) in its core economic strategy (this is discussed below in Section 10.5). He argues that new attitudes are required from policy-makers that place greater emphasis upon environmental, health and animal welfare considerations. In effect, he makes a case for government to engage more actively with the discourse of sustainable development.

10.2 The discourse of sustainable development

10.2.1 Beyond Brundtland

The two previous examples present a rather one-sided (negative) illustration of the link between productivist agriculture and environmental problems. However, there are alternative relationships proposed between production and environment that offer positive feedbacks. An emphasis upon these alternatives can be seen in the opposition to the general shift from national to international to global policies, as displayed by lobby groups at major international conferences in Seattle and

Genoa in recent years. Moreover, within these alternative scenarios other processes are presented that are antithetical to globalisation, for example what Buttel (1992) refers to as the processes of greening and environmentalism.

Greening is the process by which modern environmentally related symbols have become increasingly prominent in social discourse. Environmentalism is the greening of institutions and institutional practices, or the trend towards environmental considerations being increasingly brought to bear in political and economic decisions in educational and scientific research institutions, geopolitics and various other spheres (Robinson, 1994c). These considerations include the growing emphasis upon sustainability within economic and rural development planning, and in the marketplace (Harper, 1992; Robinson, 1994d), though it should be noted that O'Riordan (1999) argues that environmentalism is a precursor to sustainability in an evolutionary process involving shifts in economic, social and political arenas.

The notion of sustainability as an important component of agricultural development is one that has moved into mainstream political thought only recently (Table 10.4) (Robinson, 2002a). Impetus has come from the United Nations 'Earth Summit' at Rio de Janeiro in 1992 and from ideas put forward by the World Commission on Environment and Development (the Brundtland Report) in 1987 (Drummond and Marsden, 1999; Robinson, 2002b). For example, economic growth that pays due regard to environmental considerations is now incorporated as a principal policy goal of the EU in Article 2 of the Treaty of the European Community, as amended by the Treaty of the European Union. Article 130R requires environmental protection to be integrated into the definition and implementation of EU policies, and more environmental policies affecting farmers have appeared as part of EU legislation (e.g. Palacios, 1998). Moreover, promotion of sustainability was a central element in the EU's Fifth Environmental Action Programme, 'Towards Sustainability', adopted in March 1992 and operating from 1993 to 2000. This programme based its actions on the concept of sustainable development as put forward in the Brundtland Report, emphasising the need to address root causes of environmental degradation before they affect economic growth and efficiency.

Various writers have noted the major paradox inherent in the term 'sustainable development', namely the combination of the contradictory ideas of limits to growth and active promotion of growth (e.g. Korten, 1994; Redclift, 1990; 1994; Robinson, 2002a; Willers, 1994). It can be argued that reconciliation of this contradiction may be sought via the twin goals of economic self-sufficiency and production for limited use, but the tensions between 'sustainable' and 'development' are generally inadequately recognised in policies promoting sustainability (Pearce et al., 1989; Rao, 2000). Nevertheless, sustainable development is now enshrined as an attainable goal in many articles of legislation in various countries, including the EU and the USA (Clinton and Gore, 1992; EPA, 1992).

Table 10.4 Sustainable development: evolution of an idea

Pre WWII	Concept of *sustainable use* in an economic sense
1964	UN International Biological Programme
1960s	The term *sustainable* is applied to project management objectives in less-developed countries
1972	UN Stockholm Conference on the Human Environment
1970s	Ideas on conserving resources evolve initially into the concept of *ecodevelopment*
1980s	The World Conservation Strategy
1987	Brundtland Report, *Our Common Future* – the term *ecologically sustainable development* appears as a central component: development that meets the needs of the present without compromising the ability of future generations to meet their own needs
1992	Earth Summit at Rio de Janeiro – Agenda 21

(*Source*: based on Robinson, 1998b)

In some sectors of agricultural production the industrial model of farming has been challenged directly by environmental regulation that refers to growing concerns over food quality, environmental pollution and sustainability. In the USA this can be seen in legislation such as the Food Security Act of 1985, and the Food, Agriculture and Conservation and Trade Act of 1990 (Helmers and Hoag, 1994), both of which promote low-input 'sustainable' agriculture. Nevertheless, Duram (1998) notes the small amount of funding devoted to research on sustainable agriculture in the USA. Funds have been allocated under the USDA's Sustainable Agriculture Research and Education Programme (SAREP) from 1988, but this represents only 0.5 per cent of the federal government's commitment to agricultural research and education.

The achievement of sustainable forms of agricultural production is now widely recognised by governments as a long-term policy objective (Billing, 1996), and many countries are now developing what they term 'sustainable agricultural strategies' as part of their national environmental and agricultural plans (for the UK see DoE/MAFF, 1995; Scottish Office, 1995; Welsh Office, 1996). To date, policies promoting sustainability have tended to pursue one of two broad themes, namely: giving environmental considerations greater weight, but balancing this against the benefits of economic development, and using environmental capacities to place a constraint on economic activity so that environmental concerns are uppermost. The first of these two has been overwhelmingly dominant to date, suggesting a less than enthusiastic embracing of environmental dimensions of sustainability by most governments.

Increasingly, academic debate has stressed the importance of incorporating the idea of natural capital within the concept of sustainability, recognising that some aspects of the environment incorporate a critical natural capital that is essential to human survival, and is irreplaceable. A key difficulty is deciding just what is 'critical' and how it should be preserved and handed onto future generations. This is at the heart of academic debates on sustainability (see Owens, 1994), and it is also embedded in the policy implementation that is underway.

The notion of natural capital is part of the framework represented in the work of Serageldin (1996; 1999), employing ideas developed at the Environment Department of the World Bank. This sees sustainability as the maintenance of stocks of capital, consisting of natural, human, social and human-made capital (Deininger and Binswanger, 1999). Hence a sustainable society nurtures or enhances these stocks. In order to maintain the natural capital there must be maintenance of social systems and human skills (e.g. Reardon, 1995). With reference to this particular framework, Warren et al. (2001) refer to 'strong sustainability' as the maintenance of as much natural capital as possible. This reflects much of the initial thinking about sustainability as it casts doubt upon the continuing ability of technology to substitute natural capital with other forms of capital, whilst basing its arguments on a concern for conserving natural resources (e.g. Lyon, 2000).

In contrast to 'strong sustainability', in the work of Hartwick (Hartwick and Olewiler, 1986) and Solow (1986) there is a 'weak sustainability' paradigm, which allows the substitution of natural capital by human-made capital provided that the total stock of capital is maintained or enhanced. This allows for processes of substitution for depleted resources. Also, Serageldin (1996) argued that the poor in particular would always have to convert some natural capital into other assets in order to survive, but that they could still maintain 'critical natural capital' or, in other words, reach a position of 'sensible sustainability'.

Whilst the ideas of the two preceding paragraphs may seem abstract, they are being put into practical form in many development projects in Developing Countries, especially through incorporation in other frameworks. For example, Warren et al. (2001) refer to projects on sustainable livelihoods (e.g. Carney, 1998; Scoones, 2001). This is illustrated by the Department of Sustainable Livelihoods within the UK's Department for International Development (DfID) which, in effect, adopts an approach based on the notion of 'sensible sustainability'. As shown in Figure 10.3, this sees rural households as combining the management of five types of capital: natural (soil, land, water, wildlife, biodiversity), economic (savings,

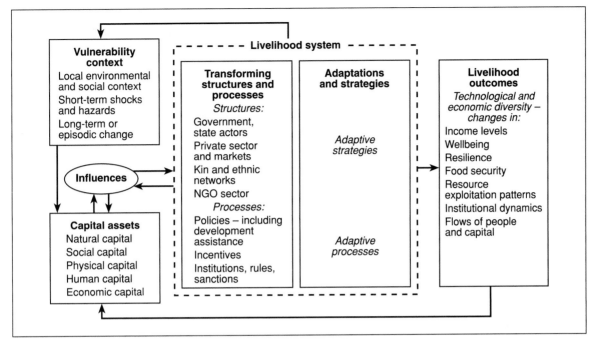

Figure 10.3 DfID's sustainable livelihoods (based on Warren *et al.*, 2001)

credit, remittances, profits), human (skill, know-ledge, ability to provide labour, health), physical (basic infrastructure and tools), and social (net-works, status, membership of informal institutions, access to more formal institutions) (Bebbington, 1999). As used in Figure 10.3, 'livelihood' is the sum of the capabilities and assets and activities required to make a living. Thus a 'sustainable livelihood' 'is one that uses the different forms of capital to maintain and enhance its capab-ilities and assets, to cope with and recover from stress and shock and to provide opportunities for the next generation' (Warren *et al.*, 2001, p. 326).

Despite the attractiveness of this conceptual-isation of sustainable livelihoods, there have been criticisms, especially regarding its use as a frame-work for policy. Warren *et al.* (2001, p. 326) refer to four criticisms:

- it is implicitly interventionist (neo-colonialist);
- it ignores or does not directly address 'processes of conflict', as instigated by those who seek to appropriate scarce resources and political power;

- because it focuses on households and individual communities this tends to overlook political and economic constraints upon rural households, and over which they have little control, e.g. policies of central government;
- there are practical problems regarding the measurement of the various capitals and the flows between them.

These criticisms are sufficient for some to have rejected the whole idea of social capital (Fine, 2000; Harriss and de Renzio, 1997). Others have used different aspects of the framework whilst noting the limitations of its practical applicability (War-ren *et al.*, 2001). Moreover, it is increasingly recog-nised that the notion of sustainable development is a process rather than a measurable end point, so that conceptual bases for sustainability need to be much more multi-faceted, that is like the 'capitals' approach but with more clearly defined criteria.

10.2.2 Sustainable agriculture

Although the term 'sustainable agriculture' has been popularised during the last 20 years (see

Altieri, 1987), its practice is as old as the origins of agriculture. Yet, its definition remains problematic (Bowler, 2002a). There has been extensive consideration of what constitutes a sustainable agricultural system, and what are the implications of the demands that such a system may have for farming in a particular area. However, it is difficult to derive a satisfactory definition that is not too simplistic and which has some practical characteristics within it that can be recognised within farming practice.

Amongst the definitional problems, O'Riordan and Cobb (1996, p. 50) refer to 'the limitation of scientific knowledge, the inability to calculate the full profit and loss account in natural resource terms, and the great divide in the public debates as to what farms are really for'. With respect to the latter, a comparison between the views of academics and farmers in the USA on what constitutes sustainable agriculture and how it might be achieved in practice revealed several statistically significant differences (Dunlap *et al.*, 1992). Farmers tended to focus on economic factors rather than environmental ones and did not recognise any need for reducing purchased inputs or energy use. This contrasted with academic views that stressed the core ecological aspects of sustainable agriculture, for example 'the ability of the agro-ecosystem to maintain productivity when subject to a stress or shock' (Conway, 1997, p. 177). The academic views also varied between academic disciplines, reflecting the need to formulate future policy on sustainability that encompasses a broad interpretation of the term.

Cobb *et al.* (1999) identify three specific problems in defining sustainable agriculture:

- The dificulty in determining what sustainable soil characteristics are, e.g. the elusive notions of soil resilience and soil health.
- The debate over the role of biodiversity.
- Failure of academic research to develop appropriate inter-disciplinary research on this topic.

Despite these difficulties, it is agreed that sustainable agriculture is primarily an approach to agriculture that balances agronomic, environmental,

economic and social optima, as implied in the definition provided by Francis and Younghusband (1990):

> Sustainable agriculture is a philosophy based on human goals and on understanding the long-term impact of our activities on the environment and on other species. Use of this philosophy guides our application of prior experience and the latest scientific advances to create integrated, resource-conserving, equitable farming systems. These systems reduce environmental degradation, maintain agricultural productivity, promote economic viability in both the short- and long-term, and maintain stable rural communities and quality of life.

As Table 10.5 suggests, this definition can be translated into general assessments of how farming might be considered to be sustainable, especially at farm level. It can be compared with the 'sustainable framework' for agricultural policy as set out by the UK government in its *Sustainable Development: The UK Strategy*, published in 1994, with four objectives:

- to produce good-quality food efficiently and inexpensively;
- to minimise resource use;
- to safeguard soil, water and air quality;
- to preserve and enhance biodiversity and landscape amenity.

In practice, this prescription is too bland and ill-defined to provide a sound basis on which to develop a strategy for achieving sustainability (Whitby and Ward, 1994). It highlights the gap between what politicians have been prepared to proscribe as a means of achieving sustainability and the content of sustainability as recognised in academic research.

Doering (1992) suggests that sustainable agriculture has four key aspects that differentiate it from most current commercial farming, with sustainability emphasising limited inputs, specific practices (e.g. organic farming) and management perspectives based on ecological and social considerations (e.g. biodynamics and permaculture) (see also Bowler, 2002a; 2002b):

- It implies less specialised farming, which often requires mixed crop and livestock

Table 10.5 Conditions to be satisfied if agricultural systems are to be sustainable

- Soil resources must not be degraded in quality through loss of soil structure or the build-up of toxic elements, nor must the depth of topsoil be reduced significantly through erosion, thereby reducing water-holding capacity
- Available water resources must be managed so that crop needs are satisfied, and excessive water must be removed through drainage or otherwise kept from inundating fields
- Biological and ecological integrity of the system must be preserved through management of plant and animal genetic resources, crop pests, nutrient cycles and animal health. Development of resistance to pesticides must be avoided
- The system must be economically viable, returning to producers an acceptable profit
- Social expectations and cultural norms must be satisfied, as well as the needs of the population with respect to food and fibre production

(*Source*: Benbrook, 1990.)

farming for reduced dependence upon purchased inputs.

- It implies that off-farm inputs should not be subsidised and that products contributing adverse environmental impacts should not receive government price support.
- It implies that farm-level decision-making should consider disadvantageous off-farm impacts of farm-based production, e.g. contamination of groundwater, removal of valued landscape features.
- It may require different types of management structure, e.g. family farms as opposed to corporate 'factory' farms.

This can be compared with Charlton's (1987) analysis of prospects for sustainable agricultural systems in the humid tropics in which he identified key practices as more effective management of fallow, fertilisation and manuring; improved tillage practices; intercropping; and agro-forestry. This topic has been taken further in work by Whiteside (1998) on smallholder farmers in southern Africa, in which he demonstrates that it is a combination of resource-conserving technologies that have the potential to achieve what he terms 'sustainable intensification'. However, these 'intermediate' technologies must be accompanied by more concerted attempts to sustain local institutions and to create an enabling external environment, e.g. through securing access to land for smallholders, research and development aimed at sustainability, and making markets work for smallholder sustainability.

Definitions of sustainable agriculture have often been narrowly focused on particular environmental, economic or social criteria rather than incorporating elements of all three. An alternative has been to identify goals for the development of sustainable rural and agricultural systems. Two examples of this are shown in Table 10.6, including the notion of a 'nested hierarchy' in which sustainability can be recognised at various different spatial scales (from farm to global). This can be compared with Buttel *et al.*'s (1992b) five major areas to consider when examining whether a system is sustainable (without a clear hierarchy): the nature of the production process, the economic and social organisation of food production, the use and management of labour, the role of scientific research and extension activities, and the organisation of marketing and distribution activities.

Whatever definiton of 'sustainable agriculture' is chosen, it is widely agreed that there is an urgent need for a greater input of ecological and environmental information in agricultural policy. At present much of the research on sustainable agriculture focuses on individual plots or fields or is performed at farm level. Yet, most of the environmental issues associated with agriculture are manifested at larger scales, and hence the need to consider measurable changes in water quality, soil quality, biodiversity and other environmental indicators at these scales (Allen *et al.*, 1991).

Table 10.6 Components of sustainable agriculture

Brklacich *et al.* (1997) recognise three components within a broader conceptualisation of agricultural sustainability:

1 *Environmental sustainability*: the capacity of an agricultural system to be reproduced in the future without unacceptable pollution, depletion or physical destruction of its natural resources and natural or semi-natural habitats
2 *Socio-economic sustainability*: the capacity of an agricultural system to provide an acceptable economic return to those employed in the productive system
3 *Productive sustainability*: the capacity of an agricultural system to supply sufficient food to support the non-farm population

Troughton (1993; 1997, p. 280), based on Lowrance (1990) extends this to a five-fold 'nested' hierarchy of sustainability:

1 *Agronomic sustainability*: the ability of the land to maintain productivity of food and fibre output for the foreseeable future. Fields
2 *Micro-economic sustainability*: the ability of farms to remain economically viable and as the basic economic and social production unit. Farms
3 *Social sustainability*: the ability of rural communities to retain their demographic and socio-economic functions on a relatively independent basis. Communities
4 *Macro-economic sustainability*: the ability of national production systems to supply domestic markets and to compete in foreign markets. Countries
5 *Ecological sustainability*: the ability of life support systems to maintain the quality of the environment while contributing to other sustainability objectives. Global

---

Table 10.7 Key questions to be answered in developing a research agenda on sustainable agricultural systems

- How can we measure sustainability at a variety of scales: farm-level, regional, national, supra-national?
- How can notions of 'sustainability' be built into agricultural policy other than by simply promoting environmentally friendly farming or landscape restoration within the overall context of an essentially economic agricultural policy?
- To what extent do existing agri-environmental measures contribute to achieving specified objectives (such as the maintenance or enhancement of biodiversity and the conservation of cultural heritage)?
- Can the implementation of environmental strategies in peripheral areas like the Lake District assist in supporting threatened traditional forms of input-extensive management? The initial evidence from the UK ESAs and Tir Cymen in Wales seems quite positive
- To what extent do selected measures contribute towards or reinforce a polarisation of the landsape into agriculturally productive areas and areas primarily producing environmental goods?
- Do certain measures assist a reduction of agricultural surpluses by promoting general extensification?
- What is the socio-conomic impact of agri-environmental measures on farms, especially in peripheral areas?
- To what extent can such measures become an integral part of the survival strategy of individual farmers?
- To what extent do they contribute to diversification and sustainability?

(*Source*: Robinson and Ghaffar, 1997)

There is clearly great scope for more work on the topic of sustainability, with a need not only to understand the operation of sustainability at different spatial scales, but what it means to the farmers themselves. By way of developing the agenda, Robinson and Ghaffar (1997) suggest a list of the key questions that need to be addressed (Table 10.7). However, it is important to consider these questions against the back-drop of ongoing changes in agricultural practice, especially in the Developed World where moves towards greater sustainability can be observed.

10.3 Towards sustainability?

Although it oversimplifies the current debates associated with agricultural sustainability, in terms of the development of moves towards sustainable agriculture it is possible to simplify the range of views into two opposing approaches:

- The ecocentric. This is an idealist view stressing the need for alternative agriculture as part of a no- or low-growth scenario for human development. This stresses a distinct alternative set of approaches to industrial-style 'modern' farming methods and therefore advocates, at best, a low-growth model of development. This approach has been associated with champions of organic and biodynamic farming, which has radical implications for changes in consumption patterns, resource allocation and utilisation, and individual lifestyles. To obtain sustainability various changes in agricultural practices are proposed, including diversified land use, integration of crops and livestock, traditional crop rotations, use of green or organic manures, nutrient recycling, low energy inputs and biological disease control. These have been applied to various models of sustainability, including organic farming, permaculture, low input-output farming, alternative agriculture, regenerative farming, biodynamic farming, and ecological farming.
- The technocentric. This is an instrumentalist view that rejects the ecocentric as being both practically and politically unrealistic. The proposed alternative is less rigorous, regarding sustainable agriculture as a contextual process that can act as a stated goal to be used to modify existing agricultural systems rather than a set of specific prescriptions. This can make it less

concerned with target setting and policy implementation, but nevertheless different types of agriculture have been proposed, such as integrated crop management, diversified agriculture, extensified agriculture and conservation agriculture. These advocate an extensive, diversified and conservation-oriented system of farming as encouraged via certain types of state regulation, e.g. limits on applications of fertilisers, imposition of minimum standards of pesticide residues in food, restrictions on types and rates of application of agrochemicals, and subsidies to promote environmentally friendly and lower input–output farming systems.

Some of the key elements within these two approaches to sustainability will now be discussed, starting with the ecocentric.

10.3.1 Organic farming

The definition of organic production given by the US Department of Agriculture (USDA) (cited in Foster and Lampkin, 2000, p. 1) is commonly used and has been widely adopted:

> A production system which avoids or largely excludes the use of synthetic compounded fertilizers, pesticides, growth regulators, and livestock feed additives. To the maximum extent feasible, organic farming systems rely upon crop rotations, crop residues, animal manures, legumes, green manures, off-farm organic wastes and aspects of biological pest control to maintain soil productivity and tilth, to supply plant nutrients, and to control insects, weeds, and other pests.

The number of farms on which production meets this definition is small, but increasing rapidly worldwide (Monk, 1998; 1999). In 1994 there were 4050 certified organic farms in the USA, nearly half in California, with organic industry sales in excess of US$2.3 billion per annum, growing by over 20 per cent per annum. In 1997 there were 5021 certified growers, and organic cropland had grown from 163,250 ha in 1992 to 344,060 ha in 1997 (representing 0.2 per cent of US cropland). About 2 per cent of US top speciality crops (lettuce, carrots, grapes and apples) are grown today under certified organic farming systems (Greene, 2000, p. 6). Nevertheless organic foods are still under 2 per cent of all US food sales. Map 10.2 shows that the largest amounts of organic cropland are in Idaho, California, North Dakota, Montana and Minnesota. The main concentrations of organic pasture and rangeland are in Alaska and Colorado. The leading areas under different types of organic production are shown in Table 10.8.

The largest certification scheme in the UK for licensing organic food production is the Soil Association Symbol Scheme, complying with EU regulations but imposing additional standards

Table 10.8 Organic production in the USA – the leading states (by area)

State	Type of organic production (acres)	State	Type of organic production (acres)
	Grain		Beans
North Dakota	53,306	Minnesota	14,060
Montana	41,287	Iowa	13,578
Minnesota	22,426	Nebraska	8,009
Idaho	21,787	Wisconsin	6,810
Nebraska	15,036	North Dakota	6,792
Colorado	15,359	South Dakota	5,435
	Oilseeds		Hay & Silage
North Dakota	12,170	Idaho	31,204
Utah	4,857	Wisconsin	13,781
California	4,411	New York	9,995
South Dakota	1,935	North Dakota	9,621
		Montana	8,795
		Minnesota	6,893
		Vermont	5,808
	Vegetables		Fruit
California	22,886	California	32,582
Washington	3,140	Arizona	4,361
Colorado	3,716	Washington	2,978
Oregon	2,345	Florida	2,625
Arizona	3,081	Colorado	1,816
Minnesota	1,684	Texas	1,344
New York	1,615	Oregon	1,231
	Organic Vegetables as % of total		Herbs/Nursery (sq ft)
Vermont	23.6	Vermont	278,710
Colorado	8.7	Pennsylvania	29,000
Utah	6.7	Texas	28,976
Maine	3.1	Arizona	20,000
Connecticut	2.8		
Arizona	2.8		
Massachusetts	2.4		
California	2.1		

(*Source:* Greene, 2000)

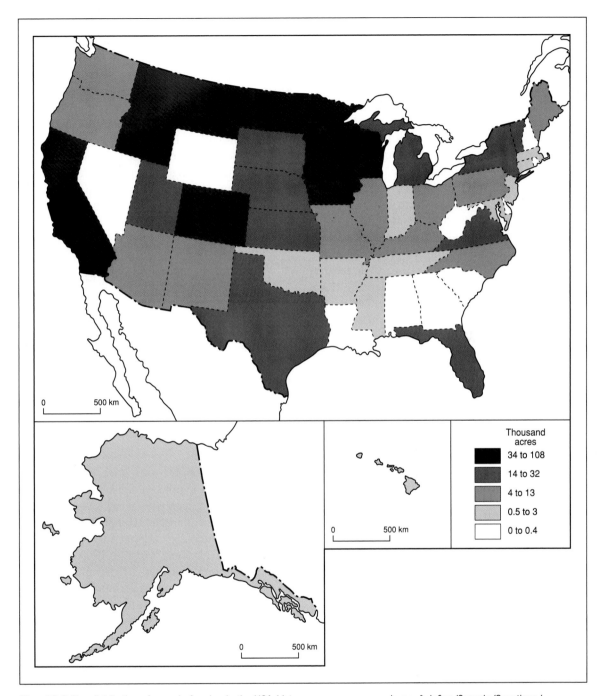

Map 10.2 The distribution of organic farming in the USA (data source: www.ers.usda.gov/briefing/Organic/Questions/ organiccroplandmap.htm)

(Marland, 1989). In the UK, sales of fruit, vegetables and other produce grown without pesticides, artificial fertilisers and other modern intensive farming methods nearly quadrupled over the five-year period 1993–8, to represent just under 2 per cent of the UK food market, or in excess of £350 million in sales and 1000 certified organic farms. However, this accounted for only 0.5 per cent of farmed land compared with 7 per cent in Austria and 3 per cent in Germany (Mintel, 1997). In 2000 there were 540,000 ha in organic production or in conversion in the UK, more than twice the area in 1999 (Soil Association, 2000), and approaching 3 per cent of all farmland. A core area of organic producers was being established in central southern England (Ilbery et al., 1999), with farmers attracted by a mixture of environmental, ethical and economic factors (Burton et al., 1999b).

In the EU, organic accreditation is underpinned by European Council Regulation 2092/91, providing specific rules for the production, inspection and labelling of products, and with implications for farming systems and the environment. Land area under organic production more than tripled in the EU between 1993 and 1998, increasing from 0.7 million ha on 29,000 holdings to 2.71 million ha on 104,000 holdings. In 1998 organic farming represented 2.1 per cent of utilisable agricultural land in the EU, though with considerable variation between countries, the highest proportions occurring in Austria (7 per cent), Finland (5 per cent) and Italy (5 per cent). Organic aid schemes in the EU have also increased considerably through the 1990s, with 25,000 agreements in Italy, 18,800 in Austria, 13,500 in Sweden and over 10,000 in Germany (Foster and Lampkin, 2000; Michelson, 2001).

Governments in many parts of the Developed World have helped to promote this advance of organic farming. For example, the UK's Organic Farming Scheme was increased by £24 million in 1999, to encourage an additional 75,000 ha of new conversion (MAFF, 2000). Between 2000 and 2006 an additional £140 million is to be allocated to organic conversion. This provides payments for conversion on the basis of £450 per ha spread over five years for arable land, £350 per ha for improved land over the same period, and £50 per ha

for unimproved land. O'Riordan and Cobb (2001) argue that in environmental terms this represents 'good value for money', but that the amounts are too low to justify either the on-farm costs of conversion or advantages associated with sustainability gains from the resulting transformation (see also Lampkin and Padel, 1994). However, their work largely dealt with a single 417 ha farm in Gloucestershire, and hence further research is required that analyses a wider range of organic enterprises (see also Wookey, 1987). In particular, in the UK, the Select Committee for Agriculture (2001) identifies the following areas as deficient in research:

- an analysis of the supply chain bottle-necks and imbalances;
- development of clear plans which respond to market demand;
- formulation of appropriate advice to farmers considering or going through the conversion process on aspects of the transition to organic production that relate to economic impacts;
- the basis for the belief that local trading schemes (such as farmers' markets, farmer box schemes) support the development of both the market and production of organic produce.

In the USA, from October 2002, producers and handlers of organic produce must be certified by a USDA-accredited certifying agent to sell, label or represent their products as '100 per cent organic', 'organic' or 'made with organic ingredients'. This is part of the National Organic Project, though there are several smaller USDA organic programmes, including export promotion, farming systems trials and financial assistance for the costs of certification. There are 23 states with certified organic animal production systems. However, meat and poultry producers faced organic labelling restrictions at times in the 1990s.

There are a number of overlapping contemporary trends that are contributing to the dynamics of organic production (McKenna et al., 2001). These include:

- increased consumer concern with food safety;
- 'greening' of the images of corporate food producers and production practices;

- formalised and wide-ranging guidelines prescribing food qualities at various organisational scales;
- movement into the marketing of organic produce by large-scale capital-intensive food producers;
- continued use of discourses incorporating 'sustainability' to define trends in production and consumption.

As these trends have evolved in recent years, organic farming has been presented as a growing arena for the intersection of the interests of capitalist modes of production and simple commodity production (Campbell and Coombes, 1999b; Whatmore, 2002). For example, TNCs have gradually increased their emphasis on food safety, 'freshness' and 'green' values as a means of enhancing their economic advantage. This can lead to economic goals taking over from more ecological values even in the context of 'organic' production. However, there is often scope for local interests to subvert global forces through exercise of various forms of control, e.g. in determining practice at the farm level or by implementation of regulatory mechanisms (Kaltoft, 1999). That is, the global imperative does not always subsume the local interest, especially where the latter may still represent large numbers of independent small producers. However, a key aspect in the expansion of organic production in the Developed World appears to be the extent to which a new food supply chain associated with organic production can be substituted for the conventional food chain. Morgan and Murdoch (2000) argue that the latter relies on intensive inputs into the food production process whereas, for organic production, farmers have to (re)learn how to farm in an ecologically benign fashion. They argue that organic agriculture resurrects local, context-dependent knowledge, which poses a challenge for the traditional agricultural extension system, farmers and retailers (e.g. Conford, 1992; Lockeretz, 1995). In contrast, the 'conventionalisation argument' asserts that organic farming is being pushed down the same path as conventional farming in terms of the penetration of agribusiness capital (Grey, 2000; Pugliese, 2001). This is substantiated by trends in California

(e.g. Buck et al., 1997), though not in New Zealand (Coombes and Campbell, 1998) or Ontario (Hall and Mogyorody, 2002) where factors operating at farm level have been important in developing small-scale independent farming organic producers.

In Europe and the USA, retailing of organic produce has often been associated with small-scale, 'local' outlets rather than chain stores such as supermarkets, and hence there has been a close link between organic farming and ethical shopping that supports particular types of production and retailing (Young and Welford, 2002). In contrast, in the UK, supermarkets have played a large role in the organic sector, starting with sales of 'organic' produce by Safeway in 1981. This has provoked concern from organic food organisations, some of which have contended that the supermarkets' approach to organic food is inimical to the underlying ethos of organic production (Clunies-Ross, 1990). Although supermarkets have initially been obliged to accept the principles of supplier standards for organic produce based on internationally accepted criteria, some are now introducing their own criteria which can be supplied to domestic suppliers. In so doing they may champion cheaper alternatives to fully 'organic' production, such as produce grown using Integrated Crop Management (ICM), which promotes natural predators and crop rotation but does not eliminate the use of pesticides and fertilisers. This is discussed further below. In 2002, nearly 90 per cent of Sainsbury's UK fresh produce supplies were being grown to ICM standards, largely through farmers in the LEAF (Linking Environment and Farming) network, which promotes 'green' farming.

10.3.2 Integrated Farming Systems/ Integrated Crop Management

Integrated Farming Systems (IFS), sometimes referred to as Integrated Crop Management (ICM), has been described as 'a whole farm policy aiming to provide the basis for efficient and profitable production which is economically viable and environmentally responsible. It integrates beneficial natural processes into modern farming practices using advanced technology and aims to minimise the environmental risks while conserving, enhancing

Table 10.9 Organic farming in the EU, Switzerland, the Czech Republic and Norway, 1998

Country	A	B	C	D	E
Austria	287,000	8.43	20,207	257,172	7.09
Belgium	11,744	0.85	421	11,299	42.81
Germany	416,518	2.43	9,209	358,332	7.19
Denmark	99,161	3.69	2,228	83,375	52.60
Spain	269,465	1.05	7,782	143,620	15.51
Finland	126,176	5.81	4,975	102,892	4.83
France	218,790	0.77	6,139	87,369	1.27
Britain	274,519	1.70	1,462	53,510	2.13
Greece	15,402	0.44	4,183	5,996	12.70
Ireland	28,704	0.66	887	15,277	1.40
Italy	785,738	5.30	42,238	466,744	20.37
Luxemburg	777	0.61	26	718	0.66
Netherlands	19,323	0.96	962	9,155	22.69
Portugal	24,902	0.65	564	11,021	1.25
Sweden	127,330	4.10	2,870	243,657	9.42
Switzerland	77,842	7.48	4,712	na	na
Czech Rep.	71,620	2.03	306	na	na
Norway	15,581	1.53	1,589	na	na
EU 15	2,706,449	2.10	104,153	1,850,137	6.54
All 18	2,871,492	2.14	110,760		

Key:
A Certified organic and in-conversion land area (ha)
B Certified organic and in-conversion land area as a percentage of utilisable agricultural area
C Certified organic and in-conversion holdings
D Agri-environment programme (2078/92) supported organic and in-conversion land area (ha)
E Agri-environment programme (2078/92) supported organic and in-conversion land as a percentage of total agri-environment programme supported land
(*Source*: Foster and Lampkin, 2000)

and recreating that which is of environmental importance' (IACPA, 1998, p. 1). It involves a set of principles and procedures to be applied, but taking account of the specific circumstances of the farm and its surroundings (BAA, 1996). The key principles of IFS/ICM are shown in Table 10.9. In summary these principles represent a technologically intensive set of production techniques which emphasise equally environment, farm incomes and food quality (Morris, 2000b; Morris and Winter 1999). Many of the elements are not new, but they are being applied in the context of modern, technologically advanced farming in an integrated fashion.

In certain respects IFS has close associations with organic farming, especially as its aim is sustainable resource use by focusing on the centrality of environmental considerations within the farming context. This represents a challenge to conventional food production practices, and it contrasts with the content of most agri-environmental schemes in which conservation measures are generally an adjunct to productivist activity (Robinson, 1994; Winter, 2000). Moreover, agri-environment measures seek to downgrade the production role of farmers by emphasising a conservationist or 'park-keeping' function whereas IFS maintains a productivist focus. Nevertheless, IFS does not seek to make such radical changes to production practices as those proposed by organic farming. In the latter, inorganic outputs are completely withdrawn whereas in IFS they are merely reduced to achieve environmental benefits and cost savings. Therefore IFS has been referred to by some as being 'a middle course' (Wibberley, 1995, p. 48).

Various research projects on the development and nature of IFS have taken place in Western Europe from the late 1970s, often focused specifically on ICM rather than mixed and/or livestock systems. Yet, despite the use of trial plots and trial farms in these projects, it has tended to be commercial interests that have dominated the dissemination of IFS and which have been developing IFS/ICM protocols with their suppliers of fresh produce. In the UK, the supermarket chain, Sainsbury's, claims that 97 per cent of its fresh fruit and vegetables are produced via ICM. However, as there is no commonly agreed ICM/IFS standard, it is difficult to corroborate such a statement.

In general it has been the commercial advantages of IFS that have been stressed in its evaluation. This might call into question its environmental dimensions, but nevertheless the fact that IFS also requires farmers to make significant changes to their existing farming practices is also recognised (e.g. BAA, 1996). Part of the problem regarding the exact nature of the changes likely to be associated with IFS is the fact that it is being defined and interpreted by various groups of people within the agri-commodity sector, especially agricultural advisors and supply trade representatives on the one hand and farmers on the other. The analysis by Morris and Winter (1999) demonstrates that these 'actors' have different conceptualisations of IFS and therefore different expectations. Their survey in western England revealed widespread lack of knowledge of IFS by farmers, whilst amongst agricultural advisors there was confusion as to whether IFS represented a return to traditional farming practices or an introduction of new technology, both aimed at protecting and enhancing the environment. It remains to be seen therefore whether IFS will become more widely disseminated amongst the farming community and, if so, what effects its adoption will have upon both the environment and farmers' balance sheets.

IFS/ICM may be compatible with some sustainability criteria. However, their primary focus is on increasing the efficiency of input use to maximise profitability, and hence many farmers using them are likely to continue using relatively intensive systems and to become progressively more specialised. In contrast to organic and low-input systems, they permit a relatively high level of pesticide use and there is no specific preference for mixed farming. Indeed, IFS/ICM farmers do not have to make environmental concerns of primary importance when determining their agronomic priorities. It can be argued therefore that integrated systems, such as IFS/ICM, address only a certain number of environmental sustainability criteria, but exclude others, notably a continued heavy reliance on external, non-renewable and fossil-fuel derived inputs. Hence, Tilzey (2000, p. 287) concludes that they simply just 'attempt to reduce in some measure the ecological "footprint" of what remains a basically unchanged configuration of intensive, agro-chemically based production'.

10.3.3 Integrated Fruit Production

Food safety barriers have been used quite frequently by the EU and Japan as a means of restricting cheap intensively produced fruit and vegetable imports from the USA (Campbell and Coombes, 1999b). These barriers include reductions in permissable applications of chemicals, banned inputs and prohibitions on items deemed likely to cause environmental damage or compromise animal welfare. Similar arguments were strongly invoked in the long-running dispute over the ban by Australia on imported apples from New Zealand, on the grounds that such imports might bring a new disease (fireblight) into Australia. Yet the legitimacy of such legislation is highly contested. For example, American scientists have claimed that bovine growth hormone has no adverse effects when meat or dairy produce are subsequently consumed by people, whereas European scientists contend that there are potential health risks both to animals being treated and to people. Similar disagreements between these two sides are manifest with regard to GM foods.

In addition to international regulation promoting 'environmentally friendly' farming practices, the changing concerns of consumers in favour of 'fresh', 'healthy' or 'green' foods are also contributing to the emergence of a 'green' protectionist barrier (e.g. McKenna, 2000a; 2000b). As a direct response to this, new systems of production have emerged in certain sectors of farming. For example, pipfruit producers in New Zealand have moved towards an Integrated Fruit Production (IFP) system, as one of their main markets, the EU, recognises this as providing 'safe' and 'environmentally grown' fruit that satisfies sanitary and phytosanitary requirements at the point of entry to the market. A related problem, though, is that the market now has multiple entry points, with individual supermarket chains becoming increasingly prescriptive in their requirements. For example, the UK supermarket chain, Tesco, announced in 1997 that IFP or similar schemes would be the minimum entry standard to gain access to their shelves, thereby removing premium payments to growers for supply under IFP (McKenna *et al.*, 2001). Thus the process of 'greening' extends to distribution chains. In the EU, 20 supermarket chains standardised and audited their environmental demands under the auspices of IFS/ICM protocols in 1996. Similarly the ISO14000 system for auditing environmental management systems is being utilised by distribution channels to guarantee food safety.

In New Zealand IFP for pipfruit was introduced in the mid-1990s as a deliberate alternative to organic production, as the latter was perceived to be associated with risk of large-scale crop failure (McKenna *et al.*, 2001). Initially, in 1995–6 six UK retailers, controlling over three-quarters of all UK sales of fresh fruit and vegetables, established the Fresh Produce Consortium, specifying that they would only sell New Zealand pipfruit if it was produced using Integrated Pest Management (IPM) systems. Such systems control pests by employing all methods consistent with economic, ecological and toxicological requirements while giving priority to natural limiting factors and economic thresholds. These systems were developed from the 1960s and have been incorporated in IFP practices for pipfruit production in parts of the EU from the late 1980s.

The details of IFP vary from country to country, but essentially it represents an integrated approach to pest and disease management, with three key aspects (see also Table 10.10):

- it encourages monitoring to determine if pest and disease thresholds have been exceeded before spraying;

- it gives preference to non-chemical controls wherever possible;
- where chemical controls are needed, it gives preference to targeted sprays such as the insect growth regulator chemicals and less toxic organic insecticides which are more specific to particular pests, thereby allowing pest predators and parasites to be effective. This replaces use of and dependence on broad-spectrum pesticides.

These characteristics of IFP differ from those of certified organic producer status principally in terms of permitted inputs. Full organic production practices prohibit synthetic chemical applications to crops, require organic certified inputs to orchard operations, such as mulch and natural fertilisers, and stipulations that post-harvest activities occur through approved packing, handling, storing and transport operators. Elemental sprays, like copper, are permitted, though some are under threat from tighter regulations proposed in some quarters. By the 1998–9 growing season 56 per cent of New Zealand's 1600 pipfruit growers were producing under IFP or were moving towards meeting this production criteria (McKenna *et al.*, 1998; 1999c).

10.4 Agri-environment schemes

One dimension of 'sustainability' that may be a key adjunct to farm-based production in the emergent third food regime is the modification of agricultural policy in favour of a broader role for farmers (Potter *et al.*, 1991; Robinson and Ilbery, 1993). This role embraces the production of 'countryside' as well as traditional assemblages of crops and livestock. In the EU this has been promoted by changes in the basis of farm support and by specific policies encouraging both environmentally friendly farming, for example the establishment of Environmentally Sensitive Areas (ESAs), and a broader aim of sustainable development (Robinson, 1994d; Whitby, 1996). Yet there are several contradictions within these policy reforms, the vital ones being whether the championing of the role of the farmer as steward of the land is really just a

Table 10.10 The principles of Integrated Farming Systems

Principle	Characteristics
Crop rotation	Promotion of soil structure and fertility and reduction of demand for agrochemicals; minimum of four different crops in rotation is recommended
Minimum soil cultivation	Provides agronomic and environmental benefits, e.g. reduced soil erosion and nitrogen volatilisation. Use of mechanical tools for weed control
Disease resistant cultivars	Enables reduced use of inputs
Modifications to sowing times	For example, later sowing times reduce pests and outbreaks of disease
Targeted application of nutrients	Saving of costs by reducing the amount of chemical applied, and provides environmental benefits, e.g. by reducing chemical contamination of groundwater
Rational use of pesticides	This can include avoidance of prophylactic spraying through crop monitoring and use of thresholds to determine the most appropriate timing of application
Management of field margins	Creates habitats for predators
Tillage systems	To favour natural control of pests, improvement of soil structure and to reduce demand for external nitrogen
Cropping sequences	Modifications to increase crop diversity
Promotion of biodiversity	Provides ecological benefits and promotion of beneficial predators. Between 3 and 5 per cent of total cropping area is usually recommended as non-agricultural vegetation

(*Source*: based on Morris and Winter, 1999)

policy for farm survival in marginal regions rather than a concerted attempt at landscape enhancement (Webster and Felton, 1993), and, second, the interpretation of 'sustainability' by policymakers.

Agri-environment policy (AEP) consists of any policy implemented by farm agencies or ministries utilising funding from agricultural support budgets, and aimed primarily at encouraging or enforcing production of environmental goods. The latter may include something quite explicitly targeted by the policy, such as hedgerow restoration, or it may refer to a more loosely stated concept, such as the generation of a desirable type of countryside. These targets may or may not be a joint product with the traditional farm outputs of food and fibre. In particular, AEP has been part of agricultural reforms in Western Europe, in which output-related support has gradually been changed in favour of area-based payments and payments for the supply of

environmental goods. Further moves in this direction have been made under the European Commission's Agenda 2000, which advocates more reductions in price support for arable crops, beef and sheepmeat (Lobley and Potter, 1998). Baldock *et al.* (1998) classified agri-environment schemes (AESs) on the basis of the nature of their focus and the nature of their targeting, as shown in Table 10.11.

Accompanying the CAP reforms in 1992 was the Agri-Environmental Regulation 2078/92, promoting wider use of agri-environmental measures in the EU. This advocated a standard model in which a contractual agreement between farmers and the state would be concluded, in which farmers agreed to produce environmental goods in return for specified payments.

The Regulation allowed for general measures, such as reduction of fertiliser and pesticide use and promotion of organic farming, which can be used on all farms, as well as measures, such as ESAs, which are more specifically targeted at a local environmental situation. Indeed, there are now some schemes that require the farmer to enter the entire farm into a management agreement. This applies to Tir Cymen in Wales and to some of the non-river valley English ESAs, such as Clun, Exmoor and the Lake District. There is a need, though, to tackle the problem that, despite the designation of substantial areas as ESAs in the UK, there are large amounts of low-intensity farmland that have no environmental policy directly affecting it (Bignal and McCracken, 1996). As a consequence, on this land there continues to be either intensification or abandonment of traditional practices, both of which are equally damaging to the landscape and to conservation of semi-natural flora and fauna. Moreover, Hart and Wilson's (1998) analysis concludes that the UK has been a reluctant participant in the Regulation and that it has largely failed to develop an expansive AEP. This is because top-down policy-making structures have hindered the incorporation of ideas from grassroots level; there has been a predominance of productivist thinking among policy-makers reducing the urge for comprehensive, holistic and well-funded AESs; an historically strong emphasis on targeted habitat protection has hampered the introduction of horizontal AESs; and the changing role of pressure groups within the AEP bargaining process has resulted in policy incrementalism rather than reform.

Using the EU's principle of subsidiarity, the details of policy design were placed in the hands of the individual member states. The Regulation allowed for a 50 per cent refund to member states on spending on AEPs, and up to 75 per cent in Objective One areas (designated with the aim of reducing deprivation in the EU's most affected areas, with per capita GDP less than 75 per cent of the EU average). Arrangements for individual countries can mean that they received less than the 50 per cent, as has been the case for the UK, which receives an EU contribution of around 17 per cent of the total costs of the schemes. Subsidiarity has also fostered significant variation in the operation of AESs in the EU. For example, in Germany and Spain the EU's agri-environment policies have been translated into measures controlled at a regional

level, so that regional interests can be paramount, sometimes to the detriment of the policies' original national intentions (Mazorra, 2000).

10.4.1 Farmers' decision-making and agri-environment schemes

The voluntary nature of many AESs within the EU has led to much research attention being directed at investigations of farmers' behaviour (participation versus non-participation) and attitudes (pro-conservation versus pro-productivism), adding to the body of knowledge first systematically studied as part of behavioural approaches within human geography in the 1960s and 1970s (e.g. Ilbery, 1978; Wolpert, 1964).

Hence a wide range of work during the last decade has added to the understanding of attitudes adopted by farmers towards the environment and the consequent effects upon their land management decisions (e.g., in a UK context, Crabb et al., 1998; Battershill and Gilg, 1997; McEarchern, 1992; Morris and Andrews, 1997; Potter and Lobley, 1996b; Robinson and Lind, 1999; Wilson, 1996). Also, AESs have often had an inherent cultural dimension, being generally sensitive to some of the idiosyncracies of local tradition integral to farming practice. Hence there are several examples of research on AESs with culturally sensitive perspectives (McEarchern, 1992; Morris and Andrews, 1997; Young et al., 1995).

Rogers (1995) argues that the basic decision-making framework focusing on the farmer can be extended to incorporate demand, supply and external factors. This enables the analytical framework to include communication channels and the role of agents of change. This is illustrated in work done by Morris et al. (2000) on the uptake of the Countryside Stewardship Arable Option Scheme in England, emphasising different steps in the adoption process:

1　Prior conditions. These are broad, contextual, external factors that may influence and shape the disposition of potential adopters of an innovation. This can include the felt needs of farmers with respect to their priorities and the balance between them; social norms regarding acceptable farming practice, and the policy environment through which regulation, incentive and advice can influence decisions.

2   Knowledge. This stage ranges from initial awareness of an innovation to knowledge and understanding of purpose and functioning. Typically, knowledge may be obtained from the mass media or from other farmers, especially by word of mouth at market or group meetings of farmers, e.g. in a co-operative or farmers' union. Mass media tend to be the most effective means of creating awareness whilst the knowledge base is then consolidated through focused literature, group meetings, personal observation, workshops and conferences.

3   Persuasion. This occurs when a farmer or potential adopter deliberately seeks information and advice and forms an attitude or develops an opinion towards the innovation regarding its potential suitability. Agricultural extension services can play an important part in this stage as can demonstration farms at which the operation of the innovation may be viewed.

4   Decision. This is where the farmer makes a detailed assessment of the feasibility and practicality of the scheme for the specific conditions pertaining on their farm. Again, direct communication between adopters and advisors is significant.

5   Adoption. This is the point at which the farmer purchases the new technology or introduces a new farming practice, perhaps involving uptake of a new government-sponsored scheme. Advice and guidance on adoption/implementation may be provided by external advisors.

6   Confirmation. The farmer makes an evaluation of their decision to adopt or not adopt in the light of consequent experiences. This can involve further input from key advisors or opportunities for sharing of experiences with others in a similar situation.

One of the major reasons for looking at the way in which farmers make decisions is the need to understand the meanings they attach to conservation and the environment. These meanings are likely to be a product of values, beliefs, knowledge systems or culture (Morris and Andrews, 1997). However, it has been found in several studies that, because farmers have 'grown up' during an era when the drive for increased production has been at the forefront of agricultural policy, the notion of efficiency and maximisation of production is central to their views on land management. They also tend to feel let down by and distrustful of the new government attitudes towards 'environmentally friendly' farming because it detracts from what they regard as their main function. Therefore, farmers and those representing the interests of conservation tend to interpret land management issues in different ways. For instance, a farmer in the uplands may consider that he or she is 'working with nature' and pursuing responsible stewardship by maximising production under difficult physical conditions (e.g. McEarchern, 1992). In contrast, a conservationist may take the view that 'improving' the land to generate increased output is causing unnecessary stress to delicate habitats that may be rare or have a high aesthetic appeal. This type of difference in perception lies at the heart of considerations regarding the successful implementation of agri-environment schemes. The difference must be better understood if conflicts between farmers and conservationists are to be reduced. Indeed, such better understanding is essential if the present voluntary nature of agri-environment schemes is to continue and for the schemes to meet their stated aims. This is stressed in work by Battershill and Gilg (1997) in south-west England, arguing that the key to adoption of AES is the attitudinal disposition of farmers – more important than structural constraints, conservation opportunities or family and financial considerations.

Wilson and Hart (2000) report findings from a survey of 1000 farm households in nine EU countries and Switzerland regarding farmers' participation in AESs, using two case study areas per country, one an intensively farmed region and one an extensively farmed region, and covering a range of different schemes as per Baldock et al.'s (1998) classification (Table 10.11).

Table 10.11 Types of agri-environment schemes (with examples)

Focus (description)	Horizontal schemes (broad and shallow)	Targeted schemes (deep and narrow)
Wide focus	OPUL (Austria) Hessian agri-environmental programme (Germany) Swiss agri-environmental programme	ESAs (UK/Denmark) Swedish agri-environmental programme Castro Verde ESA (Portugal)
More specific focus	Prime a l'herbe (France) Portuguese agri-environmental programme Scheme for grassland protection (Spain)	Bocage Avesnois (France) Scheme for the reduction of nitrate pollution, Larissa (Greece) Scheme for the arable steppes of Castilla y Leon (Spain) Countryside Stewardship (England) Organic farming (Greece)

(*Sources:* Baldock *et al.*, 1998, Wilson and Hart, 2000)

Table 10.12 Co-financeable expenditure, expenditure ratio and scheme premia (selection) for agri-environmental policies in nine EU countries, 1993–7

Country	Total co-financeable AEP expenditure 1993–7 (million)	AEP expenditure (%) 1993–7	Expenditure ratio (per ha UAA) 1993–7	Organic farming premia (per ha) 1996	Landscape protection (per ha) 1996
Austria	1553	32.9	0.450	350–750	38–575
Germany	1294	27.4	0.076	135–867	up to 750
France	1018	21.5	0.034	106–712	up to 500
Sweden	252	5.3	0.076	100–250	50–450
Portugal	197	4.2	0.051	no data	no data
UK	192	4.1	0.011	8–60	10–300
Spain	167	3.5	0.006	85–450	55–350
Denmark	38	0.8	0.014	100–150	no data
Greece	15	0.3	0.004	100–200	420
Total	4726	100			

(*Source:* Buller, 2000, pp. 229–30 and 241)

The survey revealed considerable variation in funding for such schemes across these countries (Buller, 2000; see also Colman, 1994) (Table 10.12). It also demonstrated the complexity of farmers' motivations for participation in the voluntary schemes. Figure 10.4a shows the key reasons stated by farmers for participation, revealing that economic considerations have been the primary driving force, followed by reasons relating to the schemes' goodness of fit with existing farm-management plans (see also Brouwer and Lowe, 1998). Conservation reasons appear as slightly less important. In terms of the classification employed (Table 10.11), the 'deeper' and more 'narrow' schemes are, then the greater likelihood that goodness of fit plays a major role in decision-making (see also Lobley and Potter, 1998). Nevertheless, Wilson and Hart

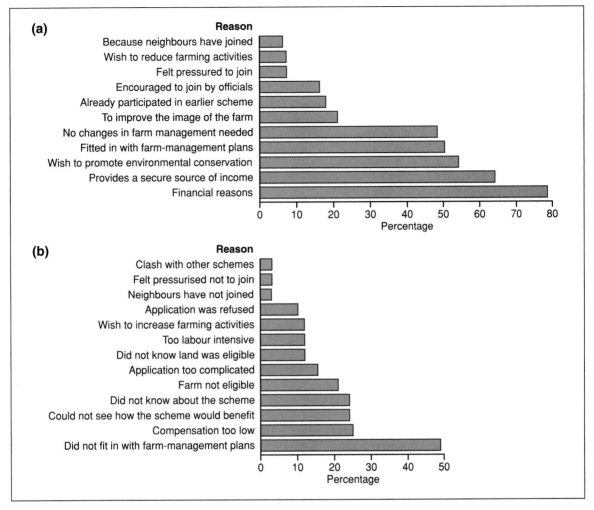

Figure 10.4 (a) Reasons for participation in agri-environment schemes; (b) Reasons for non-participation in agri-environment schemes (based on Wilson and Hart, 2000)

(2000, p. 2167) also argue that, contrary to Whitby's (1994) earlier findings, there is an increasing acknowledgement by EU farmers of the environmental benefits from participation in AESs. This was more pronounced in the northern member states, with some suggestion of a northern versus Mediterranean divide, perhaps reflecting how policy officials market the schemes to farmers and the type of farmer recruited (Wilson et al., 1999).

This 'divide' is elaborated in evaluations of agri-environmental measures under EU Regulation 2078, which suggest that in many upland and mountainous areas of southern Europe AESs have been unsuccessful for a variety of reasons. Galdos et al. (2000) highlight five in particular:

- the 'gap' between policy concerns and local realities;
- poor dissemination at a local level;
- lack of confidence amongst both local authorities and farmers that they can carry out the programme;
- the finacial aid involved does not sufficiently compensate for the amount of work and services involved;

- there is competition from support measures that the farmers find more attractive.

Reasons for non-participation of farmers (Figure 10.4b) primarily reflected practical considerations, such as lack of compatibility with existing farm management plans, financial reasons and lack of information about AES. This may lend support to Morris and Potter's (1995) notions of non-participants being split into conditional and resistant non-adopters (see below). Many studies have reported financial reasons as a deterrent to participation in UK AESs (e.g. Lobley and Potter, 1998; Wilson, 1997b). This may reflect the low level of AES payments in the UK compared with that in many other EU countries, where financial considerations seem less of a constraint for farmers. For example, in the UK, farmers complying with the requirement to set aside at least 10 per cent of their cropland have been reluctant to find more land for agri-environment schemes without higher payments by way of compensation (Potter, 1996, p. 176).

A direct comparison can be made between Wilson and Hart's typology of farmers based on participation in AESs and Morris and Potter's (1995) survey of 101 farmers in an ESA in southeast England. This included 54 per cent who were participants in the scheme and 46 per cent non-participants. They compiled a participation spectrum that classified respondents into four different groups:

- Active adopters (52% of adopters) – strongly motivated by environmental commitment.
- Passive adopters (48% of adopters), who had taken part mainly for financial reasons.
- Conditional non-adopters (37% of non-adopters), who might consider participation if a particular constraining factor such as an aspect of scheme design were to be relieved.
- Resistant non-adopters (63% of non-adopters) who were adamant in their self-exclusion.

This compares with Wilson and Hart's typology:

- Scheme enthusiasts (65% of participants), who were more likely to see scheme

objectives as financial, depended largely on the farm for income and saw maintenance of the family farming tradition as important. They had been 'converted' towards conservation-oriented beliefs, but combined financial and conservation-oriented responses. This group was dominant in highly targeted schemes.
- Neutral adopters (35% of participants) were 'neutral' about both conservation and the financial imperative for entering the schemes. They participated in the schemes for pragmatic reasons rather than for conservation. A larger number were apparent in areas where AESs were 'sold' to farmers as income support schemes rather than as conservation schemes.
- Uninterested non-adopters (54% of non-participants) did not have a high regard for the conservation objectives of the schemes and disagreed with legislative measures to control farmers' environmental management practices. More crucially they generally did not depend on their farm for income and hence were less likely to be attracted to AESs for financial reasons.
- Profit-maximising non-adopters (46% of non-participants) were farmers who felt that profit-maximisation would be compromised by joining an AES. They felt that AESs could not compensate them for potential income losses incurred by being in a scheme. Mediterranean countries have the highest proportion of this type of farmer.

Wilson (1996) developed this classification into a participation matrix according to attitude: from utilitarians to conservationists, and ESA eligibility: from high to low. This study concluded that conditional adopters could be converted to active or passive adopters by targeted changes in scheme design. However, these 'targeted changes' depend on better understanding of farmer motivation for participating in AESs. For example, Lobley and Potter (1998), in a study in south-east England, found that adopters of the ESA scheme were predominantly motivated by financial gain whereas Countryside Stewardship Scheme (CSS) adopters demonstrated predominantly conservation motives. They classified adopters into two groups: stewards, who justified participation in terms of potential conservation benefits in ways which corresponded closely to the conservation objectives of the schemes; and compliers, who focused primarily on financial gains from the scheme, provided that requirements of scheme membership fitted reasonably well with their existing farming system and their commercial imperative. They also classified non-adopters into two groups: potential enrollers, who had been put off joining the schemes by a range of personal, farm and scheme factors; and resistors, whose opposition to the philosophy of interventionist agri-environment schemes was often reinforced by limited knowledge of the schemes (Morris *et al.*, 2000, p. 2).

Wilson and Hart also considered a range of variables influencing participation. They found that AESs tend to be more suitable to relatively large farms, as in many regions AESs as an income support measure may not sufficiently benefit small family farms most in need of 'green' subsidies for farm survival (see also Wilson, 1997c). Per hectare payment methods in most AESs may disproportionately benefit larger farms, though this varies considerably depending upon type-of-farming practised. In particular, grassland protection schemes have been more beneficial to large farms, and hence extensive grassland farms are more likely to participate in schemes than intensive livestock and arable farms. This partly reflects the high degree to which AESs in the EU are targeted at upland livestock areas (Brouwer and Lowe, 1998). In general, extensive grassland farms do not have to reduce stocking rates substantially in order to be eligible for schemes. In contrast, for intensive farms the required changes can often be quite substantial. This tends to differentiate between arable and pastoral areas, with arable farmers tending not to enter AESs for financial reasons or, if they do participate, doing so as 'neutral adopters'.

Wynn *et al.*'s (2001) analysis of the decision by farmers to enter the ESA scheme in the ten Scottish ESAs revealed a number of generic factors as important in explaining the entry decision. Non-entrants were less aware of and less informed about the scheme than entrants. The probability of entry was increased where the scheme prescription fitted the farm situation and the costs of compliance were low. Key factors accelerating entry to the scheme were an interest in conservation, more adequate supply of information and the existing provision of extensive systems of production. There was the suggestion that, through the Tier 1 management payments, the scheme was being used primarily as an income source with little commitment to conservation (see also Whitby, 1994). As with research on the uptake of woodland incentives (Crabtree *et al.*, 1998), there was no evidence to support the contention that succession and the path of business development were key determinants of environmental change on farms, as suggested in other areas by Potter (1997; Potter and Lobley, 1996a; 1996b) and by Ward and Lowe (1994).

Research on AESs has highlighted the relationship between participation and farm-household characteristics (e.g. Brotherton, 1990; 1991a). In general, participants are more likely to be owner-occupiers who have completed secondary school (and perhaps have post-school qualifications), with a strong reliance on the farm as the prime source of income. The presence of a successor does not seem to be a significant variable, with strong regional and type-of-scheme variation being more important (Potter and Lobley, 1992; 1996a; Wilson and Hart, 2000, p. 2177). Of greater importance has been the quality of information provided to farmers by extension services (Cundliffe, 2000). Wilson and Hart (2000, pp. 2178–80) distinguish between three types of information delivery:

- Well-funded extension services with substantial powers, playing a crucial role in enrolling large numbers of farmers into schemes, e.g. Austria and Switzerland.
- Some provision of information to farmers, but with an emphasis on farmers being proactive in obtaining information, e.g. the UK, France, Sweden and Germany.
- Poorly funded extension services where 'other channels' such as private agronomists may be significant for farmers, e.g. the Mediterranean countries.

Lack of information tends to be a crucial factor in both non-participation and delays in joining AESs by farmers quite favourably disposed to joining a scheme.

In the UK, several schemes have been introduced, all operating on the basis of the farmers' voluntary participation in return for payment (Van Huylenbroeck and Whitby, 1999). In 1992/3 the costs of payments to farmers, running costs and monitoring costs were £33 million. This had risen to £86 million in 1996/7. Of the latter sum, 43 per cent were consumed by the ESAs scheme, with 44 designated areas. Yet these policies account for only a very small amount of the overall spending on agricultural support: 2.5 per cent of total CAP expenditure in the UK in (1995/6) and approximately 3 per cent for the EU as a whole.

10.4.2 Environmentally Sensitive Areas

First introduced in the UK's 1986 Agriculture Act and applied subsequently in various forms throughout the EU (e.g. Baldock *et al.*, 1990; Primdahl and Hansen, 1993; Whitby, 1994; Wilson, G. A., 1994; 1995), Environmentally Sensitive Areas (ESAs) cover 18 per cent of the UK in 44 separate designated areas (Map 10.3). The ESAs are areas of high conservation value in which famers can agree voluntarily to participate in the adoption of environmentally friendly farming practices and receive payment for so doing over a five-year period. Payments vary according to the type of farming and the physical character of the ESA, but the general pattern has been a combination of flat-rate payments and item payments for carrying out particular tasks (Brotherton, 1991b; Skerratt, 1994; Skerratt and Dent, 1996).

Extensions to the scheme after the initial designation of 22 areas were predicated on the official view that offering voluntary rewards to farmers for participation in the production of countryside, rather than just the traditional output of crops and livestock, is an acceptable way of promoting desirable changes to farm management practice (Wilson, 1997c). The expansion that came after the first five years of the scheme was based in part on a growth in the uptake rates and the amount of land in the ESAs entered into management agreements. For example, by the end of 1995 in the 22 English ESAs 43 per cent of eligible landowners

had entered 42 per cent of their land into management agreements. In Scotland at this time 910 farmers had registered 28.1 per cent of their eligible area in ESA agreements (Edgell and Badger, 1996, p. 32). However, there has been variation between the ESAs, with a broad contrast between those in upland and lowland areas. Many of the schemes in the lowlands have less than 50 per cent of the farmland in the designated areas entered into agreements. Despite the lower figures for the Scottish ESAs as a whole, some of those designated first had high participation rates and higher proportions of whole farms entered. For example, in the Breadalbane ESA in Tayside, in the mid-1990s three-quarters of farmers were participating and around 80 per cent of the agricultural land was covered by management agreements (Robinson, 1994).

Higher take-up rates and entry of land to some of the ESA schemes in Scotland and other parts of upland Britain are also related to the different role played by ESAs in the uplands in general (Morris and Robinson, 1996). This role is not only one of attempting to act as a curb to destructive practices in areas of agricultural intensification, but as maintainers of traditional farming landscapes, often in locations where there is little economic alternative to land-based activity (Gaskell and Tanner, 1991). So in the uplands the focus has been upon encouraging farmers to undertake tasks that they had perhaps been neglecting because of the high costs of labour involved, such as rebuilding stone walls, replanting woodland, and active conservation tasks such as the protection of rough grazing, protection of unimproved, unenclosed land in valleys and general restraint from overstocking. The rates of uptake in the uplands suggest that farmers have been willing to accept payments for undertaking tasks that they wanted to implement on their farm but had not done so through lack of money. The receipt of the requisite money has led to more labour being hired to carry out various conservation tasks and, in some cases, has added to overall farm incomes (Froud, 1994).

The motivations of farmers entering the ESA management agreements are not necessarily those in which conservation or stewardship of the land has been uppermost (Skerratt, 1994; Skerratt and

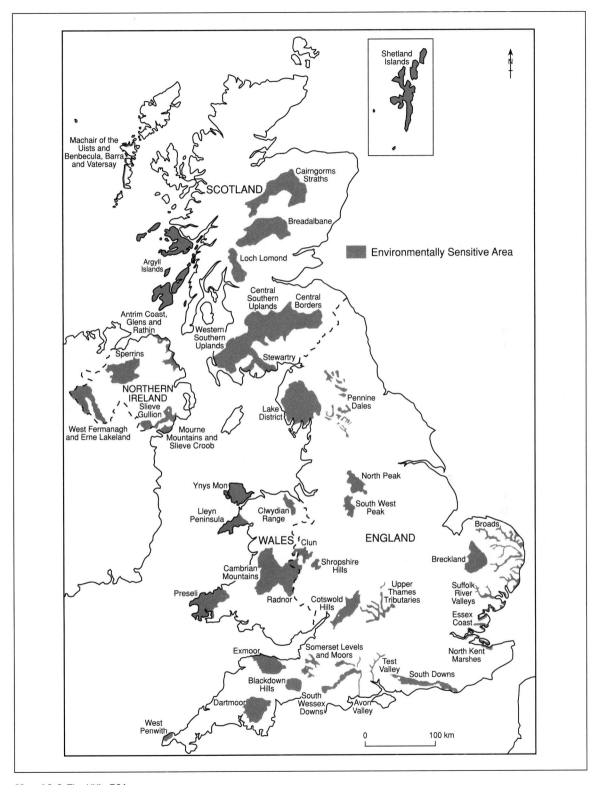

Map 10.3 The UK's ESAs

Dent, 1996), though farmers with conservation-oriented attitudes and a desire to pursue a 'traditional' approach to farming appear to be important factors in determining whether farmers will enter into management agreements (Battershill and Gilg, 1996; MacFarlane, 2000a; 2000b; Morris and Potter, 1995; Wilson, 1996). This may be especially important on farms of marginal ESA eligibility where conservation-oriented attitudes may encourage participation if monetary gain from a management agreement is likely to be quite limited (Wilson and Hart, 2001).

Assessments of the conservation value of ESAs have varied considerably. Some have claimed that farmers are being paid to do what they would do in any case, and therefore the ESA management agreements are having little impact on the environment (Baldock *et al.*, 1990; Colman, 1994). Others have stressed particular benefits arising from the scheme (Garrod *et al.*, 1994; Garrod and Willis, 1995), including increased incomes per unit area (Froud, 1994).

The ESAs now form an integral part of UK and EU policies for sustainable rural development. However, the value of their contribution to sustainability is questionable. Morris and Potter (1995), for example, argue that agri-environment policies will only be transitory in their effect and are unlikely to influence long-term attitudes of farmers. Their survey of farmers with environmentally friendly management agreements in south-east England did not suggest that current policies were likely to promote a change in farming practices that would contribute substantially to developing sustainable agriculture. The transitory nature of such policies must certainly be considered. If the money is withdrawn, will farmers simply revert to practices designed to increase output if that is the only way to maintain their incomes? Or will there be a further reduction in the numbers of farmers, especially in upland areas where some of the agri-environment policies are concentrated?

10.4.3 The Countryside Stewardship Scheme

The Countryside Stewardship Scheme (CSS) was introduced in England in 1991, giving farmers outside ESAs opportunities to follow environmentally friendly practices, on a voluntary basis, if they have particular types of land deemed worthy of special protection and management. In 1995 the Government's White Paper on *Rural England* placed strong emphasis on the CSS as the main incentive for management of the countryside outside ESAs (DoE/MAFF, 1995). It has been described as a market approach to conservation (Bishop and Phillips, 1993) because of its reliance upon landowners competing to offer environmental services and only receiving payment if they are deemed to be offering value for money. CSS was managed by the Countryside Commission until 1996, during which time 92,500 ha were enrolled in management agreements concluded under the scheme. In 1996 the Ministry of Agriculture, Fisheries and Food (MAFF) took over its administration. The budget for the CSS was increased to £23 million per annum in 1998.

A sister scheme, Tir Cymen, administered by the Countryside Council for Wales, was introduced in 1993 (Banks and Marsden, 2000) and more recently a similar one, the Countryside Premium Scheme, has been launched in Scotland (Egdell, 2000). In all of these schemes management agreements entitle farmers and landowners to claim financial payments that are attached to particular items of work. Some examples of payments within the CSS are shown in Table 10.13.

Within the management agreements in the CSS there are four major elements:

- Land management measures supported by annual maintenance payments. These specify ways in which current land management practices, such as ploughing, grazing and mowing, are to be continued or adopted to produce conservation objectives.
- Supplements for additional work. These enable land mangement measures to be achieved, for example cutting back docks and thistles prior to introduction of grazing animals.
- Public/educational access, under which agreements for new access provision may be negotiated.
- Capital items. One-off payments, for instance to reinstate a length of hedge-bank or dry

Table 10.13 Summary of items of work, codes and payment in the Countryside Stewardship Scheme

Annual management: annual payments are for ten years unless otherwise stated.
Those items with a code prefixed 'U' are upland items and are only available on LFA land.

Item	Code	Payment
Managing grassland		
Lowland hay meadows	H1	£115/ha/year
Upland hay meadows	UH11	£1550/ha/year
Lowland and culm pastures	P1	£85/ha/year
plus for old pasture <3 ha	GRP	£30/ha/year
Upland in-bye pasture	UP1	£60/ha/year
Upland rough grazing pastures (enclosed up to 20 ha)	UP2	£45/ha/year
Upland rough grazing (over 20 ha)	UP3	£20/ha/year
Chalk and limestone grassland	P4	£60/ha/year
Upland limestone grassland	UP4	£60/ha/year
Managing fen and reedbeds		
Managing fen	F	£l00/ha/year
Managing reedbed	R	£l00/ha/year
Recreating grassland on cultivated land	R1	£280/ha/year
Supplement for former set-aside land (of high environmental value)	RIX	£50/ha/year
Supplements (the following are available on grassland [including those associated with historic landscapes], fen, reedbeds and grassland re-creation items)		
Grassland supplements	GX	£40/ha/year (for up to 5 years)
Supplement for raised water levels	GW	£60/ha/year
Supplement for use of native seed	GS	£250/ha/year
Managing lowland heath		
Maintaining existing lowland heath	LHl	£20/ha/year
Enhancing existing lowland heath	LH2	£50/ha/year
Re-creating heath	LH3	£275/ha/year

(*Source:* Simpson and Robinson, 2001)

stone wall or fence against livestock, can be negotiated (Morris and Young, 1997a; 1997b).

The scheme is partly funded from the UK's budget allocation in the EU's CAP and the remainder is from the UK's domestic agricultural budget. As with other AESs in the UK, participation is voluntary and is administered by means of a management agreement. Initially these agreements ran for a period of five years, before being increased to ten years when the MAFF took over its administration. The scheme is not restricted to farmland, but nearly all of the land currently under CSS management is agricultural. This includes land under institutional ownership, for example the National Trust and local authorities entering land into CSS management agreements.

The CSS has both national objectives and regional targets. The national objectives relate to the conservation and restoration of landscape, archaeological and historical features and wildlife habitats. At its inception, increasing public access provision was seen as a high priority but the importance of this as a condition of the management agreement has diminished with the change in responsibility from the Countryside Commission to the MAFF.

Regional targets are selected by the MAFF in association with the Countryside Commission, English Nature, English Heritage, local wildlife trusts and other relevant environmental non-governmental organisations (NGOs). The targets may include habitats that are clearly recognised as being of national or international significance, but also include ones of significant local importance but which could be overlooked in a review conducted at national or international level. Types of land eligible are chalk and limestone grassland, lowland heath, waterside, coast, uplands, historic landscapes, old orchards, unimproved areas of old meadow and pasture on neutral and acid soils, community forests, peri-urban countryside, and traditional boundaries including damaged and neglected hedges, walls, banks and ditches. Emphasis is placed upon applications that are deemed to offer positive changes and where land has a special conservation interest or adjoins such areas (Countryside Commission, 1995).

In terms of take-up, at the end of 1998 143,055 ha had been enrolled in 8614 management agreements (an average of 16.61 ha per agreement). As shown in Map 10.4, the largest numbers of agreements have tended to be in the uplands, with concentrations in the south-west. The highest densities of agreement per unit area of farmland have also been in upland areas and the west, notably the Peak District, parts of Devon, the Isle of Wight, Cumbria and west Cornwall. In general, the highest uptake rates occur in the National Parks, if those which are also designated as ESAs are excluded. Agreements focusing on field boundaries, waterside land and old meadows and pasture have been most popular (See Table 10.14).

There are several features of the scheme that distinguish it from other AESs in the UK, such as the longer-running ESAs scheme (Evans and Morris, 1997). Three aspects are notable:

- it is the first large-scale AES in the UK to 'seek to buy environmental and public access goods from farmers and other land managers on a targeted and discretionary basis' (Harrison-Mayfield et al., 1998);
- it is not targeted at designated agricultural regions, but is available throughout the country, with the exception of the ESAs;
- it contains an element of competitive tendering, based on the quality of the proposed objectives by individual applicants (Potter, 1998).

The broad objective of the scheme is to conserve and create landscape and wildlife features within specified landscape types and to enhance public access to the countryside. Particular emphasis has been placed on applications offering 'positive changes' and where land has 'special conservation interest or adjoins such areas' (Countryside Commission, 1995), with four basic elements to any CS agreement (Morris and Young, 1997b):

- Land management measures specify the way in which current land management practices (such as controlling the timing and intensity of grazing, stocking rates and feeding; restricting ploughing and the application of agri-chemicals, and refraining from modifying drainage) are to be continued or adapted to produce conservation objectives.
- Supplements for additional work required in order that land management measures can be achieved (e.g. cutting to control dock and thistle before reintroducing grazing) are available.
- New public access and educational access can be negotiated.
- One-off capital items (such as restoring dry stone walls) can be financed.

In their study of the uptake of CSS in Cheshire, Morris and Young (1997b) found that the vast majority of CS agreements were on owner-occupied mixed and livestock holdings, with few on horticultural and dairy holdings, i.e. CSS was more prevalent on those holdings farmed more

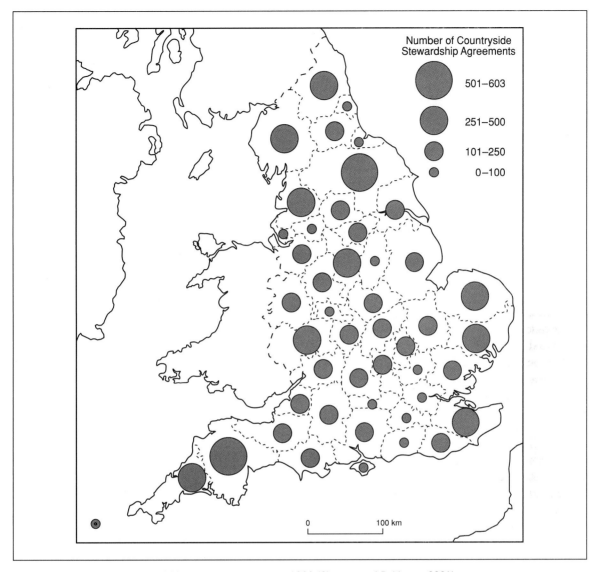

Map 10.4 Take-up rates for the CSS in England, to summer 1999 (Simpson and Robinson, 2001)

extensively. Related to this was the tendency for larger farms to take-up CSS. The evaluation of CSS in Cheshire was generally positive: 'in terms of conserving particular habitats and landscapes, CSS is undoubtedly achieving what it set out to do' (p. 315), with benefits accruing not only from the conservation of habitat and landscape but also from the creation of new jobs and a positive impact upon farm incomes (CEAS Consultants Ltd, 1996). In Cheshire this was largely through

agreements for managing lowland hay meadows, lowland grazed pasture and for supporting additional work on grassland in the form of the initial stages of conservation work. However, the analysis suggested that 'CSS is unlikely to achieve the production of widespread environmentally beneficial farming' because only a small proportion of farmland was being affected, with just small, targeted pockets of land involved rather than the wider countryside as a whole.

Table 10.14 Number of CSS agreements by landscape type, to May 1999

Landscape type	Number of agreements	%
Access	14	0.1
Arable margins	488	5.7
Chalk and limestone grassland	963	11.2
Coast	301	3.5
Countryside around towns	104	1.2
Field boundaries	2102	24.4
Historic landscapes and features	526	6.1
Lowland heath	429	5.0
Old meadows and pasture	1102	12.8
Old orchards	227	2.6
Uplands	759	8.8
Waterside land	1599	18.6
Total	8614	100.0
Access Agreements		
Open Access	791	35.6
Linear – footpaths	608	27.4
Bridleways	122	5.5
Disabled access	31	1.4
Educational access	669	30.1
Total	2221	100.0

(*Source*: Simpson and Robinson, 2001)

Barriers to uptake were evident, both geographically and related to farm characteristics. There was also often a 'halo effect' where parts of a farm were managed for conservation, but the remainder was still used intensively (Morris and Young, 1997b, p. 315). Despite the variety of landscape types included in the CSS targets, there are large areas of the countryside that are unlikely to be entered into CSS. Hence achievement of the broader environmental goals expressed for the scheme may be impossible.

Some aspects of CSS have been disappointing, notably the uptake of the Arable Field Margins option, especially in areas dominated by large-scale farming. This option includes creation of uncropped and grass field margins, in-field banks providing refuge for insects, especially beetles, and conservation headlands that are cropped using limited chemical inputs. In 1997 a new Arable Stewardship scheme was introduced on a pilot basis, aimed at improving wildlife biodiversity on arable farmland. There are five options within this scheme: overwintered unploughed stubbles, spring cereal crops undersown into grass or stubbles, insecticide-restricted crop margins, grass field margins and in-field conservation strips for habitat, including seed mixtures conducive to wildlife. Combinations of these options can produce income of £500 to £600 per ha equivalent.

From work by Simpson and Robinson (2001) in Devon and Hampshire, it is clear that the uptake of the CSS is closely associated with input from NGOs. In particular, local wildlife and conservation organisations and the local/regional representatives of national bodies are playing a significant role in promoting the possibilities offered by CSS. There is the suggestion that there are some significant regional variations that offer lessons for policy-makers in developing better strategies for increasing take-up rates of agri-environment measures. In Devon, the CSS has been taken up by around one-third of all farmers in the two areas targeted by the scheme. This is having an impact upon the maintenance of traditional grassland management practices, though there is some evidence to suggest that several participants would have pursued such practices even without CSS payments. Nevertheless, in other cases, the payments have encouraged farmers to consider the environmental benefits of certain practices whilst providing valuable revenue.

In terms of examining the links between different policy aims as applied to farming, consideration of the CSS lends weight to the much stated argument that this agri-environment policy is not particularly well integrated with price support policy within the CAP or with other aspects of rural development. Latent potential for such integration exists, but remains to be realised, in part because of the limited amount of funding accorded CSS. This problem of funding is widespread throughout AESs in the EU. If AESs on their own are equated with a move to post-productive farming systems, then it is clearly premature to speak of having entered a post-productivist era in agriculture, though broader conceptions of this transition may give more credence to the emergence of a significant change both in farming practices and policy measures (e.g. Marsden, 1998b).

With respect to a move towards a stronger ecological basis for farming, the CSS provides opportunities for environmental gains to be incorporated into agreed farm management plans. However, such plans have not always prevented farmers from reducing inputs on the land entered into CSS agreements whilst intensifying on the rest of their land (the halo effect), and hence the championing of 'whole farm' approaches to AESs (Morris and Cobb, 1993). Furthermore, in general, the continued presence of policies fostering increased production has restricted the ecological benefits of the environmentally friendly policies. Hence Bowler (1992a) concludes that policies like CSS and ESAs are more correctly described as promoting 'alternative agriculture' rather than sustainable agriculture, as they do not meet many of the latter's basic criteria. Moreover, even in areas targeted by the CSS, there is a significant proportion of non-participants. Hence it is difficult to argue with the contention that on many farms the essential practices of a largely agri-industrial farming system remain in tact. Any threat to these practices awaits further legislation to reduce price supports and to extend possibilities for different types of farming system to fluorish.

10.4.4 Further examples of agri-environment schemes

Similar AESs to the two highlighted above have been operated outside the EU. For example, before accession to the EU, environmental payments were a prime source of subsidised support for Finnish agriculture (Hyttinen and Kola, 1995). The largest subsidies were given to farmers in areas where they were not eligible for other financial support, primarily the areas of most intensive arable production in the south and west of the country. Here, on a voluntary basis, farmers could participate in a scheme promoting the extension of environmentally friendly farming practices, including cessation of cultivation on three-metre wide strips of land between fields and rivers/lakes; 30 per cent of a farm's field area to be unprepared and allowed to regenerate during winter; restrictions on the use of fertilisers; other measures to reduce pollution of watercourses; and maintenance of landscape, cultural heritage and biodiversity. The support for this programme was nearly equivalent to the price support Finland receives under the CAP and so contrasts sharply with the total level of payments under the ESAs scheme.

Finland also has a closer integration of farming and forestry than most countries, with long-term policies promoting 'greening' through extensions of farm woodland. One of the first and most extensive farm afforestation policies has operated since 1969. The origins of this policy lie in the success of productivist farm policies operating since the Second World War, whereby in the late 1960s over 400,000 ha of farmland was deemed surplus to requirements. The subsequent legislation introduced a field afforestation programme (Selby and Petajisto, 1994). However, this had little impact upon the area regarded as surplus to agricultural requirements, and eventually a strengthened programme was introduced in the early 1990s when an area of around 800,000 ha, equivalent to nearly one-third of the agricultural area and 40 per cent of the field area, was regarded as surplus to requirements given the desired 100 per cent self-sufficiency (Selby, 1994). The regional impacts of the two programmes have been quite varied. The greatest farm afforestation has occurred in the most rural areas suffering from poor social conditons and limited economic growth. Impacts on the most productive arable areas have been much weaker, though local-level variations can be observed, associated with stoniness, waterlogging, low soil fertility and outlying locations, all of which favour conversion to forestry (Mather, 1998, p. 114). This reinforces the long-term view of farm forestry as being associated with marginal land, but it also reflects the fact that the financial incentives for farm afforestation have tended to be relatively less attractive on the better farmland (Bullock *et al.*, 1994; Gasson and Hill, 1990). This suggests that there are inherent contradictions in the policy between, on the one hand, wishing to cut production, and, on the other, endeavouring to do so merely by encouraging land (and primarily not the most productive) to be afforested.

In other European countries, farm afforestation policies have also met with relatively little success in recent years, partly reflecting the negative attitudes to forestry held by many farmers (e.g. Scambler, 1989). For example, afforestation rates in the UK, France and Italy have fallen (Matthews, 1994), and there has been an absolute decrease in the area of farm woodland in France in favour of an increasing degree of agricultural specialisation (Cavailhes and Normandin, 1993; Cinotti, 1992). The planting that has occurred in France is mostly on unused scrubland and relatively agriculturally unproductive marginal land in the south of the country. Lloyd *et al.* (1995) report that farmers are discouraged from converting farmland to woods because of poor financial returns, the lack of flexibility of land under a timber crop and the increased external regulation associated with timber production (see also Williams *et al.*, 1994).

The voluntary approach in AESs and farm afforestation programmes has also been pursued in a number of initiatives operating in different parts of Canada, most of which owe their existence to the desire of conservationists to promote protection of natural habitats. Hilts (1997) lists eight such schemes:

- the Island Nature Trust in Prince Edward Island;
- corporate conservation agreements in Nova Scotia;
- local programmes operated by community groups in Quebec;
- landowner agreements negotiated by the Muskoka Heritage Foundation in Ontario;
- the Natural Heritage Stewardship Programme in Ontario. This has involved voluntary agreement from 2000 landowners to protect approximately 24,000 ha of woodland and wetlands (van Hemessen *et al.*, 1994).
- landowner agreements to conserve wildlife in the three Prairie provinces;
- stewardship agreements with farmers in the Fraser delta and with landowners around the Cowichan estuary in British Columbia;
- the Environmental Farm Plan (EFP) programme in Ontario, which is amongst the most comprehensive in the world and which is run by farmers themselves. This owes its origins to a mixture of farmers' genuine concern for environmental well-being and a fear of being forced by government-introduced regulation into different types of farm management practices to achieve greater sustainability. Introduced by a Farm Coalition, farmers are asked to evaluate their operation using 23 worksheets relating to topics such as pest control, use of water, disposal of wastes and cropping practices. This is administered at the county level by the Ontario Soil and Crop Association, with workshops provided for the participants. These encourage farmers who have completed worksheets to pursue actions that will minimise environmental risks, usually on the basis of self-identified actions (Bidgood, 1994). However, there is a need for more investigation of the outcomes of this scheme before environmental benefits can be assessed (Robinson, 2003).

In Australia a National Landcare Program was launched in 1989, since when over 2500 community Landcare groups have been formed. These involve representatives of around one-third of Australian farm businesses, typically constituting a high proportion of landholders in an area. The groups are voluntary, small-scale, cooperative, self-help and community based. The basic concept incorporates notions of healthy environments, participatory democracy and local-level action. The activities of the groups involve the organisation of field days and other information-sharing activities, property and catchment planning and projects to establish field trials and demonstration sites. In some respects its farm base represents a different type of farm regulation to that imposed by the state, but without really challenging the further intensification of agriculture. Hence, Landcare helps to deal with some of the negative environmental externalities of agricultural production rather than proposing any fundamental reassessment of farming systems (Lockie, 1998). Major impacts have been on tree planting, though perhaps with a questionable impact on problems like dryland salinity. Indeed, it appears that Landcare's promotion of environmental stewardship has been greatest in terms of having an effect on how farmers became more aware of how their actions might affect other farmers. As with several other related schemes and projects there is much need for research focusing on the directly attributable environmental impacts.

There remain new areas of research on AESs both in Europe and North America to be tapped (see Winter, 1996b).

- Those schemes that have received less attention, especially those without spatially delimited boundaries, such as the agri-environmental aspects of EU Objective 5b-funded schemes (Ward and McNicholas, 1998).
- The need to consider broader objectives of biodiversity sustainability and policy reform as well as specific aspects of environmental value associated with particular schemes.
- Exploration of the way in which AESs are mediated, e.g. through investigations of the regulatory nature of the agencies involved (Cooper, 1998). Questions have also been raised about the value of paying farmers for things they might do anyway: referred to as policy deadweight.
- The changing nature of AESs. For example, as part of ongoing reforms, the EU's 1999 Rural Development Regulation (RDR), EC 1257/1999, provided opportunities to widen the basis of AESs, to devise new measures to protect and support rural economies, to extend income-generating opportunities for new cropping, and to help farmers diversify into related economic enterprises (CEC, 1997). One outcome should be further decline in payments for production and a growth in redirected investments in AESs, woodland management, processing and marketing of new products, higher compensatory allowance for less favoured areas, the planting and managing of energy-providing crops, and training programmes to widen the skills of the rural labour force (MAFF, 2000). National initiatives are being produced that are closely related to the RDR, such as the UK's *A better quality life* (DETR, 1999) which focuses on promotion of sustainable development, with the application of sustainability indicators. The RDR is a continuation of the CAP reform process in which farming is regarded as fulfilling various functions, of which production is just one.
- The need to examine possible links between AESs and various aspects of sustainability. The sustainability of schemes using financial inducements to engage otherwise disinterested farmers is questionable. However, there have been arguments that schemes which secure 'good practice' on traditional farms are just as valid and worth supporting as the prevention of 'poor practice' on conventional farms. This may be especially true when traditional farms and the rural economy of which they are an important part are under pressure (Battershill and Gilg, 1996; Harrison-Mayfield *et al.*, 1998).

With reference to this last point, the achievement of sustainability is highly complex and most AESs can be said to have promoted sustainability only in terms of giving environmental considerations greater weight, whilst generally balancing these against the economic benefits of maintaining productivist agriculture (Whitby, 2000). Indeed, throughout the Developed World there remains a strong imbalance betwen the monetary rewards for producing food from the land as compared with producing countryside or being 'environmentally friendly'. For example, in England and Wales the expenditure of the DEFRA on agri-environment policy is less than one-twentieth of its overall budget.

In Germany regionally based AESs have been more significant than in other EU countries. The regions (*länder*) have acted as a regional filter for AESs, with 91 schemes in operation in the early 1990s (Wilson, G. A., 1994; 1995). Some of these schemes have applied across regions, but the one attracting the most attention, as a template for a new generation of AESs, has been the MEKA scheme in Baden-Wurrtemburg, with a range of environmental actions supported. Financial incentives are operated to encourage voluntary participation, with maxima of £10,000 per annum and DM550 per ha. These subsidies are paid to farmers on the basis of a complex points system, and can be applied to whole farms as well as portions. High participation rates have been recorded and a series of environmental gains from the scheme's operation (Wilson and Wilson, 2001). This can be compared with the experience in Switzerland where there are also a range of regional schemes for counties (*kantone*) and districts (*Gemeinden*) as well as national-level variants (Wilson *et al.*, 1996). There are a mixture of production subsidies with environmental prescriptions (at higher rates than paid in the EU), and environmental payments. There is evidence that the schemes have been most well received in the less physically well-endowed areas.

10.5 Genetic modification

If the range of policies discussed above in Sections 10.3 and 10.4 represent different degrees of an approach to agriculture that has ecocentric tendencies then the opposite, technocentric, approach is represented by the production of genetically modified (GM) foods. There are advocates of this approach who contend that GM crops, livestock and foods are a much more significant development, in that they offer the prospect of increased production on a large scale, thereby having the potential to dramatically increase the world's supply of certain crops and possibly also of some livestock products. However, there have also been concerted protests regarding the development of GM foods and related biotechnology. Some of this has illustrated the sharp divide in opinion that exists over biotechnology's role within the agri-food system (e.g. Kneen, 1999).

By 2000 roughly 28 million ha of GM crops were being grown worldwide, a fifteen-fold increase in just five years (Middleton, 1999b, p. 53). The five principal GM crops, soybeans, maize, cotton, rapeseed and potatoes, are grown commercially in nine countries, of which France, Spain and South Africa are the most recent adopters, having grown GM crops for the first time in 1998. The USA has nearly three-quarters of the total global area devoted to GM crops. In 1999 half of the 29 million ha devoted to soybeans in the USA were planted with GM herbicide-resistant seeds, intended to give easier control of weeds, less tillage and reductions in soil erosion.

In terms of recent biotechnical developments it has been this emergence of GM foods that has offered the most divisive views of the future of the marriage between farming and technological advances. The rapidly expanding biotechnology corporations, such as Monsanto, Hoescht and Calgene, are claiming that GM crops have the potential to eradicate global food shortages, thereby providing global food security for the poorest parts of the world. Yet strong opposition to these claims has been mounted in several Developing Countries and amongst some environmental lobby groups in the Developed World, especially in the UK. Opposition to GM crops in the UK has focused on the issue relating to the probability that the integrity of plant species may be compromised as engineered crop acreage increases and pollen-mediated gene flow affects native plants which are congenes for the transgenic crops. Another concern is the potential reduction of biodiversity as farming becomes more reliant upon heavy use of herbicides.

Attempts to organise concerted opposition in Developing Countries have appeared following the March 1999 meeting of scientists and activists in Delhi, at which the term 'biodevastation' was coined in relation to the threat posed by GM crops. The International Convention on Biodiversity includes reference to the need for an international protocol on biosafety to ensure safe handling and transfer of GM organisms between countries. Meanwhile in India 1500 organisations have campaigned against Monsanto's Bollgard cotton on the grounds that it can destroy beneficial species and can create superior pests (Power, 2001, p. 285).

Monsanto is the agricultural subsidiary of Pharmacia, a transnational pharmaceutical company with dual origins in the USA and Switzerland. Monsanto itself began as a chemical firm in St. Louis, Missouri, in 1901. The firm expanded its production of herbicides post-1945 so that today glyphosate herbicides like Roundup account for around one-sixth of the company's annual sales and half the operating income (Tokar, 2001). It produces a range of biotechnology products including bovine growth hormone and various GM crops. The latter are closely linked to its Roundup product via the development of herbicide-tolerant plants, notably 'Roundup ready' soybeans. The company claims that this will ultimately reduce herbicide use whilst others assert that herbicide-tolerant crop varieties are more likely to increase farmers' dependence on herbicides. Increased use of herbicides may occur if weeds that emerge once the original herbicide has dispersed are treated with further applications of herbicides.

Genetic engineering technology involves the insertion of genes with known characteristics and/or products into a strain of plant or animal previously lacking the desired trait. For genetically engineered food crops, the introduced genes most frequently have included those that make new or higher amounts of essential amino acids, produce certain enzymes used in photosynthesis, or make proteins that confer resistance to pests. This technology has built upon pioneering work in the 1920s by a British medical officer, Frederick Griffith, and

in the 1940s by Oswald Avery and co-workers at the Rockefeller Institute in New York, and has utilised the knowledge of DNA gained in the 1950s by Nobel laureates James Watson and Francis Crick. The successful insertion of novel genes into a plant host transforms this host into a GM organism and it is said to have been genetically engineered.

This is an 'advance' upon the plant breeding methods employed during the Green Revolution of the 1950s and 1960s when new strains of wheat and rice were developed producing dwarf crops that, under favourable growing conditions, gave high yields. Amongst the key developments of genetic modification have been successful introduction of genes that create tolerance for herbicides, the creation of inactive plant genes thereby removing undesirable characteristics, and the introduction of transforming agents as plant viruses to create a desirable product, e.g. making the plant more palatable, nutritious or combative against disease.

Another definition of 'genetically modified' refers to food produced from plants or animals that have had their genes changed by scientists in the laboratory rather than farmers in the field. There is, though, an essential continuity of current plant and animal manipulations with the thousands of years of selective breeding. Ten thousand years ago the first domestications involved deliberate selections and breeding of wild plants and animals, which lead to genetic changes (Mannion, 1999). Modern genetic engineering is able to go much further than conventional breeding and operates much faster. Extending 'traditional' plant and animal breeding techniques through GM means that scientists can now introduce virtually any gene into any organism in attempts to generate desirable properties. For example, one of the early examples of GM foods was the FlavrSavr tomato. This had its ripening gene modified so that the tomato softens more slowly and so stays longer on the vine to develop a fuller taste. It has a longer shelf life and so may reduce wastage. Also developed at the same time in the early 1990s were Monsanto's Roundup Ready soybeans. These were genetically modified to survive being sprayed with Roundup herbicide that is applied to a field to kill weeds. These two examples are significantly different from 'traditional' selective breeding methods because they have been created by new technology that permits scientists to isolate, move and modify genes, enabling manipulation of the 'blueprint' of any living organism in a precise and detailed fashion.

A problematic characteristic of the development of GM crops has been the dominance of just a few major chemical corporations, including Dow, Du Pont and Monsanto. These have expanded their markets by creating crops that will become dependent on the chemicals developed by their own company. A good example is Monsanto's creation of plants to be resistant to their high-selling Roundup herbicide. These plants are intended to produce crops requiring less reliance on pesticides and with greater intrinsic disease resistance. However, there has tended to be a dominance of new transgenic crops engineered for herbicide tolerance rather than for any intrinsic improvement in crop food quality or pest resistance (Lappé and Bailey, 1999, p. 35). Indeed, two-thirds of all transgenic food crops to date are being engineered for herbicide tolerance. Whilst this may be problematic, a broader concern relates to the fundamental difference between bio-engineered and conventional crops. Both the old and new methods can alter the genetic composition of crops, but only 'traditional' plant breeding methods ensure a degree of uniformity from generation to generation. In contrast, there is no assurance of this occurring with genetic engineering technology. Hence, for example, a field of cotton with genes resistant to a particular toxin, may enable only resistant plants to survive to reproduce. This type of highly undesirable potential outcome is one of the central arguments made by those concerned that there has been insufficient consideration of the long-term consequences of this technology.

This concern has failed to prevent the rapid spread of certain GM crops. For example, Monsanto claimed that by early 1998 half of the American grain industry was using genetically engineered seed. To increase the rate and extent of spread of this seed, Monsanto has acquired or controls eight of the major seed companies: Asgrow, Delta and Pine Land, DeKalb, Gargiulo

Tomato, Hartz, Holden, Naturemark, and Stoneville Pedigreed.

Another vital element in the rapid spread of GM technology is the relationship between farmers and the company. Lappé and Bailey (1999, p. 39) fear that many farming operations will become almost totally dependent on Monsanto for both the chemicals and seed needed each year. Another concern is that, with pesticide-secreting crops, the presence of a bacterial toxin (intended to control damage from pests) throughout a plant's life-cycle may promote the development of resistant strains of the pests. Hence the US Environmental Protection Agency (EPA) now requires US cotton growers who use GM varieties secreting a toxin derived from *Bacillus thuringiensis* (Bt) to plant refuges of up to 40 per cent non-Bt cotton to prevent natural applications of Bt bacteria becoming ineffective as widespread resistance to Bt occurs.

Many contend that genetic engineering marks a revolutionary point of departure for agriculture whereby 'traditional' selection and growing methods are giving way to 'fast-track' programmes developed by the biotechnology industry. The new technology is turning some farmers away from integrated pest management and plant protein improvement programmes to reliance on chemical treatments. The new technology is permitting substitution of chemicals for labour and the potential for the same or higher yields at a lower cost.

To date, genetic engineeering technology has not showed signs of redressing the problems of the Green Revolution, in that it requires an expensive infrastructure to support the transgenetic crops. Moreover, so far, few of the engineered crops have achieved significant increases in yield. Indeed, future increases in the development of HYVs seem more likely to arise from traditional breeding methods rather than transgenics.

Some advocates of GM crops argue that they can be a means to reduce intensity of agrichemical usage. There are three main areas of concern regarding these arguments (Tilzey, 2000, p. 287):

- If crops are guaranteed to be herbicide- and insect-resistant, farmers can spray more broad-spectrum weed killers like Roundup, with the consequent eradication of all non-crop infield and field-edge flora. This is likely to further reduce the variety of farmland wildlife by elimination of wild plants and serious reductions in farmland insects and birds.

- GM organisms may inter-breed with wild relatives to produce aggressive herbicide-resistant super-weeds. Evidence of contamination, up to a distance of 200 m, between genes from GM crops and other crops, including weeds, is emerging. This is prompting fears that some weeds could become resistant to herbicides.
- If GM seed passes on insect resistance to wild cousins then insects dependent on wild plants might be denied their only food source. This may impact upon organisms higher up the food chain. GM technology is likely to further reinforce agricultural specialisation and dependence of farmers upon external inputs, and therefore upon the agro-chemical companies who supply the package of GM seeds and pesticide.

With public concern over GM crops mounting, in October 1998 the UK government introduced a year-long ban on the introduction of all GM commercial crops. This halted development on several herbicide-resistant crops until further research on their ecological effects was completed. Meanwhile there was a voluntary three-year ban on insecticidal GM crops by the biotechnology industry. These crops have been shown to kill beneficial insects such as lacewings. Concerns have been expressed in various quarters regarding the effects of GM crops on the declining populations of farmland birds, such as the grey partridge, linnet and sky-lark, which could be affected by the plants' eradication of weeds and insects or through the GM crops interbreeding with wild species to become 'superweeds' or poison wild species. Nevertheless, the UK government seems broadly in favour of proceeding to commercial scale development of GM crops, and hence restrictions on GM crop production in the UK are likely to be scaled down in the near future.

In summary, the debates relating to GM foods include ethical and environmental considerations in addition to concerns over the role of companies developing GM crops and the outcomes of the differential adoption of GM methods that may further exacerbate differences between the Developed and Developing Worlds. The environmental concerns relate to the unknown 'knock-on' effects of planting GM crops. For example, by continually exposing pests to a toxin (in a GM crop) the pest may eventually develop a resistance to the toxin. GM plants or micro-organisms may have a competitive advantage that could disturb natural ecology if they were released into the wild. Hence there are concerns that pollen from GM crops may infiltrate plants growing nearby to produce 'super-weeds' that have a similar resistance to pests or diseases as the original GM crop. There may be unforeseen consequences comparable with the impacts of exotic insects and other organisms upon environments in which they may compete effectively with native species to the detriment of the latter. It has been shown that pollen can travel up to 20 times further than the 50 to 200 m 'exclusion zones' placed around GM trial crops in the UK. One control on possible cross-pollination from GM crops is to use GM genes that produce sterile seeds. However, there are both doubts about the effectiveness of this technology and concerns that this would mean that farmers would have to purchase fresh seeds each year, so giving greater power to the company selling the GM seeds. The role of GM crops in contributing to further diminutions in genetic diversity is another concern, especially as human modifications of the environment are rapidly diminishing that diversity.

10.6 Conclusion

This book was completed as the first trappings of the relentless march of another season of Christmas festivities appeared in the UK's shops. One of the first headlines about food shopping at Christmas referred to a central element in the multi-faceted 'farm crisis' that has been analysed throughout the book. Consumers were warned that their Christmas turkey would be produced from a rapidly dwindling band of turkey growers. Several newspapers reported that the number of growers had been cut in half in just five years, largely because an increasing proportion of the market was being taken by supermarkets who only purchased turkeys through contracts with large suppliers. Further evidence of this trend was available from the long-running radio 'soap' of village life, 'The Archers', where Ambridge's small producer of turkeys for the Christmas market, the Grundys of Grange Farm, were no longer in production having lost their farm through inability to meet their monthly rental.

Throughout the Developed World the fate of small turkey-producers has been shared by many small farmers, greatly reducing the overall number of farms and leading some observers to refer to the process as the demise of family farming. However, as the multi-faceted research of agricultural geographers has shown, this proposed outcome is an over-simplification of a much more complex set of restructuring processes. These have been associated with distinctive geographical outcomes including greater regional specialisation, local characteristics of global phenomena, and a mixture of new and familiar geopolitical frameworks affecting the whole agri-food system.

The contribution of geographers to the understanding of these processes has been greatly enhanced in the past two decades by the emergence of research extending across the breadth of the agri-food system rather than just being concerned with farm-based production. This broader focus has been accompanied by significant theoretical advances in the form of engagement with regulation theory, food regimes and new considerations of the agrarian question. Sweeping generalisations within this theorising, notably the proposed productivist-post-productivist transition, have been subjected to closer critical scrutiny in recent years, with two distinctive strands of thinking apparent. Dominant in the early 1990s, political economy emphasised structural controls on production and has provided insights to the operation of globalising processes. However, the diversity of production and consumption practices, often dependent on distinctive individual decision-making, restricts the explanatory power of structural forces and

has encouraged work looking beyond structural controls. This work has moved past the traditional approaches from behavioural geography, introducing new methods, e.g. actor-network theory, and an increasingly significant cultural dimension closely mirroring the post-modernist 'cultural turn' seen elsewhere in the social sciences. The examination of agri-food networks has been an important element in this research as has research dealing explicitly with the prevailing political economy linked to macro-level regulators of policy such as the EU's CAP.

The combination of different theories and methods has enabled agricultural geography to enter a new phase, dominated by concerns over globalising tendencies and with a focus extending well beyond the farmgate. Nevertheless, particular attention has been given to certain farm-based trends, especially the contrast between increased specialisation and diversification. In both cases, rather than the traditional mapping of spatial distribution patterns, it has been the examination of both internal and external causal factors that has gained the most attention. These factors have ranged from micro-level factors operating within the farm household to the role of TNCs and international trade agreements. These international and trade dimensions have then linked research on globalisation and restructuring in the Developed World to the experiences of the Developing World. Some of the contrasts between North and South could not be more stark.

To emphasise the differentials between North and South this book deliberately referred to the 'other side' of globalisation and focused on the continuing presence of malnutrition and starvation in parts of the tropics. Again, it should be emphasised that our understanding of the North-South divide has been enhanced by consideration of processes affecting the entire agri-food chain and not just analysis operating at the farm level. Nevertheless, it is clear from the growing body of research on agriculture in sub-Saharan Africa that a better understanding has been obtained of how subsistence production operates and how farmers respond to the opportunities presented by the commercial market. In developing this better understanding there have been a variety of studies utilising theories from different philosophies and applications at different spatial scales. Ultimately, though, the significance emerges of Yarwood's (2002, p. 13) assertion that 'decisions about which crops to grow or which farming practices to follow are influenced more by government . . . policy than local farming conditions or local market forces'. He was writing in the context of the EU, but his remarks could equally well apply to the tropics and to the multi-faceted impress of 'government': local, national, bi-lateral and multi-lateral trade agreements, the WTO and the IMF, TNCs and aid agencies. It is the complex interplay between these regulators of the agri-food system that can be seen to have shaped the trajectory and impacts of the so-called 'Green Revolution' as one of the prime solutions to the world food problem. More local regulators have also been discussed here in the form of land reforms, widely applied in Latin America for example.

Much of the work done on the impacts of globalisation and restructuring on the agri-food system has concentrated upon the impacts in the Developed World. It tends to be a different set of researchers who look at the impacts in the Developing World, and this has meant that separate worlds of academia correspond closely to their separate geographical foci. For example, African specialists look closely at issues of hunger and starvation and the production end of agri-export systems but the consumption end is largely the province of a different specialist. Moreover, relatively few agricultural geographers have applied themselves to farming systems in the Developing World – if they do, then they wear another 'hat' and call themselves development specialists, generally drawing upon not only different literature but also different theories from researchers looking at agriculture in the North. This disjuncture between various academic camps is apparent when considering the topics tackled here in Chapters 7 and 8, bringing together literature on some of the impacts of globalisation in Developing Countries and then examining aspects of the 'crises' of hunger and starvation there.

In addressing the 'other side' of globalisation the limited space did not permit an exhaustive consideration of development theory, but instead the focus was placed explicitly at farm-level responses to the ever increasing impress of commercial forces. These responses can be seen in the decline in subsistence production and the growing importance of the duality between systems of production geared to export and those focused on subsistence or for the local market. This duality can be traced to the impact of colonialism upon indigenous farming systems, but today the effect of colonialism's successor, globalisation, is multi-faceted and complex. Some of the main ramifications are addressed here citing the example of the so-called banana wars between the USA and the EU. Other aspects of the 'other side' to be addressed were the changing role of women in tropical agriculture, the so-called agricultural transition and the expansion of exports of high-value foods from the tropics to the rich North.

The last two chapters of the book dealt with two broad topics that have played a central role in agricultural geography: one with a long history – land use competition, and one with a huge surge of interest in recent years – environmental dimensions of agricultural production. The former addressed issues relating to the protection of agricultural land and the changing role of agriculture in society, both in terms of the diminishing proportion of the population in the Developed World living on the land and the need for the vast and rapidly growing cities of the Developing World to have their food produced close to the point of consumption. These considerations encompass the changing societal attitudes to agriculture, which were then pursued in some detail in the concluding chapter.

Returning to the Christmas turkey production, there is growing evidence that some parts of society care not only about the cost of that turkey but also how it was produced and what impact upon the environment the associated production system had. So it is appropriate that the final chapter dealt with the growing debates centred on issues of sustainability, the promotion of environmentally friendly farming, and the conflicts between ecocentric views of the future and technocentric views typified by those championing the development of genetically modified (GM) foods. The sustainability discourse has permeated thinking across a wide range of issues, but has become especially significant in ongoing debates about the future of agriculture and the policies required to meet multiple demands, especially in terms of production needs versus environmental protection.

Different aspects of the sustainability discourse were examined in this book, notably the attempts to define and recognise sustainable agriculture, the growth of organic production, the promotion of extensification, the growing concern about the health implications of industrial-style farming and food processing, and the debate relating to GM foods. These issues were left to last, not because they were deemed to be the least important, but quite the opposite, they are the issues that seem to be most likely to be of greatest public concern to the general public of the Developed World at the start of the twenty-first century. They are issues for which crucial questions remain to be addressed and which perhaps will form the basis for close consideration by government and academe alike (Robinson and Harris, 2003):

- Is it sensible to ship food such long distances (incurring large amounts of so-called 'food miles') if it can be produced locally?
- Have the full environmental implications (e.g. transport, pollution tax) been taken into consideration when calculating prices?
- On the other hand, is it sensible to subsidise farmers in Europe and North America (arguably, in part, to maintain the appearance of the landscape and countryside) if food can be produced more cheaply in other parts of the world?
- Is it more environmentally friendly to consume locally (industrially) produced foodstuffs, or imported organically produced foods?
- Are there standards (organic production, ethical trade, labour conditions, environmental effects) that should be adhered to in food production, at whatever cost to the consumer, or is low-cost food the priority, at whatever environmental or social cost?

References

Abraham J (1991) *Food and development: the political economy of hunger*. Kogan Page Ltd, London

Adams W M, Mortimore M (1996) Farmers, risk and environment in Northeast Nigeria. *Geography* **81**: 400–3

Adams W M, Mortimore M J (1997) Agricultural intensification and flexibility in Nigerian Sahel. *Geographical Journal* **163**: 150–60

Adeemy M S (1968) Types of farming in North Wales. *Journal of Agricultural Economics* **19**: 301–16

Adger W N, Whitby M (1993) Natural resource accounting in the land-use sector: theory and practice. *European Review of Agricultural Economics* **20**: 77–97

Adger W N, Pettenella D, Whitby M (eds) (1997) *Climate change mitigation and European land use policies*. CAB International, Wallingford

AFT (American Farmland Trust) (1990) *Saving the farm: a handbook for conserving agricultural land*. AFT Western Office, Davis, CA

Aglietta M (1974) *A theory of capitalist regulation*. New Left Books, London

Agocs P, Agocs S (1994) Too little, too late: the agricultural policy of Hungary's post-Communist government. *Journal of Rural Studies* **11**: 117–30

Aitchison J W (1992) Farm types and agricultural regions. In: Bowler I R (ed.) *The geography of agriculture in developed market economies*. Longman, London, pp. 109–33

Aitchison J W, Aubrey P (1982) Part-time farming in Wales: a typological study. *Transactions of the Institute of British Geographers* new series, **7**: 88–97

Alamgir M (1982) *Famine in south Asia*. Oxford University Press, Oxford

Alamgir M, Arora P (1991) *Providing food security for all*. New York University Press, New York

Alanen I (1999) Agricultural policy and the struggle over the destiny of collective farms in Estonia. *Sociologia Ruralis* **39**: 431–58

Albrecht D E, Murdock S H (1984) Toward a human ecological perspective on part-time farming. *Rural Sociology* **49**: 389–411

Alexandratos N (ed.) (1995) *World agriculture: towards 2010*. FAO, Rome and Wiley, Chichester and New York

Allan W (1965) *The African husbandman*. Oliver and Boyd, Edinburgh

Allanson P, Murdoch J, Garrod G, Lowe P (1995) Sustainability and the rural economy: an evolutionary perspective. *Environment and Planning A* **27**: 1797–814

Allen B J, Ballard C (2001) Beyond intensification? Reconsidering agricultural transformations. *Asia Pacific Viewpoint* **42**: 157–62

Allen P, van Dusen D, Lundy L, Gliessman S (1991) Integrating social, environmental and economic issues in sustainable agriculture. *American Journal of Alternative Agriculture* **6**: 34–9

Allen J, Thompson G (1997) Think global, then think again – economic globalization in context. *Area* **29**: 213–27

Alston M (ed.) (1991) *Family farming – Australia and New Zealand*. Centre for Rural Social Research, School of Humanities and Social Sciences, Charles Stuart University, Riverina

Alston M (1995a) Women and their work on Australian farms. *Rural Sociology* **60**: 21–32

Alston M (1995b) *Women on the land: the hidden heart of rural Australia*. University of New South Wales Press, Sydney

Alston M, Wilkinson J (1998) Australian farm women – shut out or fenced in? The lack of women in agricultural leadership. *Sociologia Ruralis* 38: 391–408

Altieri M A (1987) *Agroecology: the scientific basis of alternative agriculture.* Westview Press, Boulder, CO

Amin A, Thrift N J (eds) (1994) *Globalization, institutions and regional development in Europe.* Oxford University Press, Oxford

Anderson K (1997) A walk on the wild side: a critical geography of domestication. *Progress in Human Geography* 21: 463–85

Ansell D J, Tranter R B (1992) *Set aside: in theory and practice.* Dept. of Agricultural Economics and Management, Centre for Agricultural Strategy, University of Reading

Appadurai A (1984) How moral is south Asia's economy? – a review article. *Journal of South Asian Studies* 43: 481–97

Appadurai A (1986) Introduction: commodification and the politics of value. In: Appadurai A (ed.) *The social life of things.* Cambridge University Press, Cambridge, pp. 1–15

Arce A, Marsden T K (1993) The social construction of international food: a new research decade. *Economic Geography* 69: 293–311

Archer J C, Lonsdale R E (1997) Geographical aspects of US farmland values and changes during the 1978–1992 period. *Journal of Rural Studies* 13: 399–413

Argent N (1997) Rural crises and local farm/non-farm business linkages: a case study of Kangaroo Island. *South Australian Geographical Journal* 96: 3–19

Arkell T (1991) The decline of pastoral nomadism in the Western Sahara. *Geography* 76: 162–6

Arkleton Trust (1992) *Rural change in Europe: farm structures and pluriactivity. Final report to the European Commission.* Arkleton Trust, Nethy Bridge, Inverness-shire

Ash R, Edmonds R L (1998) Land resources, environment and agricultural production. *The China Quarterly* 156: 836–79

Atkins P J (1987) The charmed circle: von Thunen and agriculture around nineteenth century London. *Geography* 72: 129–39

Atkins P J, Bowler I R (2001) *Food and society: economy, culture, geography.* Arnold, London

Austin R B (1978) Actual and potential yields of wheat and barley in the United Kingdom. *ADAS Quarterly Review* 29: 76–87

Avery B W (1969) Problems of soil classification. In: Sheals J G (ed.) *The soil ecosystem.* Systematics Association, London, pp. 9–17

Avery B W (1973) Soil classification in the Soil Survey of England and Wales. *Journal of Soil Science* 24: 324–38

Avery B W (1980) Soil classification for England and Wales. *Soil Survey Technical Monographs,* no. 14

Avery B W (1990) *Soils in the British Isles.* Soil Survey, Harpenden

BAA (British Agrochemicals Association) (1996) *Integrated crop management.* BAA, Peterborough

Baethgen W, Magrin G (1995) Assessing the impacts of winter crop production in Uruguay and Argentina using crop simulation models. In: Rosenzweig C, Allen L, Harper L, Hollinger S, Jones J (eds) (1995) *Climate change and agriculture: analysis of potential international impacts.* American Society of Agronomy Special Publications, no. 59, pp. 207–28

Baker O E (1926) Agricultural regions of North America. *Economic Geography* 2: 459–94

Baker B B, Hanson J D, Bourdon R M, Eckert J B (1993) The potential effects of climate change on ecosystem pressures and cattle production on US rangelands. *Climatic Change* 25: 97–107

Baldock D, Conder D (eds) (1987) *Removing land from agriculture.* CPRE, Institute for European Environmental Policy

Baldock D, Cox C, Lowe P, Winter M (1990) Environmentally Sensitive Areas: incrementalism or reform? *Journal of Rural Studies* 6: 143–62

Baldock D, Mitchell K, vonMeyer H, Beaufoy G (1998) *Assessment of the environmental impact of certain agricultural measures.* Institute for European Environmental Policy, London

Baldwin M, Kellogg C E, Thorp J (1938) *Soil classification, Soils and men: United States Department of Agriculture Yearbook,* 979–1001

Banks J, Marsden T K (2000) Integrating agri-enviromental policy, farming systems and rural development, Tir Cymen in Wales. *Sociologia Ruralis* 40: 466–80

Banse M, Munch W, Tangermann S (2000) The implications of European Union accession for Central and Eastern European agricultural markets, trade, government budgets and the macroeconomy. In: Hartell J G, Swinnen J F M (eds) *Agriculture and East-West European integration.* Ashgate, Aldershot, pp. 1–32

Barlett P (1993) *American dreams, rural realities.* University of North Carolina Press, Chapel Hill

Barnes T J (1990) Analytical political economy: a geographical introduction. *Environment and Planning A* **22**: 993–1006

Barnett V, Payne R (1995) *Agricultural sustainability.* Wiley, Chichester

Barr C, Benefield C, Bunce B, Rinsdale H, Whitaker M (1986) *Landscape changes in Britain.* Institute of Terrestrial Ecology, Grange-over-Sands

Barrett H R (1995) Women in Africa: the neglected dimension in development. *Geography* **80**: 215–24

Barrett H R, Browne A (1991) Environmental and economic sustainability: women's horticultural production in the Gambia. *Geography* **76**: 241–8

Barrett H R, Browne A (1996) Export horticultural production in Sub-Saharan Africa: the incorporation of The Gambia. *Geography* **81**: 47–56

Barrett H R, Ilbery B W, Browne A W, Binns A (1999) Globalization and the changing networks of food supply: the importation of fresh horticultural produce from Kenya into the UK. *Transactions of the Institute of British Geographers* new series, **24**: 159–74

Bartlett C A, Ghoshal S (1989) *Managing across borders: the transnational solution.* Harvard Business School Press, Boston

Bascom J (2000) Revisiting the rural revolution in East Carolina. *Geographical Review* **90**: 432–45

Bateman D, Ray C (1994) Farm pluriactivity and rural policy: some evidence from Wales. *Journal of Rural Studies* **10**: 1–13

Batterbury S P J (1997) Alternative affiliations and the politics of overseas field research: some reflections. In: Robson E, Willis K (eds) *Fieldwork in developing countries: a rough guide.* Developing Areas Research Group, Institute of

British Geographers, London, monograph no. 9, pp. 85–112

Batterbury S P J (2001) Landscapes of diversity: a local political ecology of livelihood diversification in south-west Niger. *Ecumene* **6**: 437–64

Batterbury S P J, Bebbington A J (1999) Environmental histories, access to resources and landscape change: an introduction. *Land Degradation and Development* **10**: 247–61

Battershill M R J, Gilg A W (1996) Traditional farming and agro-environment policy in South-west England: back to the future? *Geoforum* **27**: 133–47

Battershill M R J, Gilg A W (1997) Socio-economic constraints and environmentally-friendly farming in the south-west of England. *Journal of Rural Studies* **7**: 91–7

Battershill M R J, Gilg A W (1998) Traditional low-intensity farming: evidence of the role of Vente Directe in supporting such farms in north-west France and some implications for conservation policy. *Journal of Rural Studies* **14**: 475–86

Bayliss-Smith T P (1982) *The ecology of agricultural systems.* Cambridge University Press, Cambridge

Bayliss-Smith T P (1984) Energy flows and agrarian change in Karnataka: the Green Revolution at micro-scale. In: Bayliss-Smith T P (ed.) *Understanding Green Revolutions.* Cambridge University Press, Cambridge, pp. 153–74

Beach E D, Boyd R, Uri N D (1995) Eliminating direct payments to farmers in the US: the impact on land values. *Regional Science Perspectives* **25**: 30–57

Beard N, Swinbank A (2001) Decoupled payments to facilitate CAP reform. *Food Policy* **26**: 121–46

Bebbington A J (1999) Capitals and capabilities: a framework for analyzing peasant viability, rural livelihoods and poverty. World Development **27**: 2021–44

Bebbington A J, Batterbury S P J (2001) Transnational livelihoods and landscapes: political ecologies of globalization. *Ecumene* **8**: 369–80

Beckerman S (1987) Swidden in Amazonia and the Amazon Rim. In: Turner B L, II, Brush S B (eds) (1987) *Comparative farming systems.* Guilford Press, New York, pp. 55–94

Bedenbaugh E J (1988) History of cropland set aside programs in the Great Plains. In: Mitchell J E (ed.) *Impacts of the Conservation Reserve Program in the Great Plains.* USDA Forest Service, Rocky Mountain Forest and Range Experiment Station, Fort Collins, Colorado, no. GTR RM-158, pp. 14–17

Bee A (2000) Globalization, grapes and gender: women's work in traditional and agro-export production in northern Chile. *Geographical Journal* **166**: 255–65

Beedell J C D, Rheman T (1999) Explaining farmers' conservation behaviour: why do farmers behave the way they do? *Journal of Environmental Management* **57**: 165–76

Bell D, Valentine G (1997) *Consuming geographies: we are what we eat.* Routledge, London

Benbrook C M (1990) Society's stake in sustainable agriculture. In: Edwards C A, Lal R, Madden P, Miller R H, House G (eds) *Sustainable agricultural systems.* Soil and Water Conservation Society, Arkeny, Iowa, pp. 37–58

Benjamin C (1994) The growing importance of diversification activities for French farm households. *Journal of Rural Studies* **10**: 331–42

Benneh G (1972) The response of farmers in northern Ghana to the introduction of mixed farming: a case study. *Geografiska Annaler* 5B, supplement **2**: 99–103

Benson V W, Witzig T J (1977) The chicken broiler industry: structure, practices, costs. *Agricultural Economics Reports*, 381. USDA, Washington DC

Bentall J, Corbridge S (1996) Urban-rural relations, demand politics and the 'new agrarianism' in northwest India: the Bharatiya Kisan Union. *Transactions of the Institute of British Geographers* new series, **21**: 27–48

Benton J (1999) *Agrarian reform in theory and practice: a study of the Lake Titicaca region of Bolivia.* Ashgate, Aldershot

Benton T (2001) One more symptom: the foot and mouth crisis in Britain. *Radical Philosophy* **110**: 7–11

Berlan-Darque M, Gasson R M (1991) Changing gender relations in agriculture: an inter-national perspective. *Journal of Rural Studies* **7**: 1–2

Berger A (2001) Agricultural development and land concentration in a central European country: a case study of Hungary. *Land Use Policy* **18**: 259–68

Bernstein H (1996) Agrarian questions then and now. *Journal of Peasant Studies* **24**: 22–49

Bernstein H, Crow B, Mackintosh M, Martin C (eds) (1990) *The food question: profits versus people.* Earthscan, London

Berry B J L (1976) The counter-urbanization process: Urban America since 1970. In: Berry B J L (ed.) *Urbanization and counter-urbanization.* Sage Publications, Beverly Hills, pp. 62–76

Berry B J L (1980) The urban problem. In: Woodruff A M (ed.) *The farm and the city: rivals or allies?* Prentice-Hall, Englewood Cliffs, NJ, pp. 37–59

Berry S (199?) Social institutions and access to resources. *Africa* **59**: 41–55

Berry W (1981) *The gift of good land: further essays cultural and agricultural.* North Point Press, San Francisco

Bessiere J (1998) Local development and heritage: traditional food and cuisine as tourist attractions in rural areas. *Sociologia Ruralis* **38**: 21–34

Bhatia B M (1991) *Famines in India.* Konark Publishers, Delhi

Bidgood M (1994) *A study of actions by Simcoe County Environmental Farm Plan participants.* Ontario Soil and Crop Improvement Association, Guelph, Ontario

Bignal E M, McCracken D I (1996) Low-intensity farming systems in the conservation of the countryside. *Journal of Applied Ecology* **33**: 413–24

Billing P (1996) Towards sustainable agriculture – the perspective of the Common Agricultural Policy in the European Union. Unpublished paper presented to the Workshop on Landscape and Nature Conservation, Stuttgart, 26.9.96

Binfield J C R, Hennessy T C (2001) Beef sector re-structuring after Agenda 2000: an Irish example. *Food Policy* **26**: 281–95

Binns A (1996) Feeding Africa's urban poor. *Geography* **81**: 380–4

Binns A, Fereday N (1996) Feeding Africa's urban poor: urban and peri-urban horticulture in Kano, Nigeria. *Geography* 81: 380–4

Binns A, Lynch K D (1998) Feeding Africa's growing cities into the 21st century: the potential of urban agriculture. *Journal of International Development* 10: 777–93

Binswanger H P, Townsend R F (2000) The growth performance of agriculture in Sub-Saharan Africa. *American Journal of Agricultural Economics* 82: 1075–86

Birch J W (1954) Observations on the delimitation of farming-type regions: with special reference to the Isle of Man. *Transactions of the Institute of British Geographers* 20: 141–58

Birch-Thomsen T, Fog B (1996) Changes within small-scale agriculture. A case study from south-western Tanzania. *Danish Journal of Geography* 96: 60–69

Bishop K, Phillips A (1993) Seven steps to the market – the development of the market-led approach to countryside conservation and recreation. *Journal of Rural Studies* 9: 315–38

Bjerke K (1991) An overview of the Agricultural Resources Conservation Program. In: Joyce L A, Mitchell J E, Skold M D (eds) (1991) *The Conservation Reserve – yesterday, today and tomorrow*. USDA Forest Service, Rocky Mountain Forest and Range Experiment Station, Fort Collins, Colorado, no. GTR RM-203, pp. 7–10

Blaikie P M (1971) Spatial organization of agriculture in some north Indian villages. *Transactions of the Institute of British Geographers* 52: 1–40

Blaikie P M (1985) *The political economy of soil erosion in developing countries*. Longman, London

Blaikie P M (1989a) Explanation and policy in land degradation and rehabilitation for developing countries. *Land Degradation and Rehabilitation* 1: 23–37

Blaikie P M (1989b) Environment and access to resources in Africa. *Africa* 59: 18–40

Blaikie P M (1995) Changing environments or changing views? A political ecology for Developing Countries. *Geography* 80: 203–14

Blaikie P M, Brookfield H (1987) *Land degradation and society*. Methuen, London

Blair A M (1987) Future landscapes on the rural-urban fringe. In: Lockhart D, Ilbery B W (eds) *The future of the British countryside*. Geo Abstracts, Norwich

Blakemore W F (1989) Bovine Spongiform Encephalopathy and scrapie: potential human hazards. *Outlook on Agriculture* 18: 165–8

Blanc M, MacKinnon N (1990) Gender relations and the family farm in Western Europe. *Journal of Rural Studies* 7: 91–7

Blanc M, Perrier-Cornet P (1993) Farm transfer and farm entry in the EC. *Sociologia Ruralis* 33: 319–35

Blandford D, De Gorter H, Harvey D (1989) Farm income support with minimal trade distortions. *Food Policy* 14: 268–73

Blasing T, Solomon A (1983) *Responses of the North American Corn Belt to climatic warming*. Carbon Dioxide Research Division, US Department of Energy, Washington DC, no. DOE/N88-004

Blunden G, Moran W, Bradly A (1997) 'Archaic' relations of production in modern agricultural systems: the example of sharemilking in New Zealand. *Environment and Planning A* 29: 1759–76

Blunden J (1977) Rural land use, Unit 15 in Open University. *Fundamentals of human geography: section II area patterns*. Open University Press, Milton Keynes

Blundy D, Vallely P (1985) *With Geldof in Africa: confronting the famine crisis*. Times Books, London

Boag B, Lawson H, Neilson R, Wright G (1994) Observations on the diversity of soil nematode fauna and weed seedbanks under different set-aside management regimes. *Aspects of Applied Biology* 40: 443–52

Board C (1965) The use of air photographs in land-use studies in South Africa and adjacent territories. *Photogrammetrica* 20: 163–70

Boardman J (1992) Agriculture and erosion in Britain. *Geography Review* 6(1): 15–19

Bohle H-G (1990) Twenty years of the 'Green Revolution' in India. *Applied Geography and Development* 37: 21–38

Bonanno A (1993) The agro-food sector and the transnational state: the case of the EC. *Political Geography* **12**: 341–60

Bonanno A, Busch L, Friedland W, Gouveia L, Mingione E (eds) (1994) *From Columbus to ConAgra: the globalization of agriculture and food.* University Press of Kansas, Lawrence

Bootsma A, Blackburn W, Stewart R, Muma R, Dumananski J (1984) Possible effects of climatic change on estimated crop yields in Canada. *LRRI Contributions, Research Branch, Agriculture Canada*, no. 83–64

Borrell B (1994) EU Bananarama III. *Policy Research Working Papers, World Bank, Washington DC*, no. 1386

Boserup E (1965) *The conditions of agricultural growth: the economics of agrarian change under population pressure.* Aldine Press, New York

Boserup E (1981) *Population and technological change – a study of long-term trends.* University of Chicago Press, Chicago

Boulay A, Robinson G M (2001) Farm diversification: lessons from dairying areas, unpublished paper presented to the Restless Ruralities Conference, University of Coventry, 5–6 July

Bouquet M (1982) Production and reproduction of family farms in south-west England. *Sociologia Ruralis* **22**: 227–44

Bourn D, Wint W (1994) Livestock, land use and agricultural intensification in sub-Saharan Africa. *Pastoral Development Network Papers, Overseas Development Institute, London*, no. 37a

Bowden L W (1965) *Diffusion of the decision to irrigate.* Research Papers, Department of Geography, University of Chicago, no. 97

Bowler I R (1979) *Government and agriculture: a spatial perspective.* Longman, London

Bowler I R (1985a) Some consequences of the industrialisation of agriculture in the European Community. In: Healey M J, Ilbery B W (eds) *The industrialisation of the countryside.* GeoBooks, Norwich, pp. 75–98

Bowler I R (1985b) *Agriculture under the Common Agricultural Policy, a geography.* Manchester University Press, Manchester

Bowler I R (1986) Intensification, concentration and specialization in agriculture: the case of the European Community. *Geography* **71**: 14–21

Bowler I R (1987) The geography of agriculture under the CAP. *Progress in Human Geography* **11**: 24–40

Bowler I R (1992a) The industrialization of agriculture. In: Bowler I R (ed.) *The geography of agriculture in developed market economies.* Longman, London, pp. 7–31

Bowler I R (1992b) 'Sustainable agriculture' as an alternative path of farm business development. In: Bowler I R, Bryant C R, Nellis D (eds) *Rural systems in transition: agriculture and environment.* CAB International, Wallingford, pp. 237–53

Bowler I R (ed.) (1992c) *The geography of agriculture in developed market economies.* Longman, London

Bowler I R (1992d) The agricultural significance of farm size and land tenure. In: Bowler I R (ed.) *The geography of agriculture in developed market economies.* Longman, London, pp. 85–108

Bowler I R (1994) The institutional regulation of uneven development: the case of poultry production in the province of Ontario. *Transactions of the Institute of British Geographers* new series, **19**: 346–58

Bowler I R (1996) *Update: Agricultural change in developed countries.* Cambridge University Press, Cambridge

Bowler I R (1999a) Recycling urban waste on farmland: an actor-network interpretation. *Applied Geography* **19**: 29–44

Bowler I R (1999b) Modelling farm diversification in regions using expert and decision support systems. *Journal of Rural Studies* **15**: 297–306

Bowler I R (1999c) Endogenous agricultural development in Western Europe. *Tijdschrift voor Economische en Social Geografie* **90**: 260–71

Bowler I R (2001) Rural alternatives. In: Daniels P W, Bradshaw M, Shaw D, Sidaway J (eds) *Human Geography: issues for the 21st century.* Pearson Education, Harlow, pp. 128–53

Bowler I R (2002a) Developing sustainable agriculture. *Geography* **87**: 205–12

Bowler I R (2002b) Sustainable farming systems. In: Bowler I R, Bryant C R, Cocklin C (eds)

The sustainability of rural systems: geographical interpretations. Kluwer, Dordrecht, pp. 169–88

Bowler I R, Ilbery B W (1987) Redefining agricultural geography. *Area*, 27–32

Bowler I R, Ilbery B W (1992) *Farm diversification in England and Wales.* Occasional Papers, Department of Geography, University of Leicester, no. 21

Bowler I R, Bryant C R, Nellis D (eds) (1992) *Rural systems in transition: agriculture and environment.* CAB International, Wallingford

Bowler I R, Clark G, Crockett A, Ilbery B W, Shaw A (1996) The development of alternative farm enterprises: a case study of family labour farms in the north Pennines of England. *Journal of Rural Studies* 12: 285–95

Bowler I R, Bryant C R, Nellis M D (eds) (1997) *Contemporary rural systems in transition. Volume 1 Agriculture and Environment.* CAB International: Wallingford

Bowler I R, Bryant C R, Cocklin C (eds) (2002) *The sustainability of rural systems: geographical interpretations.* Kluwer, Dordrecht

Boyd D S, Foody G M, Curran P J, Lucas R M, Honzak M (1996) An assessment of radiance in Landsat TM middle and thermal infrared wavebands for the detection of tropical forest regeneration. *International Journal of Remote Sensing* 17: 249–61

Boyd W, Watts M J (1997) Agro-industrial just-in-time: the chicken industry and postwar American capitalism. In: Goodman D, Watts M J (eds) *Globalising food: agrarian questions and global restructuring.* Routledge, London and New York, pp. 192–225

Boyer R (1990) *The regulation school: a critical introduction.* Columbia University Press, New York

Boyer R (1991) The transformation of modern capitalism, by the light of the regulation approach and other political economy theories. *Cepremap Papers*, 9134, Paris

Boyer R, Drache D (1996) *States against markets: the limits of globalisation.* Routledge, London

Brandth B (1995) Rural masculinity in transition: gender images in tractor advertisements. *Journal of Rural Studies* 11: 123–33

Brandth B, Hauger M S (1997) Rural women, feminism and the politics of identity. *Sociologia Ruralis* 37: 325–44

Breathnach P (2000) The evolution of the spatial structure of the Irish dairy processing industry. *Irish Geography* 33: 166–84

Breathnach P, Kenney M (1997) The impact of Irish dairy industry rationalization on the sustainability of small farming communities. In: Byron R, Walsh J, Breathnach P (eds) *Sustainable development on the North Atlantic margin.* Ashgate, Aldershot, pp. 323–38

Brenner R, Glick M (1991) The regulation approach: theory and history. *New Left Review* **188**: 45–119

Brierley J S (1992) A study of land redistribution and the demise of Granada's estate farming system 1940–1988. *Journal of Rural Studies* 9: 67–84

Briggs D J, Courtney F M (1989) *Agriculture and environment: the physical geography of temperate agricultural systems.* Longman, Harlow

Brklacich M, Bryant C, Smit B (1990) Review and appraisal of concepts of sustainable food production systems. *Environmental Management* **15**: 1–14

Brklacich M, Curran P (1994) Climate change and agricultural potential in the Mackenzie Basin. In: Cohen S (ed.) *Mackenzie Basin impact study interim report no 2.* Proceedings of the 6th biennial AES/DIAND Meeting on northern climate, Environment Canada, Downsview, Ontario, pp. 459–64

Brklacich M, McNabb D, Bryant C, Dumanski J (1997) Adaptability of agricultural systems to global climate change: a Renfrew County, Ontario, Canada pilot study. In: Ilbery B W, Chiotti Q, Rickard T (eds) *Agricultural restructuring and sustainability: a geographical perspective.* CAB International, Wallingford, pp. 185–200

Broadway M (1997) Alberta bound: Canada's beef industry. *Geography* **82**: 377–80

Broadway M J (2000) Planning for change in small towns or trying to avoid the slaughterhouse blues. *Journal of Rural Studies* 16: 37–46

Broadway M J (2001) Bad to the bone: the social costs of beef packing's move to rural Alberta.

In: Epp R, Whitson D (eds) *Writing off the rural West: globalization, governments and the transformation of rural communities*. University of Alberta Press and Parkland Institute, Edmonton, pp. 39–52

Broadway M J (2002) The British slaughtering industry: a dying business? *Geography* 87: 268–80

Broadway M J, Ward T (1990) Recent changes in the structures and location of the US meatpacking industry. *Geography* 76: 76–9

Brookfield H C (1972) Intensification and disintensification in Pacific agriculture: a theoretical approach. *Pacific Viewpoint* 13: 30–48

Brookfield H C (1984) Intensification revisited. *Pacific Viewpoint* 25: 15–44

Brookfield H C (2001) Intensification and alternative approaches to agricultural change. *Asia Pacific Viewpoint* 42: 181–96

Brookfield H A, Hart D (1971) *Melanesia: a geographical interpretation of an island world*. Methuen, London

Brotherton I (1990) Initial participation in UK set-aside and ESA schemes. *Planning Outlook* 33: 46–61

Brotherton I (1991a) What limits participation in ESAs? *Journal of Environmental Management* 32: 241–9

Brotherton I (1991b) The cost of conservation: a comparison of ESA and SSSI payment rates. *Environment and Planning A* 23: 1183–95

Brouwer F (2000) *CAP regimes and the European countryside*. CAB International, Wallingford

Brouwer F, Lowe P (eds) (1998) *CAP and the rural environment in transition: a panorama of national perspectives*. Wangeningen Pers, Wangeningen

Brouwer F, Lowe P (eds) (2000) *CAP regimes and the European countryside*. CAB International, Wallingford

Brown E (1994) Agro-exports and Sandinismo: the political economy of social transformation in Nicaragua (1979–1990), unpublished PhD thesis, University of Edinburgh

Brown L (1995) *Who will feed China?: wake-up call for a small planet*. WW Norton & Co., New York

Brown L A (1981) *Innovation diffusion: a new perspective*. Methuen, New York

Brown L H (1957) Development and farm planning in the African areas of Kenya. *East African Agricultural Journal*, October, 67–73

Brown M B (1993) *Fair trade. Reform and realities in the international trading system*. Zed Books, London

Browne A W, Harris P J C, Hofny-Collins A H, Pasiecznik N, Wallace R R (2000) Organic production and ethical trade: definition, practice and links. *Food Policy* 25: 69–90

Bryceson D F (1990) *Food insecurity and the social division of labour in Tanzania, 1919–85*. Macmillan, London

Bryceson D F (1996) De-agrarianization and rural employment in sub-Saharan Africa: a sectoral perspective. *World Development* 24: 91–111

Bryant C R (1992) Farming at the urban fringe. In: Bowler I R (ed.) *The geography of agriculture in developed market economies*. Longman, London, pp. 275–304

Bryant C R (2002) Urban and rural interactions and rural community renewal. In: Bowler I R, Bryant C R, Cocklin C (eds) *The sustainability of rural systems: geographical interpretations*. Kluwer, Dordrecht, pp. 247–70

Bryant C R, Johnston T R R (1992) *Agriculture in the city's countryside*. Belhaven Press, London

Bryant C R, Russwurm L H, McLellan A G (1982) *The city's countryside: land and its management in the rural–urban fringe*. Longman, London and New York

Bryant C R, Smit B, Brklacich M, Johnston T R, Smithers J, Chiotti Q, Singh B (2000) Adaptation in Canadian agriculture to climatic variability and change. *Climatic Change* 45: 181–201

Bryden J (1994) Prospects for rural areas in an enlarged Europe. *Journal of Rural Studies* 10: 387–94

Bryden J, Fuller A, MacKinnon N (1992) Part-time farming: a note on definitions – a further comment. *Journal of Agricultural Economics* 43: 109–10

Bryson J, Henry N, Keeble D, Martin R (eds) (1999) *The economic geography reader*. Wiley, Chichester and New York

Bryson W (1993) Britain's hedgerows. *National Geographic* 182: 94–117

Buck D C, Getz C, Guthman J (1997) From farm to table: the organic vegetable commodity chain at Northern California. *Sociologia Ruralis* 37: 3–20

Buller H (2000) Regulation 2078: patterns of implementation. In: Buller H, Wilson G A, Holl A (eds) *Agri-environmental policy in the European Union*. Ashgate, Aldershot, pp. 219–54

Buller H, Brives H (2000) France. In: Buller H, Wilson G A, Holl A (eds) *Agri-environmental policy in the European Union*. Ashgate, Aldershot, pp. 9–30

Buller H, Hoggart K (eds) (2001) *Agricultural transformation, food and the environment*. Ashgate, Basingstoke

Buller H, Wilson G A, Holl A (eds) (2000) *Agri-environmental policy in the European Union*. Ashgate, Aldershot

Bullock C H, Macmillan D C, Crabtree J R (1994) New perspectives on agroforestry in lowland Britain. *Land Use Policy* 11: 223–33

Bunce M F (1991) Local planning and the role of rural land in metropolitan regions: the example of the Toronto area. In: Van Oort G. *et al.* (eds) *Limits to rural land use*. Pudoc, Wageningen, pp. 113–22

Bunce M F (1998) Thirty years of farmland preservation in North America: discourses and ideologies of a movement. *Journal of Rural Studies* 14: 233–48

Burch D, Rickson R E, Lawrence G E (eds) (1996) *Globalization and agri-food restructuring*. Avebury, Aldershot

Burch D, Lawrence G, Rickson R E, Goss J (eds) (1998) Australasia's food and farming in a globalised economy: recent developments and future prospects. *Monash Publications in Geography and Environmental Science*, no. 50

Burch D, Goss J, Lawrence G E (eds) (1999) *Restructuring global and regional agricultures: transformations in Australia*. Ashgate, Basingstoke

Burg M, Endeveld M (eds) (1994) *Women on family farms: gender research, EC policies and new perspectives*. Circle for European Rural Studies, Wageningen Agricultural University, Wageningen

Burger A (1994) *The agriculture of the world*. Avebury, Aldershot

Burger A (1998) Land valuation and land rents in Hungary. *Land Use Policy* 15: 191–201

Burger A (2001) Agricultural development and land concentration in a central European country: a case study of Hungary. *Land Use Policy* 18: 259–68

Burt S, Sparks L (2001) The implications of Wal-Mart's takeover of Asda. *Environment and Planning A* 33: 1463–87

Burtin J (1987) *The Common Agricultural Policy and its reform*. Office for Official Publications of the European Communities, Luxembourg, 4th edition

Burton M, Rigby D, Young T (1999) Analysis of the determinants of adoption of organic horticultural techniques in the United Kingdom. *Journal of Agricultuiral Economics* 50: 48–63

Burton R (1998) The role of farmer self-identity in agricultural decision-making in the Marston Vale Community Forest. Unpublished PhD thesis, Department of Geography, De Montfort University

Busch L, Bonanno A, Lacy W B (1989) Science, technology and the restructuring of agriculture. *Sociologia Ruralis* 29: 118–30

Buttel F H (1992) Environmentalism – origins, processes and implications for rural social change. *Rural Sociology* 57: 1–27

Buttel F H (1994) Agricultural change, rural society and the state in the late 20th century. In: Symes D, Jansen A (ed.) *Agricultural restructuring and rural change in Europe*. Studies in Sociology, Agricultural University, Wageningen, pp. 13–31

Buttel F H (1996) Theoretical issues in global agri-food restructuring. In: Burch D, Rickson R E, Lawrence G E (eds) *Globalization and agri-food restructuring*. Avebury, Aldershot, pp. 17–44

Buttel F H (2000) Ecological modernization as social theory. *Geoforum* 31: 57–65

Buttel F H (2001) Some reflections on late 20th century agrarian political economy. *Sociologia Ruralis* 41: 165–81

Buttel F H, Goodman D (1989) Class, state, technology and international food regimes. *Sociologia Ruralis* **29**: 86–92

Buttel F H, Newby H (eds) (1980) *The rural sociology of advanced societies.* Allenheld, Osmun & Co., Montclair, NJ

Buttel F H, Hawkins A P, Power A G (1990a) From Limits to Growth to Global Change – constraints and contradictions in the evolution of environmental science and ideology. *Global Environmental Change* **1**: 57–66

Buttel F H, Larson O F, Gillespie Jr G W (1990b) *The sociology of agriculture.* Greenwood Press, Westport, CT

Byerlee D, Eicher C K (eds) (1997) *Africa's emerging maize revolution.* Lynne Rienner Publishers, London

Byers T J (1982) Agrarian transition and the agrarian question. In: Harriss J C (ed.) *Rural development: theories of peasant economy and agrarian change.* Hutchinson, London, pp. 82–93

Byrne R J (1996) Field boundaries in Anglesey, Gwynedd. *Landscape Research* **21**: 189–94

Cairncross F (1997) *The death of distance: how the communications revolution will change our lives.* Orion Business Books, London

Campagne P, Carrere G, Valceschini E (1990) Three agricultural regions of France: three types of pluriactivity. *Journal of Rural Studies* **4**: 415–22

Campbell H, Coombes B (1999a) New Zealand's organic food exports: current interpretations and new directions in research. In: Burch D, Goss J, Lawrence G E (eds) *Restructuring global and regional agricultures: transformations in Australia.* Ashgate, Basingstoke, pp. 61–74

Campbell H, Coombes B (1999b) Green protectionism and the exporting of organic fresh fruit and vegetables from New Zealand: crisis experiments in the breakdown of Fordist trade and agricultural policies. *Rural Sociology* **64**: 302–19

Carlyle W J (1991) Rural change in the Prairies. In: Robinson G M (ed.) *A social geography of Canada: essays in honour of J. Wreford Watson.* Dundurn Press, Toronto, pp. 330–58

Carlyle W J (1997) The decline of summer fallow on the Canadian Prairies. *Canadian Geographer* **41**: 267–80

Carlyle W J (1999) The Great Plains and the Prairies. In: Boal F W, Royle S A (eds) *North America: a geographical mosaic.* Edward Arnold, London, pp. 257–64

Carlyle W J (2002) Cropping patterns in the Canadian Prairies: thirty years of change. *Geographical Journal* **168**: 97–116

Carney D (ed.) (1998) *Sustainable rural livelihoods: what contribution can we make?* Department for International Development, London

Carney J, Watts M J (1991) Disciplining women? *Signs* **16**: 651–81

Carr S, Tait J (1991) Differences in the attitudes of farmers and conservationists and their implications. *Journal of Environmental Management* **32**: 281–94

Carson R (1962) *Silent Spring.* Houghton Mifflin, Boston

Carter H (1981) *The study of urban geography.* Edward Arnold, London, 3rd edition

Carter S (1999) Multiple business ownership in the farm sector: assessing the enterprise and employment contributions of farmers in Cambridgeshire. *Journal of Rural Studies* **15**: 417–30

Caskie P (2000) Back to basics: household food production in Russia. *Journal of Agricultural Economics* **51**: 196–209

Castells M (1977) *The urban question.* Edward Arnold, London

Castells M (1996) *The rise of the network society.* Blackwell, Oxford

Castree N (1996) Birds, mice and geography: Marxism and dialectics. *Transactions of the Institute of British Geographers* new series, **21**: 342–62

Cavailhes J, Normandin D (1993) Deprise agricole et boisement: etat des lieux, enjeux et perspectives dans cadre de la reforme de la PAC. *Revue Forestiere Francaise* **45**: 465–81

CEAS Consultants Ltd (1996) *Socio-economic effects of the Countryside Stewardship Scheme. Final Report.* CEAS Consultants Ltd., Wye College, Ashford

CEC (Commission of the European Communities) (1985) *A future for Community agriculture: Community guidelines following the consultation in connection with the Green Paper.* CEC, Luxembourg

CEC (1986) *Agriculture and environment: management agreements in four countries of the European Communities.* CEC, Brussels, report EUR10783

CEC (1997) *Agenda (2000) Volume 1. For a stronger and wider Union.* DOC 97/6, CEC, Brussels

CEC (2002) *European agriculture entering the 21st century.* Director General for Agriculture, CEC, Brussels

Chaleard J L (1996) Les mutations de l'agriculture commerciale en Afrique de l'Ouest. *Annales de Geographie* 592: 563–83

Chambers R (1997) *Whose reality counts? Putting the first last.* Intermediate Technology Publications, London

Chaplin S P (2000) Farm-based recreation in England and Wales. Unpublished PhD thesis, University College Worcester

Chapman G P (2002) The Green Revolution. In: Desai V, Potter R B (eds) *The companion to development studies.* Arnold, London, pp. 155–9

Chapman K (2001) Spreading fertilizers: the internationalization of an industry. *Geography* 86: 11–22

Charlton C A (1987) Problems and prospects for sustainable agricultural systems in the humid tropics. *Applied Geography* 7: 153–74

Chayanov A V (1966) *The theory of peasant economy.* Richard D. Irwin, Homewood, Illinois, translated by D. Thorner

Chen M (1991) *Coping with seasonality and drought.* Sage, New Delhi

Chen X (1991) Sea-level changes since the early 1920s from the long records of the two tidal gauges in Shanghai. *Journal of Coastal Research* 7: 787–99

Chen X, Zong Y (1999) Major impacts of sea-level rise on agriculture in the Yangtze delta area around Shanghai. *Applied Geography* 19: 69–84

Chiotti Q P (1998) An assessment of the regional impacts and opportunities from climate change in Canada. *Canadian Geographer* 42: 380–93

Chiotti Q P, Johnson T R (1995) Extending the boundaries of climate change research: a discussion on agriculture. *Journal of Rural Studies* 11: 335–50

Chiotti Q P, Johnston T, Smit B, Ebel B (1997) Agricultural response to climate change: a preliminary investigation of farm-level adaptation in southern Alberta. In: Ilbery B W, Chiotti Q, Rickard T (eds) *Agricultural restructuring and sustainability: a geographical perspective.* CAB International, Wallingford, pp. 201–18

Chisholm M A (1962) *Rural settlement and land use.* Hutchinson, London

Chisholm M (1985) The Development Commission's employment programmes in rural England. In: Healey M, Ilbery B W (eds) *The industrialisation of the countryside.* Geo Books, Norwich, pp. 279–92

Chul-Kyoo K, Curry J (1993) Fordism, flexible specialization and agri-industrial restructuring: the case of the US broiler industry. *Sociologia Ruralis* 33: 61–80

Chung H, Veeck G (1999) Pessimism and pragmatism: agricultural trade liberalization from the perspective of South Korean farmers. *Asia Pacific Viewpoint* 40: 271–84

Cinotti B (1992) Les agriculteurs et leurs forêts. *Revue Forestiere Francaise* 44: 356–64

Clark G (1984) The meaning of agricultural regions. *Scottish Geographical Magazine* 100: 34–44

Clark G (1992) Data sources for studying agriculture. In: Bowler I R (ed.) *The geography of agriculture in developed market economies.* Longman, London, pp. 32–55

Clark G (1999) Land-use conflict at the urban fringe. In: Pacione M (ed.) *Applied Geography: principles and practice.* Routledge, London and New York, pp. 301–8

Clark G, Gordon D (1980) Sampling for farm studies in Geography. *Geography* 65: 101–6

Clark G, Bowler I R, Crockett A, Ilbery B W, Shaw A (1997a) Rural re-regulation and institutional sustainability: a case study of alternative farming systems in England. In: Ilbery B W, Chiotti Q, Rickard T (eds) *Agricultural restructuring and sustainability: a geographical perspective*. CAB International, Wallingford, pp. 117–34

Clark G, Bowler I R, Shaw A, Crockett A, Ilbery B W (1997b) Institutions, alternative farming systems and local re-regulation. *Environment and Planning A* **29**: 731–45

Clark G L (1992) 'Real' regulation: the administrative state. *Environment and Planning A* **24**: 615–27

Clark J, Jones A, Potter C, Lobley M (1997) Conceptualising the evolution of the European Union's agri-environment policy: a discourse approach. *Environment and Planning A* **29**: 1869–85

Cleary M (1989) *Peasants, politicians and producers*. Cambridge University Press, Cambridge

Clinton W, Gore A (1992) *Putting people first: how we can all change America*. Time Books, New York

Cloke P J (1996) Looking through European eyes? A re-evaluation of agricultural deregulation in New Zealand. *Sociologia Ruralis* **36**: 305–30

Cloke P J, Le Heron R B (1994) Agricultural deregulation: the case of New Zealand. In: Lowe P, Marsden T K, Whatmore S J (eds) *Regulating agriculture*. David Fulton, London, pp. 104–26

Cloke P J, Thrift N J (1987) Intra-class conflict in rural areas. *Journal of Rural Studies* **3**: 321–33

Cloke P J, Thrift N J (1990) Class change and conflict in rural areas. In: Marsden T K, Lowe P, Whatmore S J (eds) (1990) *Rural restructuring*. David Fulton, London, pp. 165–81

Cloke P J, Le Heron R B, Roche M M (1990) Towards a geography of political economy perspective on rural change: the example of New Zealand. *Geografiska Annaler* **72B**: 13–27

Clout H D (1988) France, pp. 98–119 in Cloke P J (ed.) *Policies and plans for rural people: an international perspective*. Unwin Hyman, London

Clunies-Ross T (1990) Organic food: swimming against the tide. In: Marsden T K, Little J (eds) *Political, social and economic perspectives on*

the international food system. Avebury, Aldershot, pp. 200–14

Cobb R, Dolman P, O'Riordan T (1999) Interpretations of sustainable agriculture in the UK. *Progress in Human Geography* **23**: 209–35

Cocklin C (1995) Agriculture, society and environment: discourses on sustainability. *International Journal of Sustainable Development and World Ecology* **2**: 240–56

Cocklin C, Gray E, Smit B (1983) Future urban growth and agricultural land in Ontario. *Applied Geography* **3**: 91–104

Coffee W J, Polese M (1988) Locational shifts in Canadian employment, 1971–1981: decentralization v congestion. *Canadian Geographer* **32**: 248–56

Cohen S J (ed.) (1994) *Mackenzie Basin impact study, interim report no. 2*. Environment Canada, Ottawa

Collins J, Lappé F M (1986) *Myths of hunger*. Food First, San Francisco

Collins L (1999) Renewable energy from energy crops: a global perspective. *Geography* **84**: 169–79

Collinson M (ed.) (2000) *A history of farming systems research*. CAB International, Wallingford

Colman D (1994) Comparative evaluation of environmental policies. In: Whitby M (ed.) *Incentives for Countryside Management: the Case of Environmentally Sensitive Areas*. CAB International: Wallingford, pp. 219–52

Colman D (2000) Inefficiencies in the UK milk quota system. *Food Policy* **25**: 1–16

Colson F, Mathsin J (2002) How could be balanced the CAP pillars for the promotion of a multi-functional Europe model. Unpublished paper for the Working Group on the future role of agriculture in European food production and environmental responsibility. Akademie fur Raumforshung und Landesplannung, Hannover

Conford P (ed.) (1992) *A future for the land. Organic practices from a global perspective*. Green Books, Devon

Constance D, Heffernan W D (1991) El complejo agroalimentario global de las aves de corral. *Agricultural y Sociedad* **60**: 63–92

Conway G R (1997) *The doubly Green Revolution: food for all in the 21st century*. Penguin Books,

London and Cornell University Press, New York

Conway G R (1999) Ecological principles in agricultural policy: but which principles? a response. *Food Policy* 24: 17–20

Conway G R, Barbier E B (1990) *After the Green Revolution: sustainable agriculture for development.* Earthscan, London

Conzen M R G (1960) Alnwick: Northumberland. A study in town plan analysis, *Transactions of the Institute of British Geographers,* 27

Cook H F, Norman C (1996) Targeting agri-environment policy: an analysis relating to the use of Geographical Information Systems. *Land Use Policy* 13: 217–28

Cook I (1994) Constructing the exotic: the case of tropical fruit. In: Allen J, Massey D B (eds) *Geographical worlds.* Oxford University Press, Oxford, pp. 137–42

Cook I, Crang P (1995) The world on a plate: culinary culture, displacement and geographical knowledges. *Journal of Material Culture* 1: 131–54

Cooke G W (1979) Some priorities for British soil science. *Journal of Soil Science* 30: 187–214

Cooke R U, Doornkamp J C (1990) *Geomorphology in environmental management – a new introduction.* Clarendon Press, Oxford, second edition

Cooke R U, Harris D R (1970) Remote sensing of the terrestrial environment – principles and progress. *Transactions of the Institute of British Geographers* 50: 1–23

Coombes B L (1997) Rurality, culture and local economic development. Unpublished PhD thesis, University of Otago, Dunedin, New Zealand

Coombes B L, Campbell H (1998) Dependent reproduction of alternative modes of agriculture: organic farming in New Zealand. *Sociologia Ruralis* 38: 127–45

Coomes O T, Barham B L (1997) Rain forest extraction and conservation in Amazonia. *Geographical Journal* 163: 180–88

Cooper N (1998) The role of street-level bureaucrats in the implementation of agri-environmental schemes. Unpublished paper presented at the RGS-IBG Annual Conference, Kingston University, 5–8 January

Coppock J T (1964a) Crop, livestock and enterprise combinations in England and Wales. *Economic Geography* 40: 65–81

Coppock J T (1964b) Post-war studies in the geography of British agriculture. *Geographical Review* 54: 409–26

Coppock J T (1968) The geography of agriculture. *Journal of Agricultural Economics* 19: 153–75

Coppock J T (1971) *An agricultural geography of Great Britain.* G. Bell & Sons, London

Coppock J T (1976a) *An agricultural atlas of England and Wales.* Faber, London, 2nd edition

Coppock J T (1976b) *An agricultural atlas of Scotland.* John Donald, Edinburgh

Corbett J (1988) Famine and household coping stategies. *World Development* 16: 1099–1112

Corbridge S (ed.) (1995) *Development studies: a reader.* Edward Arnold, London

Corby H D L (1941) Changes being brought about by the introduction of mixed farming. *Farm and Forest* 2(supplement 3): 106–9

Cornelius W A, Myhre D (eds) (1998) *The transformation of rural Mexico: reforming the ejido sector.* Center for US–Mexican Studies, University of California, San Diego

Countryside Commission/Nature Conservancy Council (CC/NCC), (1989) *The Countryside Premium for set aside land.* Countryside Commission, Cheltenham, CCP267

Countryside Commission (1993) *The Countryside Premium for set aside land: monitoring and evaluation 1989–1992.* Countryside Commission, Cheltenham

Countryside Commission (1995) *Countryside Stewardship monitoring and evaluation: third interim report.* Countryside Commission, Cheltenham

Courtenay P P (1981) *Plantation agriculture.* Bell & Hyman, London

Courtenay P P (1990) Malaysian village regrouping policy and an example from Melacca. *Geography* 75: 128–34

Cox G, Lowe P, Winter M (1989) The farm crisis in Britain. In: Goodman D, Redclift M (eds) (1989) *The international farm crisis.* Macmillan, Basingstoke, pp. 113–34

Cox K R (ed.) (1997) *Spaces of globalization: reasserting the power of the local.* Guilford Press, New York

Crabb J, Firbank L, Winter M, Parham C, Dauven A (1998) Set-aside landscapes: farmer perceptions and practices in England. *Landscape Research* **23**: 237–54

Crabtree J R, Chalmers N, Barron N J (1998) Information for policy design: modeling participation in a Farm Woodland Incentive Scheme, *Journal of Agricultural Economics* **49**: 306–20

Craighill A, Goldsmith E (1994) What future for set aside? *Ecos* **15**(3/4): 58–62

CRM (Committee on Resources and Man), National Academy of Sciences (1969) *Resources and man.* W H Freeman & Co., San Francisco

Cromwell E (1996) *Governments, farmers and seeds in a changing Africa.* CAB International, Wallingford

Cross M (1987) Solving the world's food problem. *New Scientist*, 16 April, p. 45

CRR (Centre for Rural Research) (1994) *Restructuring the agro-food system: global processes and national responses.* CRR, University of Trondheim

Cundliffe I (2000) Agri-environment schemes: taking the message to farmers. *Landscape Research* **25**: 375–7

Curtis D, Hubbard M, Shepherd A (eds) (1988) *Preventing famine: policies and prospects for Africa.* Routledge, London

Curtis L F, Courtney F M, Trudgill S (1976) *Soils in the British Isles.* Longman, London and New York

Cwiertka K, Walraven B (eds) (2002) *Asian food: the global and the local.* Curzon, Richmond, Surrey

Dalecki M, Coughenor C (1992) Agrarianism in American society. *Rural Sociology* **57**: 48–64

Damianos D, Skuras D (1996) Farm business and the development of AFEs: an empirical analysis of Greece. *Journal of Rural Studies* **12**: 273–84

Damianos D, Dimara E, Hassapoyannes K, Skuras D (1998) *Greek agriculture in a changing international environment.* Ashgate, Aldershot

Danbom D (1991) Romantic agrarianism in twentieth century America. *Agricultural History* **65**: 1–12

Daniels T L (1986) Hobby farming in America: rural development or threat to commercial agriculture? *Journal of Rural Studies* **2**: 31–40

Daniels T L (1999) *When city and country collide: managing growth in the metropolitan fringe.* Island Press, Washington, DC

Daniels P W (2001) The geography of the economy. In: Daniels P W, Bradshaw M, Shaw D, Sidaway J (eds) *Human Geography: issues for the 21st century.* Pearson Education, Harlow, pp. 305–41

Darque M B (1988) The division of labour and decision-making in farming couples: power and negotiation. *Sociologia Ruralis* **28**: 271–92

Daskalopuglou I, Petrou A (2002) Utilising a farm typology to identify potential adopters of alternative farming activities in Greek agriculture. *Journal of Rural Studies* **18**: 95–104

Davis J E (1980) Capitalist agricultural development and the exploitation of the propertied laborer. In: Buttel F H, Newby H (eds) *The rural sociology of advanced societies.* Allenheld, Osmun & Co., Montclair, NJ, pp. 133–53

Davis J N, Mack N, Kirke A (1997) New perspectives on farm incomes. *Journal of Rural Studies* **13**: 57–64

Dawson J, Burt S (1998) European retailing: dynamics, restructuring and development issues. In: Pinder D (ed.) *The new Europe: economy, society and environment.* John Wiley & Sons, Chichester, pp. 157–76

Dean W (1995) *With Broadax and firebrand: the destruction of the Brazilian Atlantic forest.* University of California Press, Berkeley

Dear M J (2001) The postmodern turn. In: Minca C (ed.) *Postmodern geography: theory and praxis.* Blackwell Publishers, Oxford and Malden, MA, pp. 1–16

De Bakker H, Schelling J (1966) *Systeem van Bodem classificatie voor Nederland. De Hogere Niveans* (with English Summary). Pudoc, Wageningen

De Haan H, Kasimis B, Redclift M (eds) (1997) *Sustainable rural development.* Ashgate, Aldershot

De Honinck R (2000) The theory and practice of frontier development: Vietnam's contribution. *Asia Pacific Viewpoint* **41**: 7–21

Deininger K, Binswanger A (1999) The evolution of the World Bank's land policy: principles, experience and future challenges. *World Bank Research Observer* 14(2): 247–76

De Janvry A (1980) *The agrarian question in Latin America.* Johns Hopkins University Press, Baltimore

De Janvry A, Gordillio G, Platteau J-P, Sadoulet E (2001) *Access to land, rural poverty and public action.* Oxford University Press, Oxford and New York

Dempsey K (1992) *A man's town: inequality between women and men in rural Australia.* Oxford University Press, Melbourne

Deng S (2000) Agricultural colonization in Lam Dong Province, Vietnam. *Asia Pacific Viewpoint* 41: 35–49

Dent J B, McGregor M J (eds) (1994) *Rural and farming systems analysis.* CAB International, Wallingford

Desai V (2002) Community participation in development. In: Desai V, Potter R B (eds) *The companion to development studies.* Arnold, London, pp. 117–21

DETR (Department of Environment, Transport and the Regions) (1999) *A better quality of life: a strategy for sustainable development in the United Kingdom.* HMSO, London

DETR (2000) *Local planning authority Green Belt statistics, England 1997.* DETR, London

Dicken P, Peck J A, Tickell A (1997) Unpacking the global. In: Lee R, Wills J (eds) *Geographies of economies.* Edward Arnold, London, pp. 158–6

Dixon J (1992) Environmentally sensitive farming – where next? *Ecos* 13(3): 15–19

Djurfeldt G, Waldenstrom C (1999) Mobility patterns of Swedish farming households. *Journal of Rural Studies* 15: 331–44

DoE/MAFF (Department of the Environment/Ministry of Agriculture, Fisheries and Food) (1995) *Rural England: a nation committed to a living countryside.* Command Paper 3016, HMSO, London

Doering O (1992) Federal policies as incentives or disincentives to ecologically sustainable agricultural systems. *Journal of Sustainable Agriculture* 2: 21–36

Dorst J (1989) *The written suburb.* University of Pennsylvania Press, Philadelphia

Downs C J (1991) EC agricultural policy and land use: milk quotas and the need for a new approach. *Land Use Policy* 8: 206–10

Drakakis-Smith D (1994) Third world cities: sustainable urban development II – population, labour and poverty. *Urban Studies* 33: 673–701

Dreze J, Sen A K (1989) *Hunger and public action.* Clarendon, London

Dreze J, Sen A K (eds) (1990) *The political economy of hunger.* Clarendon, London, 2 vols

Drummond I, Marsden T K (1995a) Regulating sustainable development. *Global Environmental Change* 5: 51–63

Drummond I, Marsden T K (1995b) A case study in unsustainability: the Barbados sugar industry, *Geography* 80: 342–54

Drummond I, Marsden T K (1999) *The condition of sustainability.* Routledge, London

Drummond I, Campbell H, Lawrence G, Symes D (2000) Contingent or structural crisis in British agriculture. *Sociologia Ruralis* 40: 111–27

Duckham A N, Masefield G B (1970) *Farming systems of the world.* Chatto and Windus, London

Dunlap R E, Beus C E, Howell R E, Waud J (1992) What is sustainable agriculture? An empirical examination of faculty and farmer definitions. *Journal of Sustainable Agriculture* 3: 5–40

Duram L A (1998) Taking a pragmatic behavioral approach to alternative agricultural research. *American Journal of Alternative Agriculture* 13: 90–5

Dwivedi O P (1986) Political science and the environment. *International Social Science Journal* 38: 377–90

Dyson T (1996) *Population and food: global trends and future prospects.* Routledge, London

Earth Summit +5, (1997) *The United Nations convention on biodiversity: a constructive response to a global problem.* Special Session of the General Assembly, New York, 21.6.97

Easterling W E, Rosenberg N, McKenney M, Jones C (1992) An introduction to the methodology, the region of study, and a historical analog of climate change. *Agricultural and Forest Meteorology* 56: 3–15

Easterling W E, Crosson P R, Rosenberg N J, McKenney M S, Katz L A, Lemon K M (1993) Agricultural impacts of and responses to climatic change in the Missouri-Iowa-Nebraska-Kansas (MINK) region. *Climatic Change* 24: 23–61

Eaton C (1998) Contract farming: structure and management in developing nations, pp. 127–44

EC (European Commission) (1994) *EC agricultural policy for the 21st century, The Economics of the CAP*. EC, Brussels, Reports and Studies, nos 4 and 5

EC (1997) *Agenda 2000: For a stronger and wider Union*. EC, Brussels

EC (1998) *Explanatory Memorandum: The future for European agriculture*. EC, Brussels

Edmond H, Crabtree R (1994) Regional variation in Scottish pluriactivity: the socio-economic context for different types of non-farming activity. *Scottish Geographical Magazine* 110: 76–84

Edmond H, Corcoran K, Crabtree B (1993) Modelling locational access to markets for pluriactivity: a study in the Grampian region of Scotland. *Journal of Rural Studies* 9: 339–49

Edwards C (1992) Changing farm enterprises. In: Bowler I R (ed.) *The geography of agriculture in developed market economies*. Longman, London, pp. 134–161

Edwards M P (1994) Rethinking social development: the search for relevance. In: Booth D (ed.) *Rethinking social development: theory, research and practice*. Longman, Harlow, pp. 279–97

Efstratoglou-Todoulou S (1990) Pluriactivity in different socio-economic contexts: a test of the push-pull hypothesis in Greek farming. *Journal of Rural Studies* 6: 407–13

Ehrlich P, Ehrlich A, Daily G (1993) Food security, population and the environment. *Population and Development Review* 19: 1–32

Egdell J (2000) Consultation on the countryside premium scheme. Creating a 'market' for information. *Journal of Rural Studies* 16: 357–66

Egdell J, Badger R (1996) A new political economy perspective on Scottish agri-environmental policy. *Scottish Agricultural Economics Review* 9: 21–33

Eicher C K, Staatz J M (1990) *Agricultural development in the Third World*. Johns Hopkins Press, Baltimore.

Eikeland S (1999) New rural pluriactivity? Household strategies and rural renewal in Norway. *Sociologia Ruralis* 39: 359–76

Eikeland S, Lie I (1999) Pluriactivity in Norway. *Journal of Rural Studies* 15: 405–16

Elliott J A (2002) Towards sustainable rural resource management in sub-Saharan Africa. *Geography* 87: 197–204

Elliott J A, Campbell M (2002) The environmental imprints and complexes of social dynamics in rural Africa: cases from Zimbabwe and Ghana. *Geoforum* 33: 221–37

Ellis F (1988) *Peasant economics*. Cambridge University Press, Cambridge

Ellis F, Sumberg J (1998) Food production, urban areas and policy responses. *World Development* 26: 213–25

Ellis S, Mellor A (1995) *Soils and environment*. Routledge, London

Elmhirst R (1997) Gender, environment and culture: a political ecology of transmigration in Indonesia, unpublished PhD thesis, Wye College, University of London

Emerson H, MacFarlane R (1995) Comparative bias between sampling frames for farm surveys. *Journal of Agricultural Economics* 46: 241–51

English Nature (1999) *Natural Areas: helping to set the regional agenda for nature*. English Nature, Peterborough

EPA (Environmental Protection Agency) (1992) *Sustainable development and the Environmental Protection Agency: report to Congress*. EPA, Washington DC, report no. 230-R-93-005

Epps R, Whitson D (eds) (2001) *Writing off the rural West: globalization, governments and the transformation of rural communities*. University of Alberta Press and Parkland Institute, Edmonton

Errington A (1985) Sampling frames for farm surveys in the United Kingdom: some alternatives. *Journal of Agricultural Economics* 36: 251–8

Errington A, Gasson R M (1996) The increasing flexibility of the farm and horticultural workforce in England and Wales. *Journal of Rural Studies* 12: 127–42

Ervin D E (1988) Set-aside programmes: using United States experience to evaluate United Kingdom proposals. *Journal of Rural Studies* 4: 181–91

Esseks J (1978) The politics of farmland preservation. In: Hadwinger D, Browne W (eds) *The new politics of food*. D.C. Heath, Lexington, MA, pp. 199–216

Esser J, Hirsch J (1989) The crisis of fordism and the dimensions of a postfordist regional and urban structure. *International Journal of Urban and Regional Research* 13: 417–37

Estudillo J P, Otsuka K (1999) Green Revolution, human capital, and off-farm employment: changing sources of income among farm households in Central Luzon, 1966–1994. *Economic Development and Cultural Change* 47: 497–523

Eurostat (1999) *Eurostat Yearbook 1998/99. A statistical eye on Europe 1987–1997*. EC, Brussels and Luxembourg

Evans L T (1998) *Feeding the ten billion: plants and population growth*. Cambridge University Press, Cambridge

Evans N J (1992) Towards an understanding of farm tourism in Britain. In: Gilg A W (ed.) *Progress in rural policy and planning*. Belhaven, London, pp. 140–4

Evans N J (1996) Evaluating recent changes to the publication of UK Agricultural Census data. *Geography* 81: 225–34

Evans N J (2000) The impact of BSE in cattle on high-nature-value conservation sites in England. In: Millward H, Beesley K, Ilbery B W, Harrington L (eds) *Agricultural and environmental sustainability in the new countryside*. St. Mary's University and Rural Research Centre, Nova Scotia Agricultural College, Truro, Nova Scotia, pp. 92–110

Evans N J, Ilbery B W (1989) A conceptual framework for investigating farm-based accommodation and tourism in Britain. *Journal of Rural Studies* 5: 257–66

Evans N J, Ilbery B W (1992a) Farm-based accommodation and the restructuring of agriculture: evidence from three English Counties. *Journal of Rural Studies* 8: 85–96

Evans N J, Ilbery B W (1992b) The distribution of farm-based accommodation in England and Wales. *Journal of the Royal Agricultural Society of England* 153: 67–80

Evans N J, Ilbery B W (1993) The pluriactivity, part-time farming and farm diversification debate. *Environment and Planning A* 25: 945–59

Evans N J, Ilbery B W (1996) Exploring the influence of farm-based pluriactivity on gender relations in capitalist agriculture. *Sociologia Ruralis* 36: 74–92

Evans N J, Morris C (1997) Towards a geography of agri-environmental policies in England and Wales. *Geoforum* 28: 189–204

Evans N J, Yarwood R (1995) Livestock and landscape. *Landscape Research* 20: 141–6

Evans N J, Yarwood R (2000) The politicization of livestock: rare breeds and countryside conservation. *Sociologia Ruralis* 40: 228–48

Evans N J, Morris C, Winter M (2002) Conceptualizing agriculture: a critique of post-productivism as the new orthodoxy. *Progress in Human Geography* 26: 313–32

Evans R (1997) Accelerated soil erosion in Britain. *Geography* 82: 149–62

Ewell P T, Merrill-Sands D (1987) Milpa in Yucatan: a long-fallow maize system and its alternatives in the Maya peasant economy. In: Turner B L, II, Brush S B (eds) *Comparative farming systems*. Guilford Press, New York, pp. 95–129

Fairhead J, Leach M (1996) *Misreading the African landscape: society and ecology in a forest-savanna mosaic*. Cambridge University Press, Cambridge

Falconer K, Ward N (2000) Using modulation to green the CAP: the UK case. *Land Use Policy* 17: 269–78

Fanfani R (1994) Agro-food districts: a new dimension for policy-making and the role of institutions. In: Centre for Rural Research, *Restructuring the agro-food system: global* processes and national responses. CRR, University of Trondheim, pp. 81–9

FAO (Food and Agriculture Organisation) (1999) *FAO symposium on agriculture, trade and food security, Geneva, 20–24 September 1999. Synthesis of case studies, X3065/E*. FAO, Rome

Farmer B H (1977) *Green Revolution? Technology and change in ricegrowing areas of Tamil Nadu and Sri Lanka*. Macmillan, London

Farns P (1991) Soil classification: the updated FAO system and soils in the UK. *Geography Review* 5(1): 27–31

Faulkner O T, Mackie J R (1936) The introduction of mixed farming in northern Nigeria. *Empire Journal of Experimental Agriculture* 4: 89–96

Fearnside P M (1993) Deforestation in the Brazilian Amazon – the effect of population and land tenure. *Ambio* 22: 537–45

Feitelson E (1999) Social norms, rationales and policies: reforming farmland protection in Israel. *Journal of Rural Studies* 15: 431–46

Fennell D A, Weaver D B (1997) Vacation farms and ecotourism in Saskatchewan, Canada. *Journal of Rural Studies* 13: 467–76

Fennell R (1979) *The Common Agricultural Policy of the European Community: its institutional and administrative organisation*. Granada, London

Fine B (1994) Toward a political economy of food. *Review of International Political Economy* 1: 579–85

Fine B (2000) *Social capital versus social theory*. Routledge, London

Fine B, Heasman M, Wright J (1996) *Consumption in the age of affluence*. Routledge, London

Fine B, Leopold E (1993) *The world of consumption*. Routledge, London and New York

Fink V S (1991) What work is real? Changing roles of farm and ranch wives in south-east Ohio. *Journal of Rural Studies* 7: 17–22

Firbank L G, Arnold H R, Eversham B C, Mountford J O, Radford G L, Telfer M G H, Treweek J R, Webb N R C, Wells T C E (1993) *Managing set-aside land for wildlife*. ITE/NERC, HMSO, London

Fischer G, Frohberg K, Parry M, Rosenzweig C (1995) Climate change and world food supply, demand and trade. In: Rosenzweig C, Allen L, Harper L, Hollinger S, Jones J (eds) *Climate change and agriculture: analysis of potential international impacts*. American Society of Agronomy Special Publications, no. 59, pp. 341–82

Flynn A, Marsden T K (1992) Food regulation in a period of agricultural retreat: the British experience. *Geoforum* 23: 85–93

Flynn A, Marsden T K (1995) Rural change, regulation, and sustainability. *Environment and Planning A* 27: 1180–92

Flynn A, Marsden T K, Ward N (1994) Retailing, the food system and the regulatory state, in Lowe *et al.* (eds) *Regulating Agriculture*. David Fulton Publishers, London

Forster P G, Maghimbi S (eds) (1999) *Agrarian economy, state and society in contemporary Tanzania*. Ashgate, Aldershot

Foster C, Lampkin N (2000) *Organic and in-conversion land area, holdings, livestock and crop production in Europe*. Report to the European Commission, FAIR3-CT96-1794, University College, Aberystwyth

Foster I, Harrison S, Clark D (1997) Soil erosion in the West Midlands: act of God or agricultural mismanagement? *Geography* 82: 231–40

Fothergill S (1986) Industrial employment and planning restraint in the London Green Belt. In: Towse R J (ed.) *Industrial/office development in areas of planning restraint – the London Green Belt*. Proceedings of a Conference held at the School of Geography, Kingston Polytechnic, 11 April, (1986) pp. 27–40

Fothergill S, Gudgin G, Kitson M, Monk S (1985) Rural industrialisation: trends and causes. In: Healey M, Ilbery B W (eds) *The industrialisation of the countryside*. Geo Books, Norwich, pp. 147–60

Found W C (1971) *A theoretical approach to rural land-use patterns*. Edward Arnold, London

Fox M W (1996) *Agricide: the hidden farm and food crisis that affects us all*. Krieger Publishing, New York

Francis C A, Younghusband G (1990) Sustainable agriculture: an overview. In Francis C A, Flora C B, King L D (eds); *Sustainable agriculture in Temperate Zones*. John Wiley and Sons: New York. 1–23

Franklin S H (1969) *The European peasantry: the final phase*. Methuen, London

Freidberg S E (2001) Gardening on the edge: the social conditions of unsustainability on an African urban periphery. *Annals of the Association of American Geographers* **91**: 349–69

Friedland W (1991) The transnationalisation of production and consumption of food and fibre: challenges for rural research. In: Aimas R, Withi N (eds) *Final futures in an international world*. University of Trondheim Press, Trondheim, Norway, pp. 115–35

Friedland W H (1994a) The new globalisation: the case of fresh produce. In: Bonanno A, Busch L, Friedland W, Gouveia L, Mingione E (eds) *From Columbus to ConAgra: the globalization of agriculture and food*. University Press of Kansas, Lawrence

Friedland W H (1994b) The global fresh fruit and vegetable system: an industrial organization analysis. In: McMichael P (ed.) *The global restructuring of agro-food systems*. Cornell University Press, Ithaca, pp. 173–89

Friedland W H (1997) 'Creating space for food' and 'agro-industrial just-in-time'. In: Goodman D, Watts M J (eds) *Globalising food: agrarian questions and global restructuring*. Routledge, London and New York, pp. 226–34

Friedland W H, Barton A E, Thomas R J (1981) *Manufacturing green gold: capital, labor and technology in the lettuce industry*. Cambridge University Press, New York

Friedmann H (1978a) World market, state, and the family farm: social bases of household production in the era of wage labor. *Comparative Studies in Society and History* **20**: 545–86

Friedmann H (1978b) Simple commodity production and wage labour on the American plains. *Journal of Peasant Studies* **6**: 71–100

Friedmann H (1980) Household production and the national economy: concepts for the analysis of agrarian formations. *Journal of Peasant Studies* **7**: 158–84

Friedmann H (1982) The political economy of food: the rise and fall of the international food order of the post-war era. *American Journal of Sociology* **88S**: 248–66

Friedmann H (1986) Family enterprise in agriculture: structural limits and political possibilities. In: Cox G, Lowe P, Winter M (eds) *Agriculture: people and politics*. Allen & Unwin, London, pp. 20–40

Friedmann H (1987) The family farm and the international food regimes. In: Shanin T (ed.) (1987) *Peasants and peasant societies: selected readings*. Basil Blackwell, Oxford, pp. 247–58

Friedmann H (1990) The origins of third world food dependence. In: Bernstein H, Crow B, Mackintosh M, Martin C (eds) *The food question: profits versus people*. Earthscan, London, pp. 13–31

Friedmann H (1993) The political economy of food. *New Left Review* **197**: 29–57

Friedmann H, McMichael P (1989) Agriculture and the state system: the rise and fall of national agricultures, 1870 to the present. *Sociologia Ruralis* **29**: 93–117

Frissel M J (ed.) (1978) *Cycling of mineral nutrients in agricultural ecosystems*. Elsevier, Amsterdam

Froud J (1994) The impact of ESAs on lowland farming. *Land Use Policy* **11**: 107–18

Frouws J, Mol A (1997) Ecological modernization theory and agricultural reform. In: Haan H, Long N (eds) *Images and realities of rural life*. Van Gorcum, Assen, pp. 220–39

Fuller A J (1990) From part-time farming to pluriactivity: a decade of change in rural Europe. *Journal of Rural Studies* **6**: 361–73

Fuller R M, Parsell R J (1990) Classification of TM imagery in the study of land use in lowland Britain: practical considerations for operational use. *International Journal of Remote Sensing* **11**: 1901–17

Fuller R M, Groom G B, Jones A R (1994) The land cover map of Great Britain: an automated classification of Landsat Thematic Mapper data. *Photogrammetric Engineering and Remote Sensing* **60**: 553–62

Fulton A L A, Clark R J (1996) Farmer decision making under contract farming in northern Tasmania. In: Burch D, Rickson R E, Lawrence G E (eds) (1996) *Globalization and agri-food restructuring*. Avebury, Aldershot, pp. 219–38

Funahashi (1996) Farming by the older generation: the exodus of young labor in Yasothon province, Thailand. *Southeast Asian Studies* **33**: 107–21

Furuseth O J (1997) Sustainability issues in the industrialization of hog production in the United States. In: Ilbery B W, Chiotti Q, Rickard T (eds) *Agricultural restructuring and sustainability: a geographical perspective*. CAB International, Wallingford, pp. 293–312

Furuseth O J, Lapping M B (eds) (1999) *Contested countryside: the rural-urban fringe in North America*. Ashgate, Aldershot

Furuseth O J, Pierce J T (1982) A comparative analysis of farmland preservation programmes in North America. *Canadian Geographer* **26**: 191–206

Gabriel Y, Lang T (1995) *The unmanageable consumer*. Sage Publications, London

Galdos R, Gutierrez J A, Icaran C, Urrestarazu E R, de Buruaga M S, Arbulu A (2000) European policies and mountain-based agriculture on the Cantabrian-Pyrennean arc. Unpublished paper presented at the Anglo-Spanish Rural Geography Symposium, Universidade de Valladolid

Gale Jr H F (1996) Age cohort analysis of the twentieth century decline in US farmer numbers. *Journal of Rural Studies* **12**: 15–26

Gallup J L, Sachs J D, Mellinger A D (1999) Geography and economic development. In: Pleskovic and Stiglitz (eds) Annual World Bank Conference on Developed Economies, 1998. World Bank, Washington DC, pp. 127–88

Ganderton P (2000) *Mastering Geography*. Macmillan, Basingstoke

Ganjanapan A (1997) The politics of environment in northern Thailand: ethnicity and highland development programmes. In: Hirsch P (ed.) *Seeing forests for trees: environment and environmentalism in Thailand*. Silkworm Books, Chiang Mai

Gant R L, Talbot P (2000) Wall posters from fieldwork. *Teaching Geography* **25**: 82–6

Garforth C, Usher R (1997) Promotion and uptake pathways for research output: a review of analytical frameworks and communication channels. *Agricultural Systems* **55**: 301–22

Garrod G, Willis K (1995) Valuing the benefits of the South Downs Environmentally Sensitive Area. *Journal of Agricultural Economics* **46**: 160–73

Garrod G, Willis K, Saunders C (1994) The benefits and costs of the Somerset Levels and Moors ESA. *Journal of Rural Studies* **10**: 131–45

Gaskell P T, Tanner M F (1991) Agricultural change and Environmentally Sensitive Areas. *Geoforum* **22**: 81–90

Gaskell P T, Winter M (1996) *Beef farming in Great Britain: farmer responses to the 1992 CAP reforms and implications of the 1996 BSE crisis*. Report to the Countryside Commission, Scottish Natural Heritage and English Nature

Gasson R M (1966) The influence of urbanization on farm ownership and practice. *Studies in Rural Land Use, Department of Agricultural Economics, Wye College, University of London*, no. 7

Gasson R M (1967) Some economic characteristics of part-time farming in Britain. *Journal of Agricultural Economics* **18**: 111–20

Gasson R M (1980) Roles of farm women in England. *Sociologia Ruralis* **20**: 165–80

Gasson R M (1984) Farm women in Europe: their needs for off-farm employment. *Sociologia Ruralis* **24**: 216–28

Gasson R M (1986) Part-time farming in England and Wales. *Journal of the Royal Agricultural Society* **147**: 34–41

Gasson R M (1987) The nature and extent of part-time farming in England and Wales. *Journal of Agricultural Economics* **38**: 175–821

Gasson R M (1988a) *The economics of part-time farming*. Longman, London

Gasson R M (1988b) Farm diversification and rural development. *Journal of Agricultural Economics* **39**: 175–82

Gasson R M (1989) Farm work by farmers' wives, *Occasional Papers, Farm Business Unit, Wye College, University of London*, no. 15

Gasson R M (1990) Part-time farming and pluriactivity. In: Britton D (ed.) *Agriculture in Britain: changing pressures and policies*. CAB International, Wallingford, pp. 161–72

Gasson R M (1991) Part-time farming: a note on definitions – comment. *Journal of Agricultural Economics* **42**: 200–01

Gasson R M (1992) Farmers' wives: their contribution to the farm business. *Journal of Agricultural Economics* **39**: 175–82

Gasson R M (1994) Gender issues in European agricultural systems. In: Dent J B, McGregor M J (eds) *Rural and farming systems analysis.* CAB International, Wallingford, pp. 236–42

Gasson R M (1998) Educational qualifications of UK farmers: a review. *Journal of Rural Studies* **14**: 487–98

Gasson R M, Errington A (1993) *The farm family business.* CAB International, Wallingford

Gasson R M, Hill P (1990) *An economic evaluation of the Farm Woodland Scheme.* Farm Business Unit, Department of Agricultural Economics, Wye College, University of London

Gasson R M, Winter M (1992) Gender relations and farm household pluriactivity. *Journal of Rural Studies* **8**: 387–97

Gasson R M, Crow G, Errington A, Hutson J, Marsden T K, Winter M (1988) The farm as a family business: a review. *Journal of Agricultural Economics* **39**: 1–41

Geertz C (1963) *Agricultural involution in Indonesia.* University of California Press, Berkeley

Gertler M E (1999) Sustainable communities and sustainable agriculture on the Prairies, In: Pierce J T, Dale A (eds) *Communities, development and sustainability across Canada.* UBC Press, Vancouver, pp. 121–39

Ghaffar A, Robinson G M (1997) Restoring the agricultural landscape: the impact of government policies in East Lothian, Scotland. *Geoforum* **28**: 205–17

Gibb R (2000) Trade not aid. *Geography Review* **13**(5): 18–22

Gibb R, Lemon A (2001) Zimbabwe in crisis. *Geography Review* **14**(5): 24–8

Giddens A (1998) *The third way.* Polity Press, Cambridge

Gilbert A, Ward P (1984) Community action by the urban poor: democratic involvement, community self-help or a means of social control? *World Development* **12**: 8–20

Gilg A W (ed.) (1992) *Restructuring the countryside.* Avebury, Aldershot

Gilg A W, Battershill M (1998) Quality farm food in Europe: a possible alternative to the industrialised food market and to current agri-environmental policies: lessons from France. *Food Policy* **23**: 25–40

Giordano M, Matzke G (2001) Classics in human geography revisited: Commentary 2. In: Watts M (1983) *Silent violence, Progress in Human Geography* **25**: 623–8

Glennie P, Thrift N J (1996) Consumption, shopping and gender. In: Wrigley N, Lowe M (eds) *Retailing, consumption and capital.* Longman, Harlow, pp. 221–38

Goebel A (1998) Process, perception and power: notes from 'participatory' research in a Zimbabwean Resettlement Area. *Development and Change* **29**: 277–305

Goldschmidt W (1978) *As you sow: three studies in the social consequences of agribusiness.* Allenheld and Osmun, Montclair, NJ

Goldsmith E, Mander J (eds) (2001) *The case against the global economy: And for a turn towards localization.* Earthscan, London

Golledge R G, Stimson R J (1997) *Spatial behavior: a geographical perspective.* Guilford Press, New York and London

Golledge R G, Timmermans H (1990) Applications of behavioural research on spatial problems, 1: cognition. *Progress in Human Geography* **14**: 57–99

Gonzalez J J, Benito C G (2001) Profession and identity. The case of family farming in Spain. *Sociologia Ruralis* **41**: 343–57

Goodenough R (1984) The great America crop surplus: 1983 solution. *Geography* **69**: 351–3

Gooding R W, Mason D C, Settle J J, Veitch N, Wyatt B K (1993) The application of GIS to the monitoring and modelling of land cover and use in the United Kingdom. In: Mather P M (ed.) *Geographic information handling – research and applications.* Wiley, Chichester and New York, pp. 185–96

Goodman D (1994) World-scale processes and agro-food systems: critique and agenda, *Working Papers, Center for the Study of Social Transformations, University of California at Santa Cruz,* no. 94–11

Goodman D (1999) Agro-food studies in the 'age of ecology': nature, corporeality, bio-politics. *Sociologia Ruralis* **39**: 17–38

Goodman D (2001) Ontology matters: the relational materiality of nature and agro-food studies. *Sociologia Ruralis* **41**: 182–200

Goodman D, DuPuis E M (2002) Knowing food and growing food: beyond the production – consumption debate in the sociology of agriculture, *Sociologia Ruralis* **42**: 5–22

Goodman D, Redclift M (1981) *From peasant to proletarian: capitalist development and agrarian transitions*. Basil Blackwell, Oxford

Goodman D, Redclift M (eds) (1989) *The international farm crisis*. Macmillan, Basingstoke

Goodman D, Redclift M (1991) *Refashioning Nature*. Blackwell, Oxford

Goodman D, Watts M J (1994) Reconfiguring the rural or fording the divide? Capitalist restructuring and the global agro-food system. *Journal of Peasant Studies* **22**: 1–49

Goodman D, Watts M J (eds) (1997) *Globalising food: agrarian questions and global restructuring*. Routledge, London and New York

Goodman D, Sorj B, Wilkinson J (1987) *From farming to biotechnology: a theory of agro-industrial development*. Basil Blackwell, New York

Goodwin M (1998) 'The governance of rural areas: some emerging research issues'. *Journal of Rural Studies* **14**: 5–12

Goodwin M, Painter J (1996) Local governance, the crises in Fordism and the changing geographies of regulation. *Transactions of the Institute of British Geographers* **21**: 635–48

Goodwin M, Painter J (1997) Concrete research, urban regimes and regulation theory. In: Lauria M (ed.) *Reconstructing urban regime theory*. Sage, Thousand Oaks, CA, pp. 13–29

Goosens F (1997) Failing innovation in the Zairian cassava production system: a comparative historical analysis. *Sustainable Development* **5**: 36–42

Gorton M (2001) Agricultural land reform in Moldova. *Land Use Policy* **18**: 269–80

Gough K (1981) *Rural society in southeast Asia*. Cambridge University Press, Cambridge

Gough M, Marrs R (1990a) A comparison of soil fertility between semi-natural and agricultural plant communities: implications for the creation of species-rich grassland on abandoned agricultural land. *Biological Conservation* **51**(2): 83–96

Gough M, Marrs R (1990b) Trends in soil chemistry and floristics associated with the establishment of a low input meadow system on an arable clay soil in Essex, England. *Biological Conservation* **52**(2): 135–46

Gould P R (1963) Man against his environment: a game theoretic framework. *Annals of the Association of American Geographers* **53**: 290–7

Grabowski R (1995) Commercialization, nonagricultural production, agricultural innovation and economic development. *Journal of Developing Areas* **30**: 41–62

Graetz R D, Pech R P, Davis A W (1988) The assessment and monitoring of sparsely-vegetated rangelands using calibrated Landsat data. *International Journal of Remote Sensing* **9**: 1201–22

Grainger A (1995) National land use morphology. *Geography* **80**: 235–46

Grant R (1993) Against the grain. Agricultural trade policies. *Political Geography* **12**(3):

Gray I, Lawrence G, Dunn T (1993) *Coping with change: Australian farmers in the 1990s*. Centre for Rural Social Research, Wagga Wagga

Gray J (2000) The CAP and the re-invention of the rural in the European Community. *Sociologia Ruralis* **40**: 30–52

Gray L (1999) Is land being degraded? A multi-scale investigation of landscape change in southwestern Burkina Faso. *Land Degradation and Development* **10**: 47–60

Green R E (1995) The decline of the corncrake *crex crex* in Britain continues. *Bird Study* **42**: 66–75

Green R M, Lucas N S, Curran P J, Foody G M (1996) Coupling remotely sensed data to an ecosystem simulation model – an example involving a coniferous plantation in upland Wales. *Global Ecology and Biogeography Letters* **5**: 192–205

Greene C R (2000) US organic farming emerges in the 1990s: adoption of certified systems. *Economic Research Service, Agriculture Information Bulletin, USDA*, no. 770

Gregor H F (1963) Industrialized drylot dairying: an overview. *Economic Geography* **39**: 299–318

Gregory D (1978) *Ideology, science and human geography.* Hutchinson, London

Gregory R D, Baillie S R (1998) Large-scale habitat use of some declining British birds. *Journal of Applied Ecology* 35: 785–99

Gregory R D, Noble D H, Campbell L C, Gibbons D W (1999) *The state of the nation's birds.* RSPB, Sandy, Bedfordshire

Grey M (2000) The industrial food stream and its alternatives in the United States: an introduction. *Human Organization* 59: 143–50

Griffin E (1973) Testing the von Thunen theory of Uruguay. *Geographical Review* 63: 500–16

Grigg D B (1992a) *The transformation of agriculture in the West.* Blackwell, Oxford and Cambridge, MA

Grigg D B (1992b) Agriculture in the world economy: an historical geography of decline. *Geography* 77: 210–22

Grigg D B (1993) The role of livestock products in world food consumption. *Scottish Geographical Magazine* 109: 66–74

Grigg D B (1995) *An introduction to agricultural geography.* Routledge, London, 2nd edition

Grigg D B (1997) The world's hunger: a review 1930–1990. *Geography* 82: 197–206

Grigg D B (1999) The changing geography of world food consumption in the second half of the twentieth century. *Geographical Journal* 165: 1–11

Grigg D B (2001) Food imports, food exports and their role in national food consumption. *Geography* 86: 171–6

Grove A T, Klein F M G (1979) *Rural Africa.* Cambridge University Press, Cambridge

Guthman J (1998) Regulating meaning, appropriating nature: the codification of Californian organic agriculture. *Antipode* 30: 135–54

Guy C (1994) *The retail development process.* Routledge, London

Guyomard H, Bureau J-C, Gohin A, Le Mouel C (2000) Impact of the 1996 FAIR Act on the Common Agricultural Policy in the World Trade Organisation context: the decoupling issue. *Food Policy* 25: 17–34

Hagerstrand T (1967) *Innovation diffusion as a spatial process.* University of Chicago Press, Chicago, translated by A. Pred

Haggett P (1975) *Geography: a modern synthesis.* Harper and Row, New York

Halfacree K H (1997) Contrasting roles for the post-productivist countryside: a postmodern perspective on counter-urbanization. In: Cloke P J, Little J K (eds) *Contested countryside cultures: otherness, marginalization and rurality.* Routledge, London, pp. 70–91

Halfacree K H (1999) A new space or spatial effacement? Alternative futures for the post-productivist countryside. In: Walford N S, Everitt J, Napton D (eds) *Reshaping the countryside: perceptions and processes of rural change.* CAB International, Wallingford, pp. 67–76

Hall A, Mogyorody V (2001) Organic farmers in Ontario: an examination of the conventionalization argument. *Sociologia Ruralis* 41: 399–422

Hall A D (1936) *The improvement of native agriculture in relation to population and public health.* Oxford University Press, London

Hall P (ed.) (1968) *Von Thunen's isolated state.* Pergamon, New York and London

Hallberg M C, Spitze R G F, Ray D E (eds) (1994) *Food, agriculture and rural policy into the twenty-first century: issues and trade-offs.* Westview Press, Boulder, CO

Halliday J (1989) Attitudes to farm diversification: results from a survey of Devon farmers. *Journal of Agricultural Economics* 40: 191–202

Hammond R (1994) A geography of overseas aid. *Geography* 79: 210–21

Haney W G, Miller L C (1991) US farm women, politics and policy. *Journal of Rural Studies* 7: 115–21

Hanley N (ed.) (1991) *Farming and the countryside: an economic analysis of external costs and benefits.* CAB International, Wallingford

Hanson L L (2001) The disappearance of the open West: individualism in the midst of agricultural restructuring. In: Epps R, Whitson D (eds) *Writing off the rural West: globalization, governments and the transformation of rural communities.* University of Alberta Press and Parkland Institute, Edmonton, pp. 165–84

Harding A E (1988) Examples of remote sensing applications for local resource planning. In:

Vaughan R A, Kirby R P (eds) *GIS and remote sensing for local resource planning*. Remote Sensing Products and Publications, Dundee, pp. 101–111

Harper S (ed.) (1992) *The greening of rural policy: international perspectives*. Plenum Press, London and New York

Harrison M (1982) Towards a practical theory of agrarian transition. In: Harriss J C (ed.) (1982) *Rural development: theories of peasant economy and agrarian change*. Hutchinson, London, pp. 399–404

Harrison M (2001) *King sugar: Jamaica, the Caribbean and the world sugar economy*. Latin America Bureau, London

Harrison M, Flynn A, Marsden T K (1997) Contested regulatory practice and the implementation of food policy: exploring the local and national interface. *Transactions of The Institute of British Geographers* new series, **22**: 473–87

Harrison-Mayfield L, Dwyer J, Brookes G (1998) The socio-economic effects of the Countryside Stewardship Scheme. *Journal of Agricultural Economics* **49**: 157–70

Harriss B (1990) The intra-family distribution of hunger in south Asia. In: Dreze J, Sen A K (eds) *The political economy of hunger*. Clarendon, London, 2 vols, pp. 351–424

Harriss J C (1981) Social organisation and irrigation: ideology, planning and practice in Sri Lanka's settlement schemes. In: Bayliss-Smith T (ed.) *Understanding green revolutions*. Cambridge University Press, Cambridge, pp. 315–36

Harriss J C (ed.) (1982) *Rural development: theories of peasant economy and agrarian change*. Hutchinson, London

Harriss J, de Renzio P (1997) Missing link or analytically missing? The concept of social capital. An introductory bibliographic essay. *Journal of International Development* **9**: 919–37

Hart G (1996) The agrarian question and industrial dispersal in South Africa: agro-industrial linkages through Asian lenses. *Journal of Peasant Studies* **23**: 245–77

Hart J F (1956) The changing distribution of sheep in Britain. *Economic Geography* **32**: 260–74

Hart J F (1975) *The look of the land*. Prentice-Hall, Englewood Cliffs, NJ

Hart J F (1976) Urban encroachment on rural areas. *Geographical Review* **66**: 1–17

Hart J F (1978) Cropland concentrations in the South. *Annals of the Association of American Geographers* **68**: 505–17

Hart J F (1980) Land use change in a Piedmont county. *Annals of the Association of American Geographers* **70**: 492–527

Hart J F (1986) Change in the Corn Belt. *Geographical Review* **76**: 51–72

Hart J F (1991) The perimetropolitan bow wave. *Geographical Review* **81**: 35–51

Hart J F (1992) Non-farm farms. *Geographical Review* **82**: 166–79

Hart J F (1998) *The rural landscape*. Johns Hopkins University Press, Baltimore and London

Hart J F (2001) Half a century of cropland change. *Geographical Review* **91**: 525–43

Hart J F, Mayda C (1998) The industrialization of livestock production in the United States. *Southeastern Geographer* **38**: 58–78

Hart K, Wilson G A (1998) UK implementation of Agri-environment Regulation 2078/92/EEC: enthusiastic supporter or reluctant participant? *Landscape Research* **23**: 255–72

Hart P W E (1978) Geographical aspects of contract farming, with special reference to the supply of crops to processing plants. *Tijdschrift voor Economische en Social Geographie* **69**: 205–15

Hart P W E (1992) Marketing agricultural produce. In: Bowler I R (ed.) *The geography of agriculture in developed market economies*. Longman, London, pp. 162–206

Hartell J G, Swinnen J F M (eds) (2000) *Agriculture and East-West European integration*. Ashgate, Aldershot

Harts-Broekhuis A, Huisman H (2001) Resettlement revisited: land reform results in resource-poor regions in Zimbabwe. *Geoforum* **32**: 285–98

Hartwick E (1998) Geographies of consumption: a commodity-chain approach. *Environment and Planning D: Society and Space* **16**: 423–37

Hartwick J M, Olewiler N D (1986) *The economics of natural resource use*. Harper and Row, New York

Haugen M S (1990) Female farmers in Norwegian agriculture – from traditional farm women to professional farmers. *Sociologia Ruralis* **30**: 197–209

Hartmann B, Boyce J (1983) *A quiet violence*. Zed Books, London

Harvey D W (1973) *Social justice and the city*. Edward Arnold, London

Harvey D W (1982) *The limits of capital*. Blackwell, Oxford

Harvey D W (1989) *The condition of postmodernity*. Blackwell, Oxford

Harvey D W (1996) *Justice, nature and the geography of difference*. Blackwell, Oxford

Harvey D W (2000) *Spaces of hope*. Edinburgh University Press, Edinburgh

Hassenein N, Kloppenburg J (1995) Where the grass grows again: knowledge exchange in the sustainable agriculture movement. *Rural Sociology* **60**: 721–40

Haugen M S (1990) Female farmers in Norwegian agriculture: from traditional farm women to professional farmers. *Sociologia Ruralis* **30**: 197–209

Hayami Y (1984) Assessment of the Green Revolution. In: Eicher C K, Staatz J M (eds) *Agricultural development in the Third World*. Johns Hopkins Press, Baltimore, pp. 389–96

Hayami Y, Ruttan V (1985) *Agricultural development: an international perspective*. Johns Hopkins University Press, Baltimore

Haynes J (1992) Making a virtue of idleness? *Ecos* **13**(3): 3–8

Hazell P B R, Ramasamy C (1991) *The Green Revolution reconsidered: the impact of high yielding rice varieties in South India*. Johns Hopkins University Press, Baltimore

Healey M, Ilbery B W (eds) (1985) *The industrialisation of the countryside*. Geo Books, Norwich

Heffernan W D, Constance D H (1994) Transnational corporations and the globalization of the food system, pp. 29–51 in Bonnanno A, Busch L, Friedland W H, Gouveia L, Minquione E (eds) *From Columbus to ConAgra: the globalization of agriculture and food*. University Press of Kansas, Lawrence

Heimlich R E, Kula O E (1991) Economics of livestock and crop production on post-CRP lands. In: Joyce L A, Mitchell J E, Skold M D (eds) *The Conservation Reserve – yesterday, today and tomorrow*. USDA Forest Service, Rocky Mountain Forest and Range Experiment Station, Fort Collins, Colorado, no. GTR RM-203, pp. 11–23

Helburn N (1957) The bases for a classification of world agriculture. *Professional Geographer* **9**: 2–7

Helmers G A, Hoag D L (1994) Sustainable agriculture. In: Hallberg M C, Spitze R G F, Ray D E (eds) *Food, agriculture and rural policy into the twenty-first century: issues and trade-offs*. Westview Press, Boulder, CO, pp. 11–134

Henderson G (1998) Nature and fictitious capital: the historical geography of an agrarian question. *Antipode* **30**: 73–118

Henderson G (2000) *California and the fictions of capital*. Oxford University Press, New York

Henshall J D (1967) Models of agricultural activity. In: Chorley R J, Haggett P (eds) *Models in geography*. Methuen, London, pp. 425–58

Herdt R W, Steiner R A (1995) Agricultural sustainability: concepts and conundrums. In: Payne and Steiner (eds), pp. 3–14

Herod A (2001) Labor internationalization and the contradictions of globalization: or why the local is sometimes still important in a global economy. *Antipode* **30**: 407–26

Herington J (1984) *The outer city*. Harper and Row, London

Higginbotham J (1997) On- and off-farm business diversification by farm households. Unpublished PhD thesis, University of Coventry, in conjunction with the National Agricultural Centre

Hill B (1993) The 'myth' of the family farm: defining the family farm and assessing its importance in the European Community. *Journal of Rural Studies* **10**: 359–70

Hill B (1999) Farm household incomes: perceptions and statistics. *Journal of Rural Studies* **15**: 345–58

Hill J, Meiger J (1988) Regional land cover and agricultural area statistics and mapping in the Département Ardeche, France, by use of Thematic Mapper. *International Journal of Remote Sensing* 9: 1573–95

Hill M, Aaronovitch S, Baldock D (1989) Non-decision-making in pollution control in Britain. *Policy and Politics* 17: 227–40

Hill R D (1997a) An agricultural transition on the Pacific Rim: explorations towards a model. In: Magee T, Watters R (eds) *Asia Pacific: new geographies of the Pacific Rim*. C Hurst & Co., New York, pp. 93–112

Hill R D (1997b) Questions for Chinese agriculture. *Land Use Policy* 14: 335–6

Hill R D (1998) Stasis and change in forty years of southeast Asian agriculture. *Singapore Journal of Tropical Geography* 19: 1–25

Hill R, Smith D L (1994) Is von Thunen alive and well? A transport-cost surface model for South Australian wheat. *Australian Geographical Studies* 32: 183–90

Hilts S G (1997) Achieving sustainability in rural land management through landowner involvement in stewardship programmes. In: Ilbery B W, Chiotti Q, Rickard T (eds) (1997) *Agricultural restructuring and sustainability: a geographical perspective*. CAB International, Wallingford, pp. 267–77

Hirtz F (1994) *Managing insecurity*. University of California Press, Berkeley

Hodge I (1996) On penguins and icebergs: the Rural White Paper and the assumptions of rural policy. *Journal of Rural Studies* 12: 1–7

Hoggart K (1990) Let's do away with rural. *Journal of Rural Studies* 6: 245–57

Hoggart K, Mendoza C (1999) African immigrant workers in Spanish agriculture. *Sociologia Ruralis* 39: 538–62

Hoggart K, Black R, Buller H (1995) *Rural Europe: identity and change*. Edward Arnold, London

Hollander G M (1995) Agro-environmental conflict and world food system theory: sugar cane in the Everglades Agricultural Area. *Journal of Rural Studies* 12: 309–18

Holloway L E (2000a) 'Hell on earth and paradise all at the same time': the production of small-holding space in the British countryside. *Area* 32: 307–16

Holloway L E (2000b) Hobby-farming in the UK: producing pleasure in the post-productivist countryside. Unpublished paper presented at the Anglo-Spanish Rural Geography Symposium, Universidade de Valladolid

Holloway L E, Ilbery B W (1996) Farmers' attitudes towards environmental change, especially climate change, and the adjustment of crop mix and farm management. *Applied Geography* 16: 159–71

Holloway L E, Ilbery B W (1997) Global warming and navy beans: decision making by farmers and food companies in the UK. *Journal of Rural Studies* 13: 343–55

Holloway L E, Kneafsey M (2000) Reading the space of the farmers' market: a case study from the UK. *Sociologia Ruralis* 40: 366–83

Holloway L E, Ilbery B W, Gray D (1995) Modelling the impact on navy beans and vining peas of temperature changes predicted from global warming. *European Journal of Agronomy* 4: 281–7

Holmes J M (2002) Diversity and change in Australia's rangelands: a post-productivist transition with a difference? *Transactions of the Institute of British Geographers* new series, 27: 362–84

Hopkins J (1998a) Commodifying the countryside: marketing myths of rurality. In: R W Butler, Hall C M, Jenkins J M (eds) *Tourism and recreation in rural areas*. Chichester: John Wiley & Sons Ltd, pp. 139–56

Hopkins J (1998b) Signs of the post-rural: marketing myths of a symbolic countryside. *Geografika Annaler* 80B: 65–82

Hossain M (1989) *Green Revolution in Bangladesh. Impact on growth and distribution of income*. University Press, Dhaka

Hossell J E, Jones P J, Marsh J S, Parry M L, Rehman T, Tranter R B (1996) The likely effects of climatic change on agricultural land use in England and Wales. *Geoforum* 27: 149–57

Hughes A (1996) Retail restructuring and the strategic significance of food retailers' own labels: a UK–USA comparison. *Environment and Planning A* 28: 2201–26

Hughes A (2000) Retailers, knowledges and changing commodity networks: the case of the cut flower trade. *Geoforum* 31: 175–90

Hulme M, Mitchell J, Ingram W, Johns T, New M, Viner D (1999) Climate change scenarios for global impacts studies. *Global Environmental Change* 9(4): 3–19

Hyttinen P, Kola J (1995) Farm forests and rural livelihood in Finland. *Journal of Rural Studies* 12: 387–96

Iacoponi L, Brunori G, Rovai M (1995) Endogenous development and the agro-industrial district. In: Van der Ploeg J, Van Dijk G (eds) *Beyond modernization: the impact of endogenous rural development.* Van Gorcum, Assen, pp. 28–69

IACPA (Integrated Arable Crop Production Alliance) (1998) *Integrated farming: agricultural research into practice.* MAFF, London

IFLS (Institut fur landliche Strukturforshung) (1999) *Implementation and effectiveness of agri-environmental schemes established under Regulation 2978/92 (Project FAIR1 CT95-274) Final Consolidated Report.* IFLS, Frankfurt

Ilbery B W (1978) Agricultural decision-making: a behavioural perspective. *Progress in Human Geography* 2: 448–66

Ilbery B W (1981) Dorset agriculture: a classification of regional types. *Transactions of the Institute of British Geographers* new series, 6: 214–27

Ilbery B W (1982) The decline in hop farming in Hereford and Worcestershire. *Area* 14: 203–12

Ilbery B W (1983a) Harvey's principles reapplied: a case study of the declining West Midlands hop industry. *Geoforum* 14: 111–23

Ilbery B W (1983b) A behavioural analysis of hop farming in Hereford and Worcestershire. *Geoforum* 14: 447–59

Ilbery B W (1983c) Goals and values of hop farmers. *Transactions of the Institute of British Geographers* new series, 8: 329–41

Ilbery B W (1984) Agricultural specialisation and farmer decision behaviour: a case study of hop farming in the West Midlands. *Tijdschrift voor Economische en Sociale Geografie* 75: 329–34

Ilbery B W (1985a) *Agricultural geography: a social and economic analysis.* Oxford University Press, Oxford

Ilbery B W (1985b) Factors affecting the structure of viticulture in England and Wales. *Area* 17: 147–54

Ilbery B W (1985c) Horticultural decline in the Vale of Evesham, 1950–1980. *Journal of Rural Studies* 1: 109–20

Ilbery B W (1987a) The development of farm diversification in the UK. Evidence from Birmingham's urban fringe. *Journal of the Royal Agricultural Society of England* 148: 21–35

Ilbery B W (1987b) A future for the farms? *Geographical Magazine,* 249–54

Ilbery B W (1988) Farm diversification and the restructuring of agriculture. *Outlook on Agriculture* 17: 35–9

Ilbery B W (1989) Farm-based recreation: a possible solution to falling farm incomes? *Journal of the Royal Agricultural Society of England* 150: 57–66

Ilbery B W (1990) Adoption of the arable set-aside scheme in England. *Geography* 75: 69–71

Ilbery B W (1991) Farm diversification as an adjustment strategy on the urban fringe of the West Midlands. *Journal of Rural Studies* 7: 207–18

Ilbery B W (1992) From Scott to ALURE – and back again. *Land Use Policy* 9: 131–42

Ilbery B W (1996) Farm-based recreation and the development of Farm Attraction Groups in England and Wales. *Geography* 81: 86–91

Ilbery B W (ed.) (1998) *The geography of rural change.* Longman, Harlow

Ilbery B W (1999) The de-intensification of European agriculture. In: Pacione M (ed.) (1999) *Applied Geography: principles and practice.* Routledge, London and New York, pp. 274–87

Ilbery B W (2001) Changing geographies of global food production In: Daniels P W, Bradshaw M, Shaw D, Sidaway J (eds) *Human Geography: issues for the 21st century.* Pearson Education, Harlow, pp. 253–73

Ilbery B W (2002) Geographical aspects of the 2001 outbreak of foot and mouth disease in the UK. *Geography* 87: 142–7

Ilbery B W, Bowler I R (1993) The Farm Diversification Grant Scheme: adoption and non-adoption in England and Wales. *Environment and Planning C* **11**: 161–70

Ilbery B W, Bowler I R (1994) International competition in agriculture and government intervention: the case of horticulture in England. *Geography* **79**: 361–6

Ilbery B W, Bowler I R (1996) Industrialisation and world agriculture. In: Douglas I, Huggett R, Robinson M (eds) *Companion Encyclopaedia of Geography*. Routledge, London, pp. 228–48

Ilbery B W, Bowler I R (1998) From agricultural productivism to post-productivism. In: Ilbery B W (ed.) *The geography of rural change*. Longman, Harlow, pp. 57–84

Ilbery B W, Evans N J (1996) Post-productive agriculture in the South Midlands. In: Bowler I (ed.) *Progress in research on rural geography*. Occasional Papers, Department of Geography, University of Leicester, no. 35, pp. 83–4

Ilbery B W, Kneafsey M (1997) Regional images and the promotion of quality products and services in the lagging regions of the European Union. Unpublished paper presented to the Third Anglo-French Rural Geography Symposium, Université de Nantes, 11–14 September

Ilbery B W, Kneafsey M (1998) Product and place: promoting quality products and services in the lagging rural regions of the European Union. *European Urban and Regional Studies* **5**: 329–41

Ilbery B W, Kneafsey M (1999) Niche markets and regional speciality food products in Europe: towards a research agenda. *Environment and Planning A* **31**: 2207–22

Ilbery B W, Kneafsey M (2000) Registering regional speciality food and drink products in the United Kingdom: the case of PDOs and PGIs. *Area* **32**: 317–26

Ilbery B W, Stiell B (1991) The uptake of the Farm Diversification Grant Scheme in England. *Geography* **76**: 259–63

Ilbery B W, Healy M, Higginbotham J, Noon D (1996) Agricultural adjustment and business diversification by farm households. *Geography* **81**: 301–10

Ilbery B W, Chiotti Q, Rickard T (eds) (1997) *Agricultural restructuring and sustainability: a geographical perspective*. CAB International, Wallingford

Ilbery B W, Bowler I R, Clark G, Crockett A, Shaw A (1998) Farm based tourism as an alternative enterprise: a case study from the northern Pennines, England. *Journal of Regional Studies* **32**: 355–64

Ilbery B W, Kneafsey M, Bowler I, Clark G (1999a) Quality products and services in the lagging rural regions of the European Union: a producer perspective. Unpublished paper presented to the International Rural Geography Symposium, Halifax and Truro, Nova Scotia, July 11–17

Ilbery B W, Holloway L, Arber R (1999b) The geography of organic farming in England and Wales in the 1990s. *Tijdschrift voor economische en social geografie* **90**: 285–95

Ilbery B W, Kneafsey M, Bowler I, Clark G (2000) Consumer perspectives on quality products and services from lagging rural regions in the European Union. Unpublished paper presented to the Anglo-Spanish Rural Geography Symposium, Universidade de Valladolid, July

IPF (Intergovernmental Policy on Forestry) (2001) *Country report – Indonesia*. IPF, New York

Ives J D (1991) The floods in Bangladesh – who is to blame? *New Scientist* **130**(1764): 34–7

Jackson G H (1994) Set aside. A responsible use of land? *Journal of the Royal Agricultural Society of England* **155**: 34–9

Jsckson J C (1969) Towards an understanding of plantation agriculture. *Area* **1**: 36–41

Jackson P (1999) Commodity cultures: the traffic in things. *Transactions of the Institute of British Geographers* **24**: 95–108

Jackson P, Holbrook B (1995) Multiple meanings: shopping and the cultural politics of identity. *Environment and Planning A* **27**: 1913–30

Jackson P, Thrift N J (1995) Geographies of consumption. In: Miller D (ed.) *Acknowledging consumption: a review of new studies*. Routledge, London, pp. 204–37

Jaeger W K (1986) *Agricultural mechanization: the economics of animal draft power in West Africa*. Westview Press, Boulder, CO

Jaffee S (1994) Exporting high-value food commodities: success stories from Developing Countries, *World Bank Discussion Papers,* 198

Jarosz L (1996) Working in the global food system: a focus for international comparative analysis. *Progress in Human Geography* 20: 41–55

Jarosz L, Qazi J (2000) The geography of Washington's world apple: global expressions in a local landscape. *Journal of Rural Studies* 16: 1–12

Jayasuriya S K, Shand R T (1986) Technological change and labour absorption in Asian agriculture: some emerging trends. *World Development* 14: 415–28

Jenny H (1941) *Factors of soil formation.* McGraw-Hill, London

Jervell A M (1999) Changing patterns of family farming and pluriactivity. *Sociologia Ruralis* 39: 100–116

Jessop B (1990) Regulation theories in retrospect and prospect. *Economy and Society* 19: 153–216

Jessop B (1994) Post-Fordism and the state. In: Amin A (ed.) *Post-Fordism: a reader.* Blackwell, Oxford, pp. 251–78

Jokisch B D (1997) From labour circulation to international migration: the case of south central Ecuador. *Yearbook of the Conference of Latin American Geographers* 23: 63–75

Johnston R J (1997) *Geography and geographers: Anglo-American human geography since 1945.* Edward Arnold, London, 5th edition

Jokinen P (2000) Europeanisation and ecological modernization: agri-environment policy and practices in Finland. In: Mol A J P, Sonnenfeld D A (eds) *Ecological modernization around the world: perspectives and critical debates.* Frank Cass, London, pp. 138–70

Jones A R (1984) Agriculture: organisation, reform and the EEC. In: Williams A (ed.) *Southern Europe transformed: political and economic change in Greece, Italy, Portugal and Spain.* Harper and Row, London, pp. 236–67

Jones A R (1989a) The role of the SAFER in agricultural restructuring: the case of Languedoc-Rousillon, France. *Land Use Policy* 6: 249–61

Jones A R (1989b) The reform of the EECs table wine sector: agricultural despecialisation in the Languedoc. *Geography* 74: 29–37

Jones A R (1990) New directions for West German agricultural policy. The example of Schleswig-Holstein. *Journal of Rural Studies* 6: 9–16

Jones A R (1991) The impact of the EC's set-aside programme: the response of farm businesses in Rendsburg-Eckenforde, Germany. *Land Use Policy* 8(2): 10824

Jones A R (1992) Set aside: the German experience. *Ecos* 13(3): 30–4

Jones A R, Clark J (2000) Of vines and policy vignettes: sectoral evolution and institutional thickness in Languedoc. *Transactions of the Institute of British Geographers* 25: 333–54

Jones A R, Fasterding F, Plankl R (1993) Farm household adjustments to the EC's set-aside policy. Evidence from Rheinland-Pflaz. *Journal of Rural Studies* 9(1): 65–80

Jones M E (1976) Topographic climates: soils, slopes and vegetation. In: Chandler T J, Gregory S (eds) *Climate of the British Isles.* Longman, London, pp. 288–306

Jones P, Hillier D (1997) Changes brewing: superpub developments in the UK. *Geography Review* 10(3): 26–8

Joseph A E, Keddie P D (1981) The diffusion of grain corn production through Southern Ontario, 1946–1971. *Canadian Geographer* 25: 333–49

Joseph A E, Keddie P D (1985) The diffusion of grain corn production through Southern Ontario, 1971–1981. *Canadian Geographer* 29: 168–72

Josling T E, Hamway D (1976) Income transfer effects of the CAP. In: Davey B *et al.* (eds) *Agriculture and the state.* Macmillan, London, pp. 180–205

Joyce L A, Mitchell J E, Skold M D (eds) (1991) *The Conservation Reserve – yesterday, today and tomorrow.* USDA Forest Service, Rocky Mountain Forest and Range Experiment Station, Fort Collins, Colorado, no. GTR RM-203

Kabeer N (1991) Gender dimensions of rural poverty: analysis from Bangladesh. *Journal of Peasant Studies* 18: 241–62

Kabeer N (1994) *Reversed realities: gender hierarchies in development thought*. Verso, London

Kalabamu F T (2000) Land tenure and management reforms in East and Southern Africa – the case of Botswana. *Land Use Policy* 17: 305–20

Kaldis P E (2002) The Greek fresh-fruit market in the framework of the Common Agricultural Policy. Unpublished PhD thesis, University of Coventry

Kaltoft P (1999) Values about nature in organic farming: practice and knowledge. *Sociologia Ruralis* 39: 39–53

Kates R W (1962) *Hazard and choice perception in flood plain management*. Research Papers, Department of Geography, University of Chicago, no. 78

Kates R W (1993) Theory, evidence, study design. In: Turner B L, Hyden G, Kates E W (eds) *Population growth and agricultural change in Africa*. University Press of Florida, Gainesville, pp. 35–52

Katz B, Liu A (2000) Moving beyond sprawl: toward a broader metropolitan agenda. *Brookings Review* 18(2): 31–4

Katz C (2001) Vagabond capitalism and the necessity of social reproduction. *Antipode* 33: 709–28

Kato T (1994) The emergence of abandoned paddy fields in Negeri Sembilan, Malaysia. *Southeast Asian Studies* 32: 145–72

Kautsky J H (1994) *Karl Kautsky: Marxism, revolution and democracy*. Transaction Publishers, New Brunswick, NJ

Kautsky K (1920) *Dictatorship of the proletariat*. National Labour Press, Manchester, translated by H J Stenning

Kautsky K (1988) *The agrarian question*. Zwan, London, 2 vols

Kay A (1998) *The reform of the Common Agricultural Policy: the case of the MacSharry reforms*. CAB International, Wallingford

Keddie P D (1983) The renewed viability of grain corn production in southern Ontario, 1961–1971 and changes in harvesting and storage methods. *Canadian Geographer* 27: 223–39

Keddie P D, Wandel J (2001) The 'second wave': the expansion of soybeans across southern Ontario, 1951–96. *Great Lakes Geographer* 8: 15–30

Keeble D D E, Owens P L, Thompson C (1983) The urban-rural manufacturing shift in the European Community. *Urban Studies* 20: 405–18

Keefe E K, Brenneman L E, Giloane W, Long A K, Moore J M, Jr Walpole N A (1987) *Hungary. A country study*. Department of Army, US Government, Washington, DC

Keeler M E, Skuras D G (1990) Land fragmentation and consolidation policies in Greek agriculture. *Geography* 76: 73–6

Kelly G D, Hill G D E (1987) Updating maps of climax vegetation cover with Landsat MSS data in Queensland, Australia. *Photogrammetric Engineering and Remote Sensing* 53: 633–7

Kelly C E, Ilbery B W (1996) Defining and examining rural diversification: a framework for analysis. *Tijdschrift voor Economische en Social Geografie* 87: 177–85

Kelly C E, Ilbery B W, Gillmor D A (1992) Farm diversification in Ireland: evidence from County Wicklow. *Irish Geography* 25: 23–32

Kelly M (2001) Sustainable rural livelihoods: a case study of Malawi. Unpublished PhD thesis, Kingston University, Kingston-upon-Thames, Surrey

Kelly P (1999) Rethinking the local in labour markets: the consequences of cultural embeddedness in a Philippine growth zone. *Singapore Journal of Tropical Geography* 20: 56–75

Kelly P F (1999) The geographies and politics of globalization. *Progress in Human Geography* 23: 379–400

Kemp D (1991) The greenhouse effect and global warming: a Canadian perspective. *Geography* 76: 121–30

Kenny M, Lobao L M, Curry J, Goe W R (1989) Mid-western agriculture in US Fordism. From the New Deal to economic restructuring. *Sociologia Ruralis* 29: 131–48

Khan M (1997) Aligarh city: urban expansion and encroachment. *Asian Profile* 25: 47–51

King J G M (1939) Mixed farming in northern Nigeria. *Empire Journal of Experimental Agriculture* 7: 271–84 and 285–98

King R L (1987) Italy. In: Clout H D (ed.) *Regional development in Western Europe*. David Fulton, London, 3rd edition, pp. 129–64

Kinsella J, Wilson S, de Jong F, Renting H (2000) Pluriactivity as a livelihood strategy in Irish farm households and its role in rural development. *Sociologia Ruralis* 40: 497–511

Kleinman P J A, Pimentel D, Bryant R B (1995) The ecological sustainability of slash-and-burn agriculture. *Ecosystems and Environment* 52: 235–49

Kneafsey M, Ilbery B W (2001) Regional images and the promotion of speciality food and drink in the West Country. *Geography* 86: 131–40

Kneafsey M, Ilbery B W, Jenkins T (2001) Exploring the dimensions of culture economies in rural West Wales. *Sociologia Ruralis* 41: 296–310

Kneen B (1999) *Farmageddon: food and the culture of biotechnology.* New Society Publishers, Gabriola Island, British Columbia

Knudsen D C (ed.) (1996) *The transition to flexibility.* Kluwer, Dordrecht

Kodras J E (1993) Shifting global strategies of US foreign food aid, 1955–90. *Political Geography* 12: 232–46

Komen M H C, Peerlings J H M (1997) The effects of the Dutch 1996 Energy Tax on agriculture. In: Adger W N, Pettenella D, Whitby M (eds) *Climate change mitigation and European land use policies.* CAB International, Wallingford, pp. 171–86

Kop D, Wallace I (1990) The wheat imports of non-traditionally wheat-producing countries. *Geography* 75: 148–52

Korten D C (1994) Sustainable development case studies. In: Meffe G K, Carroll D R (eds) *Principles of conservation biology.* Sinauer Associates Inc., Sunderland, MA, pp. 32–49

Kostov P, Lingard J (2002) Subsistence farming in transactual economies: lessons from Bulgaria. *Journal of Rural Studies* 18: 83–94

Kothari U (1990) Women's work and rural transformation in India: a study from Gujerat, unpublished PhD thesis, University of Edinburgh

Kreuger A O, Schiff M, Valdés A (eds) (1992) *The political economy of agricultural pricing policy.* John Hopkins University Press for the World Bank, Baltimore, 4 vols

Kritzinger A, Vorster J (1996) Women farm workers on South African deciduous fruit farms: gender roles and the structuring of work. *Journal of Rural Studies* 12: 339–52

Kulikoff A (1992) *The agrarian origins of American capitalism.* University Press of Virginia, Charlottesville, VA

Kuznesof S, Tregear A, Moxey A (1997) Regional foods: a consumer perspective. *British Food Journal* 99: 199–206

Lampkin N H, Padel S (1994) *The economics of organic farming.* CAB International, Wallingford

Landau A (2001) The agricultural negotiations in the World Trade Organisation: the same old story? *Journal of Common Market Studies* 39: 913–25

Lane C (1998) African herders and aboriginal Australians unite. *IIED Drylands Programme (Haramata), Publications*, no. 33

Lang T (1992) *Food fit for the world? How the GATT food trade talks challenge public health, the environment and the citizen. A discussion paper.* The SAFE Alliance, London, and the Public Health Alliance, Birmingham

Lang T (1998) BSE and CJD: recent developments. In: Ratzan S (ed.) *The next cow crisis: health and the public good.* UCL Press, London, pp. 67–85

Lang T, Barling D, Caracher M (2001) Food, social policy and the environment: towards a new model. *Social Policy and Administration* 35: 538–58

Langhorne R (2001) *The coming of globalization: its evolution and contemporary consequences.* Palgrave, Basingstoke and New York

Lant C L, Roberts R S (1990) Greenbelts in the Cornbelt: riparian wetlands, intensification values and market failure. *Environment and Planning A* 22: 1375–88

Lappé M, Bailey R (1999) *Against the grain: the genetic transformation of global agriculture.* Earthscan, London

Lash S, Urry J (1994) *Economies of signs and space: after organised capitalism.* Polity Press, Cambridge

Law J (1992) Notes on the theory of the actor-network: ordering, strategy and heterogeneity. *Systems Practice* 5: 379–93

Lawrence G (1987) *Capitalism and the country-side: the rural crisis in Australia.* Pluto Press, Sydney

Lawrence G, Knuttila M, Gray I (2001) Globalization, neo-liberalism and rural decline: Australia and Canada. In: Epps R, Whitson D (eds) *Writing off the rural West: globalization, governments and the transformation of rural communities.* University of Alberta Press and Parkland Institute, Edmonton, pp. 89–108

Laws G, Harper S (1992) Rural ageing: perspectives from the US and the UK. In: Bowler I R, Bryant C R, Nellis D (eds) *Rural systems in transition: agriculture and environment.* CAB International, Wallingford, pp. 96–109

Lawson H M, Wright G, Bacon E, Yeoman D (1994) Impact of set-aside management strategies on the soil seedbank and weed flora at Woburn. *Aspects of Applied Biology* **40**: 453–9

Leach M, Mearns R, Scoones I, Community based sustainable development: consensus or conflict? *IDS Bulletin*, 28

Leaf M J (1987) Intensification in peasant farming: Punjab in the Green Revolution. In: Turner B L, II, Brush S B (eds) *Comparative farming systems.* Guilford Press, New York, pp. 248–75

Leborgne D, Lipietz A (1988) New technologies, new modes of regulation: some spatial implications. *Environment and Planning D: Society and Space* **6**: 263–80

Lee R (2000) *Transactions of the Institute of British Geographers* new series

Le Heron R B (1988) Food and fibre production under capitalism: a conceptual agenda. *Progress in Human Geography* **12**: 409–30

Le Heron R B (1993) *Globalized agriculture: political choice.* Pergamon Press, Oxford

Le Heron R B (2002) Globalisation, food regimes and rural networks. In: Bowler I R, Bryant C R, Cocklin C (eds) *The sustainability of rural systems: geographical interpretations.* Kluwer, Dordrecht, pp. 81–96

Le Heron R B, Roche M M (1995) A fresh place in food's space. *Area* **27**: 23–33

Lehman T (1992) Public values, private lands: origins and ironies of farmland preservation in Congress. *Agricultural History* **66**: 257–72

Lehman T (1995) *Public values, private lands.* University of North Carolina Press, Chapel Hill

Lele U, Stone S (1989) *Population pressure, environment and agricultural intensification: variations on the Boserup hypothesis.* MADIA Discussion Papers. World Bank, Washington DC, no. 4

Lem W (1988) Household production and reproduction in rural Languedoc: social relations of petty commodity production in Murviel-les-Beziers. *Journal of Peasant Studies* **15**: 500–29

Lenin V I (1982) The differentiation of the peasantry. In: Harriss J C (ed.) *Rural development: theories of peasant economy and agrarian change.* Hutchinson, London, pp. 130–8

Leopold A (1949) *A Sand County almanac.* Oxford University Press, New York

Lewis G J, Maund D J (1976) The urbanisation of the countryside: a framework for analysis. *Geografiska Annaler* **58B**: 17–27

Lewis N, Moran W, Cocklin C (2002) Restructuring, regulation and sustainability In: Bowler I R, Bryant C R, Cocklin C (eds) *The sustainability of rural systems: geographical interpretations.* Kluwer, Dordrecht, pp. 97–122

Liepins R (1995) Women in agriculture: advocates for a gendered sustainable agriculture. *Australian Geographer* **26**: 118–26

Liepins R (1996) Reading agricultural power: media as sites and processes in the construction of meaning. *New Zealand Geographer* **52**: 3–10

Liepins R (1998a) 'Women with broad vision': nature and gender in the environmental activism of Australia's 'Women in agriculture' movement. *Environment and Planning A* **30**: 1179–86

Liepins R (1998b) The gendering of farming and agricultural politics: a matter of discourse and power. *Australian Geographer* **29**: 371–88

Liepins R (1998c) Fields of action: Australian women's agricultural activism in the 1990s. *Rural Sociology* **63**: 128–56

Liepins R, Bradshaw B (1999) Neo-liberal agricultural discourse in New Zealand: economy, culture and politics linked. *Sociologia Ruralis* **39**: 563–82

Lighthall D R, Roberts R S (1995) Towards an alternative logic of technological change:

insights from Corn Belt agriculture. *Journal of Rural Studies* 12: 319–34

Lim J N-W, Douglas I (1998) The impact of cash cropping on shifting cultivation in Sabah, Malaysia. *Asia Pacific Viewpoint* 39: 315–26

Lipietz A (1992) *Towards a new economic order*. Polity Press, Cambridge

Lipton M (1982) Why poor people stay poor. In: Harriss J C (ed.) *Rural development: theories of peasant economy and agrarian change*. Hutchinson, London, pp. 66–81

Lipton M (1989) *New seeds, poor people*. Unwin Hyman, London

Liverman D M (1990) Vulnerability to global environmental change. In: Kasperson R, Kates R (eds) *Understanding global environmental change: the contributions of risk analysis and management*. Clark University, Worcester, MA, pp. 27–44

Liverman D M (1991) The regional impact of global warming in Mexico: uncertainty, vulnerability and response. In: Schmandt J, Clarkson J (eds) *The regions and global warming: impacts and response strategies*. Oxford University Press, New York, pp. 44–68

Lloyd T, Watkins C, Williams D (1995) Turning farmers into foresters via market liberalisation. *Journal of Agricultural Economics* 46: 361–70

Lobley M, Potter C (1998) Environmental stewardship in UK agriculture: a comparison of the Environmentally Sensitive Areas Programme and the Countryside Stewardship Scheme in south-eastern England. *Geoforum* 29: 41–52

Loboda J, Rog Z, Tykkylainen M (1998) Market forces and community development in rural Poland. In: Neil C, Tykkylainen M (eds) *Local economic development: a geographical comparison of rural community restructuring*. United Nations University Press, Tokyo, pp. 97–124

Lockeretz W (1995) Organic farming in Massachusetts: an alternative approach to agriculture in an urbanized state. *Journal of Soil and Water Conservation* 50: 663–7

Lockie S (1998) Landcare and the state: 'action at a distance'. In: Burch D, Lawrence G, Rickson R E, Goss J (eds) *Australasia's food and farming in a globalised economy: recent developments*

and future prospects. *Monash Publications in Geography and Environmental Science*, no. 50, pp. 15–28

Lockie S, Kitto S (2000) Beyond the farm gate: production-consumption networks and agri-food networks. *Sociologia Ruralis* 40: 3–19

Logan B I, Moseley W G (2002) The political ecology of poverty alleviation in Zimbabwe's Communal Areas Management Program for Indigenous Resources (CAMPFIRE). *Geoforum* 33: 1–14

Long N, Long A (1992) *Battlefields of knowledge: the interlocking of theory and practice in social research and development*. Routledge, London

Lovering J (1989) The restructuring debate. In: Peet R J, Thrift N J (eds) *New models in geography: the political economy perspective, volume 1*. Unwin Hyman, London, pp. 198–223

Lowe P, Brouwer F (2000) Agenda 2000: a wasted opportunity? In: Brouwer F, Lowe P (eds) *CAP and the rural environment in transition: a panorama of national perspectives*. Wangeningen Pers, Wangeningen, pp. 321–34

Lowe P, Ward N (1998) Regional policy, CAP reform and rural development in Britain: the challenge for New Labour. *Journal of Regional Studies* 32: 469–74

Lowe P, Buller H, Ward N (2002) Setting the next agenda? British and French approaches to the second pillar of the Common Agricultural Policy. *Journal of Rural Studies* 18: 1–18

Lowe P, Marsden T K, Whatmore S J (eds) (1994) *Regulating agriculture*. David Fulton, London

Lowe P, Murdoch J, Marsden T K, Munton R J C, Flynn A (1993) Regulating the new rural spaces: the uneven development of land. *Journal of Rural Studies* 9: 205–22

Low N (1995) Regulation theory, global competition among cities and capital embeddedness. *Urban Policy and Research* 13: 205–21

Lowrance J (1990) Research approaches for ecological sustainability. *Journal of Soil and Water Conservation* 45: 51–7

Lund P (1991) Part-time farming: a note on definitions. *Journal of Agricultural Economics* 42: 196–9

Lundberg M, Milanovic B (2000) The truth about global inequality. *Financial Times*, 25 February

Lynch K D (1994) Urban fruit and vegetable supply in Dar es Salaam. *Geographical Journal* **160**: 307–18

Lynch K D (1995) Sustainable urban food supply for Africa. *Sustainable Development* **3**: 80–88

Lynch K D (1999) Commercial horticulture in rural Tanzania – an analysis of key influences. *Geoforum* **30**: 171–83

Lynch K D, Binns A, Olofin E (2001) Urban agriculture under threat: the land security question in Kano, Nigeria. *Cities* **18**: 159–71

Lyon F (2000) Trust, networks and norms: the creation of social capital in agricultural economies in Ghana. *World Development* **28**: 663–81

Lyons K (1996) Agro-industrialization and social change within the Australian context: a case study of the fast food industry. In: Burch D, Rickson R E, Lawrence G E (eds) *Globalization and agri-food restructuring*. Avebury, Aldershot, pp. 239–50

McDonald R B, Hall F G (1980) Global crop forecasting. *Science* **208**: 670–79

McEarchern C (1992) Farmers and conservation: conflict and accommodation in farming politics. *Journal of Rural Studies* **8**: 159–71

MacFarlane R (1996) Modelling the interaction of economic and socio-behavioural factors in the prediction of farm adjustment. *Journal of Rural Studies* **12**: 365–74

MacFarlane R (2000a) Building blocks or stumbling blocks? Landscape ecology and farm-level participation in agri-environment policy. *Landscape Research* **25**: 321–32

MacFarlane R (2000b) Achieving whole-landscape management across multiple land management units: a case study from the Lake District ESA. *Landscape Research* **25**: 229–54

McHenry H (1996) Farming and environmental discourses: a study of the depiction of environmental issues in a German farming newspaper. *Journal of Rural Studies* **12**: 375–86

McInerney J, Turner M, Hollingham M (1989) *Diversification in the use of farm resources.* Reports, Department of Agricultural Economics, University of Exeter, no. 232

McIntire J, Gryseels G (1987) Crop–livestock interactions in sub-Saharan Africa and their implications for farming systems research. *Experimental Agriculture* **23**(supplement 3): 235–43

MacKenzie F (1992a) The politics of partnership: farm women and farm land, Ontario. In: Bowler I R, Bryant C R, Nellis D (eds) *Rural systems in transition: agriculture and environment*. CAB International, Wallingford, pp. 85–95

MacKenzie F (1992b) The worse it got the more we laughed: a discourse of resistances among farmers of Eastern Ontario. *Environment and Planning D: Society and Space* **10**: 691–713

McKenna M K L (2000a) Can rural voices affect rural choices? Deregulation in New Zealand's apple industry. *Sociologia Ruralis* **40**: 366–83

McKenna M K L (2000b) Can rural voices affect rural choices? Restructuring and (de)regulation in New Zealand's apple industry. In: Millward H, Beesley K, Ilbery B W, Harrington L (eds) (2000) *Agricultural and environmental sustainability in the new countryside*. St. Mary's University and Rural Research Centre, Nova Scotia Agricultural College, Truro, Nova Scotia, pp. 52–70

McKenna M, Murray W (1999) Jungle law in the orchard: comparing 'globalisation' in the New Zealand and Chilean apple industries. Unpublished paper presented to the New Zealand Geographical Society Conference, Massey University, Palmerston North, New Zealand, July 5–8

McKenna M K L, Roche M M, Le Heron R B (1998) Sustaining the fruits of labour: a comparative localities analysis of the Integrated Fruit Production programme in New Zealand's apple industry. *Journal of Rural Studies* **14**: 393–410

McKenna M K L, Roche M M, Le Heron R B (1999a) H J Heinz and global gardens: creating quality, leverage and localities. *International Journal of Sociology of Agriculture and Food* **8**: 35–51

McKenna M K L, Campbell H, Roche M (1999b) It's not easy being green: 'food safety' practices in New Zealand's apple industry. *Australian Geographical Studies* **37**: 63–81

McKenna M K L, Roche M, Mansvelt J, Berg L (1999c) Core issues in New Zealand's apple industry: global-local challenges. *Geography* 84: 275–81

McKenna M K L, Le Heron R B, Roche M M (2001) Living local, growing global: renegotiating the export production regime in New Zealand's pipfruit sector. *Geoforum* 32: 157–66

MacKinnon N, Bryden J, Bell C, Fuller A, Spearman M (1991) Pluriactivity, structural change and farm household vulnerability in Western Europe. *Sociologia Ruralis* 31: 58–71

McMichael P (1992a) Tensions between national and international control of the world food order: contours of a new food regime. *Sociological Perspectives* 35: 343–65

McMichael P (1992b) Agro-food restructuring in the Pacific Rim: a comparative-international perspective on Japan, South Korea, the United States, Australia and Thailand. In: Palat R (ed.) *Pacific Asia and the future of the world-system*. Greenwood, Westport, pp. 103–16

McMichael P (ed.) (1992c) *Food systems and agrarian change in the late 20th century*. Cornell University Press, Ithaca

McMichael P (1993) World food system restructuring under a GATT regime. *Political Geography* 12: 198–214

McMichael P (ed.) (1994) *The global restructuring of agro-food systems*. Cornell University Press, Ithaca

McMichael P (1996) Globalisation: myths and realities. *Rural Sociology* 61: 25–56

McMichael P (1999) Virtual capitalism and agri-food restructuring. In: Burch D, Goss J, Lawrence G E (eds) *Restructuring global and regional agricultures: transformations in Australia*. Ashgate, Basingstoke, pp. 3–22

McMichael P (2000) *Development and social change: a global perspective*. Sage, Pine Forge, CA

McMichael P, Myhre I (1991) 'Global regulation versus the nation state: agricultural food systems and the new politics of capital. *Capital and Class* 43: 83–105

Macmillan G J (1993) Gold mining and land-use change in the Brazilian Amazon. Unpublished PhD thesis, University of Edinburgh

Macmillan G J (1994) *At the end of the rainbow? Gold, land and society in the Brazilian Amazon*. Earthscan Publications, London

Macmillan W (1991) Famine: the unnatural disaster. *Geography Review* 5(1): 18–23

MacNeish R S (1992) *The origins of agriculture and settled life*. University of Oklahoma Press, Norman and London

McQuaig J, Manning E (1982) *Agricultural land use change in Canada: process and consequence*. Lands Directorate, Environment Canada, Ottawa

McReynolds S A (1998) Agricultural labour and agrarian reforms in El Salvador: social benefit or economic burden. *Journal of Rural Studies* 14: 459–74

McWilliams C (1939) *Factories in the fields*. Little Brown, Boston

Madeley J (2000) *Hungry for trade: how the poor pay for trade*. Zed Books, London

MAFF (Ministry of Agriculture Fisheries and Food) (1993) *Agriculture and England's environment: set-aside management, a consultation document*. MAFF, London

MAFF (1999) *Incidence of BSE – monthly statistics*. HMSO, London

MAFF (2000) *England Rural Development Plan 2000–2006*. MAFF, London

Maizels A (1997) *Commodity supply management by producing countries: a case-study of the tropics*. Clarendon Press, Oxford

Mann S A (1990) *Agrarian capitalism in theory and practice*. University of North Carolina Press, Chapel Hill

Mann S A, Dickinson J M (1978) Obstacles to the development of a capitalist agriculture. *Journal of Peasant Studies* 5: 466–81

Mannion A M (1995a) *Agriculture and environmental change: temporal and spatial dimensions*. John Wiley and Sons, Chichester

Mannion A M (1995b) The origins of agriculture: an appraisal. *Geographical Papers, Department of Geography, University of Reading*, no. 117

Mannion A M (1997) Agriculture and land transformation: present trends and future prospects. *Outlook on Agriculture* 26: 151–8

Mannion A M (1998) Land transformation: trends, prospects and challenges. *Geographical Papers, University of Reading*, no. 125

Mannion A M (1999) Domestication and the origins of agriculture: an appraisal. *Progress in Physical Geography*

Mannion A M, Bowlby S (eds) (1992) *Environmental issues in the 1990s*. John Wiley and Sons, Chichester

Marchand M H, Parpart J L (1995) *Feminism, postmodernism, development*. Routledge, London

Marland A (1989) An overview of organic farming in the United Kingdom. *Outlook on Agriculture* 18: 24–7

Marden P (1992) 'Real' regulation reconsidered. *Environment and Planning A* 24: 751–67

Marsden T K (1985) Capitalist farming and the farm family: a case study. *Sociology* 18: 205–24

Marsden T K (1988) Exploring political economy approaches in agriculture. *Area* 20: 315–21

Marsden T K (1989) Restructuring rurality – from order to disorder in agrarian political economy. *Sociologia Ruralis* 29: 312–17

Marsden T K (1990) Towards the political economy of pluriactivity. *Journal of Rural Studies* 6: 375–82

Marsden T K (1992) Exploring a rural sociology for the Fordist transition: incorporating social relations into economic restructuring. *Sociologia Ruralis* 32: 209–30

Marsden T K (1996) Rural geography trend report: the social and political bases of rural restructuring. *Progress in Human Geography* 20: 246–58

Marsden T K (1997) Creating space for food: the distinctiveness of recent agrarian development In: Goodman D, Watts M J (eds) *Globalising food: agrarian questions and global restructuring*. Routledge, London and New York, pp. 169–91

Marsden T K (1998a) Economic perspectives. In: Ilbery B W (ed.) (1998) *The geography of rural change*. Longman, Harlow, pp. 13–30

Marsden T K (1998b) Agriculture beyond the treadmill? Issues for policy, theory and research practice. *Progress in Human Geography* 22: 265–75

Marsden T K (1999a) Beyond agriculture? Toward sustainable modernization. In: Redclift M, Lekakis J N, Zanias G P (eds) *Agriculture and world trade liberalization: socio-environmental perspectives on the Common Agricultural Policy*. CAB International, Wallingford, pp. 238–59

Marsden T K (1999b) Rural futures: the consumption countryside and its regulation. *Sociologia Ruralis* 39: 501–20

Marsden T K (2000a) Food matters and the matter of food: towards a new food governance? *Sociologia Ruralis* 40: 20–9

Marsden T K (2000b) The condition of rural sustainability, unpublished paper presented at the Annual Conference of the RGS-IBG, University of Sussex, Brighton, January 4–8

Marsden T K, Arce A (1995) Constructing quality: emerging food networks in the rural transition. *Environment and Planning A* 27: 1261–79

Marsden T K, Little J (eds) (1990) *Political, social and economic perspectives on the international food system*. Avebury, Aldershot

Marsden T K, Wrigley N (1995) Regulation, retailing and consumption. *Environment and Planning A* 27: 1899–1912

Marsden T K, Wrigley N (1996) Retailing, the food system and the regulatory state. In: Wrigley N, Lowe M S (eds) *Retailing, consumption and capital: towards the new retail geography*. Longman, London, pp. 33–47

Marsden T K, Munton R J C, Whatmore S J, Little J K (1986a) Towards a political economy of capitalist agriculture: a British perspective. *International Journal of Urban and Regional Research* 10: 498–521

Marsden T K, Whatmore S J, Munton R J C, Little J K (1986b) The restructuring process and economic centrality in capitalist agriculture. *Journal of Rural Studies* 2: 271–80

Marsden T K, Whatmore S J, Munton R J C (1987) Uneven development and the restructuring process in British agriculture: a preliminary exposition. *Journal of Rural Studies* 3: 297–308

Marsden T K, Munton R J C, Whatmore S J, Little J K (1989) Strategies for coping in capitalist agriculture: an examination of the responses of farm families in British agriculture. *Geoforum* 20: 1–14

Marsden T K, Lowe P, Whatmore S J (eds) (1990) *Rural restructuring*. David Fulton Publishers, London

Marsden T K, Lowe P, Whatmore S J (eds) (1992) *Labour and locality*. David Fulton Publishers, London

Marsden T K, Murdoch J, Lowe P, Munton R J C, Flynn A (1993) *Constructing the country-side: restructuring rural areas*. UCL Press, London

Marsden T K, Munton R J C, Ward N, Whatmore S J (1996) Agricultural geography and the political economy approach: a review. *Economic Geography* 72: 361–76

Marsden T K, Flynn A, Harrison M (2000) *Consuming interests: the social provision of foods*. University of London Press, London

Marsh J (1982) *Back to the land*. Quartet Books, London

Marshall E J P, West T M, Winston L (1994) Extending field boundary habitats to enhance farmland wildlife and improve crop and environment protection. *Aspects of Applied Biology* 40: 387–410

Martin S, Tricker M (1989) The impacts of rural employment initiatives on local labour markets. *Proceedings of the British – Dutch Symposium on Rural Geography*, KNAG, Amsterdam, Nederlandse Geografische Studies, no. 92

Marx K (1971) *A contribution to the critique of political economy*. Lawrence and Wishart, London

Marx K (1999) *Capital: a new abridgement*. Oxford University Press, Oxford

Massey D B (1996) *Spatial divisions of labour: social structures and the geography of production*. Macmillan, London and Basingstoke, 2nd edition

Mather A S (1998) The changing role of forests. In: Ilbery B W (ed.) *The geography of rural change*. Longman, Harlow, pp. 106–27

Mather C (1999) Agro-commodity chains, market power and territory: re-regulating South African citrus exports in the 1990s. *Geoforum* 30: 61–70

Mathijs E, Meszaros S (1997) Privatisation and restructuring of Hungarian agriculture. In: Swinnen J F M, Buckwell A, Mathijs E (eds) *Agricultural privatisation, land reform and farm restructuring in Central and Eastern Europe*. Ashgate, Aldershot, pp. 161–88

Matless D (1998) *Landscape and Englishness*. Reaktion Books, London

Matthews J D (1994) Implementing forest policy in the lowlands of Britain. *Forestry* 67: 1–12

May J (1992) Ranching and redistribution after Apartheid. *Journal of Rural Studies* 8: 257–68

May J (1996) 'A little taste of something more exotic': the imaginative geographies of every-day life. *Geography* 81: 57–64

Mayfield L H (1996) The local economic impact of small farms: a spatial analysis. *Tijdshrift voor economishe en social geografie* 87: 387–98

Maxwell D (1994) The household logic of urban farming in Kampala. In: Egziabher A G (ed.) *Cities feeding people: an examination of urban agriculture in East Africa*. International Development Research Centre, Ottawa, pp. 47–66

Mazorra A P (2000) Analysis of the evolution of farmers' early retirement policy in Spain. The case of Castilla and Leon. *Land Use Policy* 17: 113–20

Meier V (1999) Cut-flower production in Colombia – a major development success story for women? *Environment and Planning A* 31: 273–89

Mendelsohn R O (1998) *The impact of climatic change on the US economy*. Cambridge University Press, Cambridge

Michalak W Z (1993) GIS and land use change analysis: integration of remotely-sensed data into GIS. *Applied Geography* 13: 28–44

Michelson J (2001) Recent development and political acceptance of organic farming in Europe. *Sociologia Ruralis* 41: 3–20

Middleton N (1999a) *The global casino*. Oxford University Press, Oxford, 2nd edition

Middleton N (1999b) All in the genes. *Geographical* 71(12): 51–5

Midmore P, Lampkin N (1989) Organic farming as an alternative to set-aside and an option for extensification. In: Dugaard A, Nielsen A H (eds) *Economic aspects of environmental regulations in agriculture*. Verlag, Kiel, pp. 267–78

Mignon C (1971) L'agriculture a temps partiel dans le department du Puy-de-Dome. *Revue d'Auvergne* 85: 1–41

Miller F A (ed.) (1991) *Agricultural policy and the environment*. Centre for Agricultural Strategy, Reading, Report no. 24

Miller L (1996) Contract farming under globally-oriented and locally-emergent agribusiness in Tasmania. In: Burch D, Rickson R E, Lawrence G E (eds) (1996) *Globalization and agri-food restructuring*. Avebury, Aldershot, pp. 203–18

Millward H, Beesley K, Ilbery B W, Harrington L (eds) (2000) *Agricultural and environmental sustainability in the new countryside*. St. Mary's University and Rural Research Centre, Nova Scotia Agricultural College, Truro, Nova Scotia

Mintel (1997) *Organic and ethical foods*. Mintel, London

Mitchell D (2000) Review essay. George Henderson's *California and the Fictions of Capital*. *Antipode* **32**: 90–8

Mitchell M, Doyle C (1996) Spatial distribution of the impact of agricultural policy reforms in rural areas. *Scottish Geographical Magazine* **112**: 76–82

Mohan G (2000) Dislocating globalisation: power, politics and global change. *Geography* **85**: 121–33

Mohan G (2002) Participatory development. In: Desai V, Potter R B (eds) *The companion to development studies*. Arnold, London, pp. 49–54

Mohan G, Stokke K (2000) Participatory development and empowerment: the dangers of localism. *Third World Quarterly* **21**: 247–68

Mohanty C T (1991) Under Western eyes: feminist scholarship and colonial discourses. In: Mohanty C T, Russo A, Torres L (eds) *Third World women and the politics of feminism*. Indiana University Press, Bloomington, pp. 51–80

Mol A J P (1997) Ecological modernization: industrial transformation and environmental reform. In: Redclift M, Woodgate G (eds) *The international handbook of environmental sociology*. Edward Elgar, London, pp. 138–49

Molnar J, Wu L (1989) Agrarianism, family farming and support for state intervention in agriculture. *Rural Sociology* **54**: 227–45

Monench M (1989) Forest degradation and the structure of biomass utilisation in an Himalayan foothill village. *Environmental Conservation* **16**: 137–46

Monk A (1998) The Australian organic basket and the global supermarket. In: Burch D, Lawrence G, Rickson R E, Goss J (eds) Australasia's food and farming in a globalised economy: recent developments and future prospects. *Monash Publications in Geography and Environmental Science*, no. 50, pp. 69–82

Monk A (1999) The organic manifesto: organic agriculture in the world food system. In: Burch D, Goss J, Lawrence G E (eds) *Restructuring global and regional agricultures: transformations in Australia*. Ashgate, Basingstoke, pp. 75–86

Monke E A, Avillez F, Pearson S, Marenco G (1998a) Evaluation of small farm agriculture. In: Monke E A, Avillez F, Pearson S (eds) *Small farm agriculture in Southern Europe: CAP reform and structural change*. Ashgate, Aldershot, pp. 7–30

Monke E A, Avillez F, Pearson S (eds) (1998b) *Small farm agriculture in Southern Europe: CAP reform and structural change*. Ashgate, Aldershot

Mooij J (1998) Food policy and politics: the political economy of the public distribution system in India. *Journal of Peasant Studies* **25**: 77–101

Mooij J (1999) Real targeting: the case of food distribution in India. *Food Policy* **24**: 49–70

Mooney H P (1988) *My own boss: class, rationality and the family farm*. Westview Press, Boulder, CO

Mooney P (1982) Labor time, production time, and capitalist development in agriculture: a reconsideration of the Mann-Dickinson thesis. *Sociologia Ruralis* **22**: 279–91

Mooney P (1987) Desperately seeking: one-dimensional Mann and Dickinson. *Rural Sociology* **52**: 286–95

Mooney S, Arthur L (1990) The impacts of climatic change on agriculture in Manitoba. *Canadian Journal of Agricultural Economics* **38**: 685–94

Moran W (1993) The wine appellation as territory in France and California. *Annals of the Association of American Geographers* **83**: 694–717

Moran W, Blunden G, Greenwood J (1993) The role of family farming in agrarian change. *Progress in Human Geography* 17: 22–42

Moran W, Blunden G, Workman M, Bradly A (1996) Family farmers, real regulation, and the experience of food regimes. *Journal of Rural Studies* 12: 245–58

Morgan K, Murdoch J (2000) Organic versus conventional agriculture: knowledge, power and innovation in the food chain. *Land Use Policy* 31: 159–74

Morgan W B (1990) Agrarian structure. In: Pacione M (ed.) *The geography of the Third World.* Routledge, London and New York, pp. 77–113

Morgan W B, Munton R J C (1971) *Agricultural geography.* Methuen, London

Morner M, Svensson T (1991) Introduction: changing rural societies in the Third World in the 19th and 20th centuries. In: Morner and Svensson (eds) *The transformation of rural society in the Third World.* Routledge, London, pp. 20–42

Morris A, Robinson G M (1996) Guest editorial: rural Scotland. *Scottish Geographical Magazine* 112: 66–9

Morris C (1999) A 'quiet revolution': Integrated Farming Systems in the transition to a more sustainable agriculture in the UK. Unpublished paper presented at the International Rural Geography Symposium, Halifax and Truro, Nova Scotia, July 11–17

Morris C (2000a) Quality assurance schemes: a new way of delivering environmental benefits in food production? *Journal of Environmental planning and Management* 43: 433–48

Morris C (2000b) A 'quiet revolution'? Integrated Farming Systems in the transition to a more sustainable agriculture in the UK. In: Millward H, Beesley K, Ilbery B W, Harrington L (eds) *Agricultural and environmental sustainability in the new countryside.* St. Mary's University and Rural Research Centre, Nova Scotia Agricultural College, Truro, Nova Scotia, pp. 71–91

Morris C, Andrews C (1997) The construction of environmental meanings within 'farming culture' in the UK: the implications for agri-environmental research. In: Ilbery B W, Chiotti Q, Rickard T (eds) *Agricultural restructuring and sustainability: a geographical perspective.* CAB International, Wallingford, pp. 87–99

Morris C, Cobb P (1993) Agriculture and conservation – the whole farm approach. *Ecos* 14 (3/4)

Morris C, Evans N J (1999) Research on the geography of agricultural change: redundant or revitalized? *Area* 31: 349–58

Morris C, Potter C (1995) Recruiting the new conservationists: farmers' adoption of agri-environmental schemes in the UK. *Journal of Rural Studies* 11: 51–63

Morris C, Winter M (1999) Integrated Farming Systems: the third way for European agriculture? *Land Use Policy*

Morris C, Young C (1997a) Never mind the environment, feel the 'quality': a discussion of the agri-environmental potential of quality assurance schemes. *North West Geographer* 3rd series, 1: 36–47

Morris C, Young C (1997b) Towards environmentally beneficial farming? An evaluation of the Countryside Stewardship scheme. *Geography* 82: 305–16

Morris C, Young C (1999) Quality assurance schemes: a new way of delivering environmental benefits in food production? *Journal of Environmental Planning and Management*

Morris C, Young C (2000) 'Seed to shelf', 'teat to table', 'barley to beer' and 'womb to tomb': discourses of food quality and quality assurance schemes in the UK. *Journal of Rural Studies* 16: 103–16

Morris J, Mills J, Crawford I M (2000) Promoting farmer uptake of agri-environment schemes: the Countryside Stewardship Arable Options Scheme. *Land Use Policy* 17: 241–54

Mortimore M (1991) Five faces of famine: the autonomous sector in the famine process. In: Bohle H-G, Cannon T, Gugo G, Ibrahim F (eds) *Famine and food security in Africa and Asia.* Naturwissenschaftliche Gessellschaft, Bayreuth, pp. 11–35

Moulaert F, Swyngedouw E A (1989) A regulation approach to the geography of flexible production systems. *Environment and Planning D: Society and Space* 7: 327–45

Muller C (1992) Tourism and farming in Lower Normandy: the example of country gites and farm accommodation. *Geographie Sociale* **12**: 249–60

Munton R J C (1981) Agricultural land use in the London Green Belt. *Town and Country Planning* **50**: 17–19

Munton R J C (1986) Greenbelts: the end of an era? *Geography*, 206–14

Munton R J C (1988) Agricultural change and technological development: tendencies apart? Unpublished paper delivered to the Institute of British Geographers Annual Conference, University of Loughborough

Munton R J C (1990) Farming families in upland Britain: options, strategies and futures. Unpublished paper delivered to the Annual Conference of the Association of American Geographers, Toronto

Munton R J C (1992) Factors of production in modern agriculture. In: Bowler I R (ed.) *The geography of agriculture in developed market economies*. Longman, London, pp. 56–84

Munton R J C, Marsden T K (1991) Dualism or diversity in family farming: patterns of change in British agriculture. *Geoforum* **22**: 105–17

Murdoch J (1995) Actor-networks and the evolution of economic forms: combining description and explanation in theories of regulation, flexible specialisation and networks. *Environment and Planning A* **27**: 731–57

Murdoch J (1998) The spaces of actor-network theory. *Geoforum* **29**: 357–74

Murdoch J (2000) Networks – a new paradigm of rural development? *Journal of Rural Studies* **16**: 407–19

Murdoch J, Clark J (1994) Sustainable knowledge. *Geoforum* **25**: 115–132

Murdoch J, Marsden T K (1994) *Resconstituting rurality*. UCL Press, London

Murdoch J, Marsden T K (1995) The spatialization of politics: local and national actor-spaces in environmental conflict. *Transactions of the Institute of British Geographers* **20**: 368–80

Murdoch J, Miele M (1999) Back to nature: changing worlds of production in the food sector. *Sociologia Ruralis* **39**: 465–83

Murdock S H, Albrecht D E (1998) An ecological investigation of agricultural patterns in the United States. In: Micklin M, Poston Jr D L (eds) *Continuities in sociological human ecology*. Plenum Press, New York and London, pp. 299–316

Murphy P (1985) *Tourism: a community approach*. Methuen, New York

Murphy A (2002) The emergence of online food retailing: a stakeholder perspective. *Tijdschrift voor Economische en Social Geografie* **93**: 47–61

Myers N (1991) *Population, resources and the environment: the critical challenges*. United Nations Population Fund, New York

NALS (National Agricultural Lands Study) (1980) *Where have all the farmlands gone?* NALS, Washington, DC

NALS (1981) *NALS final report*. Soil Conservation Service, USDA, Washington, DC

Naples N (1994) Contradictions in agrarian ideology: restructuring gender, race-ethnicity, and class. *Rural Sociology* **59**: 110–35

Nash E F (1965) A policy for agriculture. In: McGrone G, Attwood E A (eds) *Agricultural policy in Britain: selected papers*. University of Wales Press, Cardiff, pp. 35–51

Naylor E L (1982) Retirement policy in French agriculture. *Journal of Agricultural Economics* **33**: 25–36

Naylor E L (1986) Milk quotas in France. *Journal of Rural Studies* **2**: 153–61

Naylor E L (1993) Milk quotas and the changing pattern of dairy farming in France. *Journal of Rural Studies* **10**: 53–64

Naylor E L (1995) Agricultural policy reforms in France. *Geography* **80**: 281–3

Nellis M D, Harrington L M B, Sheeley J (1997) Policy, sustainability and scale: the US Conservation Reserve Program. In: Ilbery B W, Chiotti Q, Rickard T (eds) *Agricultural restructuring and sustainability: a geographical perspective*. CAB International, Wallingford, pp. 219–31

Netting R M (1993) *Smallholders, householders: farm families and the ecology of intensive, sustainable agriculture*. Stanford University Press, Stanford

North R, Gorman T (1990) *Chickengate: an independent analysis of the salmonella in eggs scare.* Health and Welfare Unit, Institute of Economic Affairs, London

NRCS (National Resources Conservation Service, USDA) (1998) *Keys to soil taxonomy, eighth edition.* National Soil Survey Centre, United States Department of Agriculture, Washington, DC

NSCGP (Netherlands Scientific Council for Government Policy) (1992) *Ground for choices: four perspectives for the rural areas in the European Community.* Report no. 42. NSCGP, The Hague

Nurse K, Sandiford W (1995) *Windward Islands bananas: challenges and options under the Single European Market.* Freidrich Ebert Stiftang, Kingston, Jamaica

Nusser S M, Goebel J J (1997) The National Resources Inventory: a long-term multi-resource monitoring program. *Environmental and Ecological Statistics* 4: 181–204

Nygard B, Storstad O (1998) De-globalization of food markets? Consumer perceptions of safe food: the case of Norway. *Sociological Ruralis* 38: 35–53

Obosu-Mensah K (1999) *Food production in urban areas: a study of urban agriculture in Accra, Ghana.* Ashgate, Aldershot

O'Brien R (1992) *Global financial integration: the end of Geography.* Royal Institute of International Affairs/Pinter Publishers, London

O'Brien P (1994) Implications for public policy, in Shertz L, Daft L (eds) *Food and agricultural markets: the quiet revolution.* National Planning Association, Washington, DC

O'Connor A (1978) *The geography of tropical African development.* Pergamon, Oxford, 2nd edition

OECD (Organisation for Economic Co-operation and Development) (1996) *Agricultural policies, markets and trade in transition economies.* OECD, Paris

O'Hare G (2002) Climate change and the temple of sustainable development. *Geography* 97: 234–46

Ohmae K (1996) *The end of the nation state: the rise of reginal economies.* Harper Collins, London

Ohlmer B, Olson K, Brehmer B (1998) Understanding farmers: decision-making processes and improving managerial assistance. *Agricultural Economics* 18: 273–90

Oldrup H (1999) Women working off the farm: reconstructing gender identity in Danish agriculture. *Sociologia Ruralis* 39: 343–58

Olofin E A, Fereday N, Ibrahim A T, Aminu Shehu M, Adamu Y (1997) *Urban and peri-urban horticulture in Kano, Nigeria.* Natural Resources Institute, Chatham

Ontario Institute of Agrologists (1975) *Foodland: preservation or starvation. Statement on Land Use Policy.* OIA, Toronto

O'Riordan T (1987) Agriculture and environmental protection. *Geography Review* 1(1): 35–40

O'Riordan T (1999) From environmentalism to sustainability. *Scottish Geographical Journal* 115: 151–66

O'Riordan T, Cobb R (1996) That elusive definition of sustainable agriculture. *Town and Country Planning*, February: 50–51

O'Riordan T, Cobb R (2001) Assessing the consequences of converting to organic agriculture. *Journal of Agricultural Economics* 52: 22–35

Orwin C S, Whetham E H (1971) *History of British agriculture, 1846–1914.* David & Charles, Newton Abbot, second edition

Osborn C T, Schnepf M, Keim R (1994) *The future use of Conservation Reserve Program acres: a national survey of farm owners and operators.* Soil and Water Conservation Society, Ankeny, Iowa

Overton M (1996) *Agricultural Revolution in England: the transformation of the agrarian economy, 1500–1850.* Cambridge University Press, Cambridge

Owens D W, Coombes M G, Gillespie A E (1986) The urban–rural shift and employment change in Britain 1917–81. In: Danson M (ed.) *Redundancy and recession: restructuring the regions?* Geo Books, Norwich, pp. 23–47

Owens S (1994) Land, limits and sustainability: a conceptual framework and some dilemmas for the planning system. *Transactions of the Institute of British Geographers* new series, **19**: 439–56

Pacione M (1989) Land use conflict in the urban fringe. *Journal of the Scottish Association of Geography Teachers* 18: 4–11

Pacione M (ed.) (1999) *Applied Geography: principles and practice*. Routledge, London and New York

Page B (2003) Agriculture. In: Sheppard E, Barnes T J (eds) *A companion to economic geography*. Blackwell, Oxford, and Malden, MA, pp. 242–56

Palacios S P I (1998) Farmers and the implementation of the EU Nitrates Directive in Spain. *Sociologia Ruralis* 38: 146–62

Parasuraman S (1995) Economic marginalisation of peasants and fishermen due to urban expansion. *Pakistan Development Review* 34: 121–38

Paxton A (1994) *The food miles report: the dangers of long-distance food transport*. SAFE Alliance, London

Parnwell M J G (2002) Agropolitan and bottom-up development. In: Desai V, Potter R B (eds) *The companion to development studies*. Arnold, London, pp. 112–17

Parry M L (1992) Agriculture as a resource system In: Bowler I R (ed.) *The geography of agriculture in developed market economies*. Longman, London, pp. 207–38

Parry M L, Livermore M (2002) Climate change, global food supply and risk of hunger. *Issues in Environmental Science and Technology* 17: 109–37

Parry M L, Carter T R, Konijn N T (eds) (1988) *The impacts of climate variations on agriculture*. Reidel, Dordrecht, 2 vols

Pearce D W, Markandya A, Barbier E (1989) *Blueprint for a green economy*. Earthscan, London

Pedder J W R (1940) Preliminary notes on cattle improvement and the possibilities for mixed farming in the middle belt and southern provinces of Nigeria. *Tropical Agriculture* 17 Supplement 3: 43–9

Peet J R (1969) The spatial expansion of commercial agriculture in the 19th century: a von Thunen interpretation. *Economic Geography* 45: 283–301

Pelupessy W (1997) *The limits of economic reform in El Salvador*. Macmillan, London

Perry M, Le Heron R B, Hayward D J, Cooper I (1997) Growing discipline through Total Quality Management in a New Zealand horticultural region. *Journal of Rural Studies* 13: 289–304

Peters J (1990) Saving farmland: how well have we done? *Planning* 56: 12–17

Pfeffer M J, Lapping M B (1994) Farmland preservation, development rights and the theory of the growth machine: the view of planners. *Journal of Rural Studies* 10: 233–48

Phillips A D M (1975) Underdraining and investment in the Midlands in the mid 19th century. In: Phillips A D M, Turton B J (eds) *Environment, Man and economic change: essays presented to S.H. Beaver*. Longman, London, pp. 253–74

Phillips E, Gray I (1995) Farming 'practice' as temporally and spatially situated intersections of biography, culture and social structure. *Australian Geographer* 26: 127–32

Phillips M (1998) Social perspectives. In: Ilbery B W (ed.) *The geography of rural change*. Longman, Harlow, pp. 31–54

Philo C (1992) Neglected rural geographies: a review. *Journal of Rural Studies* 8: 193–207

Pierce J T (1990) *The food resource*. Longman Scientific & Technical, Harlow

Pierce J T (1992) The policy agenda for sustainable agriculture. In: Bowler I R, Bryant C R, Nellis M D (eds), *Contemporary rural systems in transition. Volume 1 Agriculture and Environment*. CAB International, Wallingford, pp. 221–36

Pierce J T, Seguin J (1993) Exclusionary agricultural zoning in British Columbia and Quebec: problems and prospects. *Progress in Rural Policy and Planning* 3: 287–310

Pimentel D, Allen J, Beers A, Guinard L, Linder R, McLaughlin P, Meer B, Musonda D, Perdue D, Poisson S, Siebert S, Stoner K, Salazar R, Hawkins A (1987) World agriculture and soil erosion. *Bioscience* 37: 277–83

Pincetl S S (1999) *Transforming California: a political history of land use and development*. John's Hopkins University Press, Baltimore, MO

Pingali P, Bigot Y, Binswanger H P (1987) *Agricultural mechanization and the evolution of farming systems in Sub-Saharan Africa*. Johns Hopkins University Press, Baltimore

Pingali P, Rajaram S (1998) *Technological opportunities for sustaining wheat productivity growth toward 2020*. 2020 Briefs, International Food Policy Research Institute, Washington, DC, no. 51

Pini B (2002) The exclusion of women from agri-political leadership: a case study of the Australian sugar industry. *Sociologia Ruralis* 42: 65–76

Piscorz W (2000) The effect of European Union accession on Poland's agricultural markets and budgetary expenditures. In: Hartell J G, Swinnen J F M (eds) *Agriculture and East-West European integration*. Ashgate, Aldershot, pp. 85–106

Platt R H (1985) The farmland conservation debate: NALS and beyond. *Professional Geographer* 37: 433–42

Potter C (1996) Environmental reform of the CAP: an analysis of the short- and long-range opportunities. In Curry N, Owen S (eds) *Changing rural policy in Britain: Planning, administration, agriculture and the environment*. Countryside and Community Press, Cheltenham, pp. 165–83

Potter C (1997) Environmental change and farm restructuring: the impact of the farm family life cycle. In: Ilbery B W, Chiotti Q, Rickard T (eds) *Agricultural restructuring and sustainability: a geographical perspective*. CAB International, Wallingford

Potter C (1998) *Against the grain: agri-environment reform in the United States and the European Union*. CAB International, Wallingford

Potter C, Burney J (2002) Agricultural multifunctionality in the World Trade Organisation – legitimate non-trade concern or disguised protectionism? *Journal of Rural Studies* 18: 35–48

Potter C, Burnham P, Edwards A, Gasson R M, Green B (1991) *The diversion of land: conservation in a period of farming contraction*. Routledge, London and New York

Potter C, Goodwin P (1998) Agricultural liberalisation in the European Union: an analysis of the implications for nature conservation. *Journal of Rural Studies* 14: 287–98

Potter C, Lobley M (1992) The conservation status and potential of elderly farmers: results from a survey in England and Wales. *Journal of Rural Studies* 8: 133–43

Potter C, Lobley M (1996a) Unbroken threads? succession and its effects on family farms in Britain. *Sociologia Ruralis* 36: 286–306

Potter C, Lobley M (1996b) The farm family life-cycle, succession paths and environmental change in Britain's countryside. *Journal of Agricultural Economics* 47: 172–90

Potter C, Lobley M (1998) Landscapes and livelihoods: environmental protection and agricultural support in the wake of Agenda (2000) *Landscape Research* 23: 223–36

Powell J M, Fernandez-Riveira S, Williams T O, Renard C (eds) (1995) *Livestock and sustainable nutrient cycling in mixed farming systems of sub-Saharan Africa*. ILCA, Addis Ababa, 2 vols

Power M (2001) Alternative geographies of global development and inequality. In: Daniels P W, Bradshaw M, Shaw D, Sidaway J (eds) *Human Geography: issues for the 21st century*. Pearson Education, Harlow, pp. 274–302

Pretty J N (1995) *Regenerating Agriculture: Policies and Practice for Sustainability and Self-Reliance*. Earthscan Publications: London

Primdahl J, Hansen B (1993) Agriculture in Environmentally Sensitive Areas: implementing the ESA measure in Denmark. *Journal of Environmental Planning and Management* 36: 231–38

Pritchard W N (1996) Shifts in food regimes, regulation and producer co-operatives: insights from the Australian and US dairy industries. *Environment and Planning A* 28: 857–75

Pritchard W N (1999) The regulation of grower-processor relations: a case study from the Australian wine industry. *Sociologia Ruralis* 39: 186–201

Pritchard W N (2000) Beyond the modern supermarket: geographical approaches to the analysis of contemporary Australian retail restructuring. *Australian Geographical Studies* 38: 204–18

Pugliese P (2001) Organic farming and sustainable rural development: a multi-faceted and promising convergence. *Sociologia Ruralis* 41: 112–30

Pywell R, Pakeman R, Parr T (1994) Opportunities for habitat reconstruction on farmland in lowland Britain. *Aspects of Applied Biology* **40**: 469–77

Qualman D (2001) Corporate hog farming: the view from the family farm. In: Epps R, Whitson D (eds) *Writing off the rural West: globalization, governments and the transformation of rural communities*. University of Alberta Press and Parkland Institute, Edmonton, pp. 21–38

Quisumbing A R, Brown I R, Feldstein H S, Haddad L, Pena C (1995) *Women: the key to food security*. International Food Policy Institute, Washington, DC

Rabinowicz E (2000) Eastward European Union enlargement and the future of the CAP. In: Hartell J G, Swinnen J F M (eds) *Agriculture and East-West European integration*. Ashgate, Aldershot, pp. 215–46

Rackham O (1990) *Trees and woodland in the British landscape. The complete history of Britain's trees, woods and hedgerows*. Orion Publishing, London

Rahman S (2000) Women's employment in Bangladesh agriculture: composition, derterminants and scope. *Journal of Rural Studies* **16**: 497–508

Rakes P (1982) The state and the peasantry in Tanzania. In: Harriss J C (ed.) (1982) *Rural development: theories of peasant economy and agrarian change*. Hutchinson, London, pp. 350–80

Ramsay S (1997) Agenda 2000: implications for Scottish farmers. *Working Papers, SAC Rural Business Development Unit, Bush Estate, Penicuik*

Ramsey D, Everitt J C (2001) Post-Crow farming in Manitoba: an analysis of the wheat and hog sectors. In: Epps R, Whitson D (eds) *Writing off the rural West: globalization, governments and the transformation of rural communities*. University of Alberta Press and Parkland Institute, Edmonton, pp. 3–20

Rangeley W R (1987) Irrigation and drainage in the world. In: Jordan W R (ed.) *Water and water policy in world food supplies*. Texas A and M University Press, College Station, TX, pp. 29–36

Rao K S (1997) Natural resource management in Himalaya – a recourse to issues and strategies. *ENVIS Monographs, G.B. Pant Institute of Himalayan Environment and Development*. Kosi-Katarmal, Almora, India

Rao P K (2000) *Sustainable development: economics and policy*. Blackwell, Oxford

Rasmussen K (1992) *An elementary introduction to satellite image processing on a PC-based system*. Institute of Geography, University of Copenhagen, Copenhagen

Raven H, Lang T, Dumonteil C (1995) *Off our trolleys? Food retailing and the hypermarket economy*. Institute for Public Policy Research, London

Ray C (1998) Culture, intellectual property rights and territorial rural development. *Sociologia Ruralis* **38**: 1–19

Reardon T (1995) Sustainability issues for agricultural-research strategies in the semi-arid tropics – focus on the Sahel. *Agricultural Systems* **48**: 345–59

Redclift M R (1994) Reflections on the 'sustainable development' debate. *International Journal of Sustainable Development and World Ecology* **1**: 3–21

Redclift M, Lekakis J, Zanias G P (eds) (1998) *Agriculture and world trade liberalisation: socio-environmental perspectives on the Common Agricultural Policy*. CAB International, Wallingford

Reij C, Waters-Bayer A (eds) (2001) *Farmer innovation in Africa*. Earthscan, London

Reij C, Scoones I, Toulmin C (1996) *Sustaining the soil: indigenous and water conservation in Africa*. Earthscan, London

Reilly J (1999) What does climate change mean for agriculture in Developing Countries? A comment on Mendelsohn and Dinar. *World Bank Research Observer* **14**(2): 295–305

Reilly W (ed.) (1973) *The use of land: a citizens' policy guide to urban growth*. Thomas Y. Crowell, New York

Renard M-C (1999) The interstices of globalization: the example of Fair Coffee. *Sociologia Ruralis* **39**: 484–500

Rennie S J (1991) Subsistence agriculture versus cash cropping: the social repercussions. *Journal of Rural Studies* 7: 5–10

Repassy H (1991) Changing gender roles in Hungarian agriculture. *Journal of Rural Studies* 7: 23–30

Reutlinger S (1999) From 'food aid' to 'aid for food': into the 21st century. *Food Policy* 24: 7–16

Rhodes V J (1998) The industrialization of hog production. In: Royer J S, Rogers R T (eds) *The industrialization of agriculture: vertical co-ordination in the US food system.* Ashgate, Aldershot, pp. 217–40

Ribaudo M O, Piper S, Schaible G D, Langner L L, Colacicco D (1989) CRP: what economic benefits? *Journal of Soil and Water Conservation* 44: 421–4

Rice R A, Vandermeer J H (1990) Climate and the geography of agriculture. In: Carroll C R, Vandermeer J H, Rosset P M (eds) *Agroecology.* McGraw Hill, New York, pp. 21–63

Rigg J (1989) The new rice technology and agrarian change: guilt by association. *Progress in Human Geography* 13: 374–99

Rigg J (1990) Developing World: the Green Revolution 25 years on. *Geography Review* 4(1): 32–4

Rigg J (1998) Rural-urban interactions, agriculture and wealth: a south-east Asian perspective. *Progress in Human Geography* 22: 497–522

Rigg J (2001) *More than the soil: rural change in southeast Asia.* Prentice Hall, Harlow

Ritson C, Harvey D R (eds) (1997) *The Common Agricultural Policy.* CAB International, Wallingford, second edition

Ritzer G (1996a) The McDonaldization thesis: is expansion inevitable? *International Sociology* 11: 291–308

Ritzer G (1996b) *The McDonaldization of society: an investigation into the changing character of contemporary social life.* Pine Forge Press, California

Robbins P (1999) Meat matters: cultural politics along the commodity chain in India. *Ecumene* 6: 399–423

Robertson C J (1930) *World sugar production and consumption: an economic geographical survey.* John Bale, London

Robinson A H, Lindberg J B, Brinkman L W (1961) A correlation and regression analysis applied to rural farm densities in the Great Plains. *Annals of the Association of American Geographers* 52: 414–25

Robinson D A (1999) Agricultural practice, climate change and the soil erosion hazard in parts of southeast England. *Applied Geography* 19: 13–28

Robinson D A, Blackman J (1990) Water erosion of arable land on the South Downs. *Geography Review* 4(1): 19–23

Robinson G M (1981) A statistical analysis of agriculture in the Vale of Evesham during the 'great agricultural depression'. *Journal of Historical Geography* 7: 37–52

Robinson G M (1983) The evolution of the horticultural industry in the Vale of Evesham. *Scottish Geographical Magazine* 99: 89–100

Robinson G M (1985a) The adoption of peanut-growing in Belize, Central America. *Singapore Journal of Tropical Geography* 6: 116–26

Robinson G M (1985b) Geographical perspectives on the world food problem. *New Zealand Journal of Geography* 80: 2–6

Robinson G M (1988a) *Agricultural change: geographical studies of British agriculture.* North British Publishing, Edinburgh

Robinson G M (1988b) Spatial changes in New Zealand's food processing industry, 1973–84. *New Zealand Geographer* 44: 69–79

Robinson G M (1989) Sugar cane in Belize: locational adjustments to a saturated market. In: Munro D M (ed.) *Conservation and development issues in Belize.* Department of Geography, University of Edinburgh, Edinburgh, pp. 152–77

Robinson G M (1991a) The environment and agricultural policy in the European Community: land use implications in the United Kingdom. *Land Use Policy* 8: 95–107

Robinson G M (1991b) Changing policies for the management of rural resources in the United Kingdom. *Scottish Association of Geography Teachers' Journal* 20: 29–39

Robinson G M (1991c) The environmental dimensions of recent agricultural policy in the United Kingdom. In: Jones G E, Robinson G M (eds)

Land use change and the environment in the European Community (Biogeography Research Group, Institute of British Geographers, London), Biogeography Monographs, no. 4, pp. 63–76

Robinson G M (1991d) The city beyond the city. In: Robinson G M (ed.) *A social geography of Canada: essays in honour of J. Wreford Watson*. Dundurn Press, Toronto, pp. 302–29

Robinson G M (1993a) Beyond MacSharry: the road to CAP reform. In: Bolsius E C A, Clark G, Groendijk J G (eds) *The retreat: rural land-use and European agriculture*. Royal Dutch Geographical Society/Department of Human Geography, University of Amsterdam, Amsterdam, pp. 28–41

Robinson G M (1993b) Trading strategies: New Zealand, the GATT, CER and trade liberalisation. *New Zealand Geographer* 49: 13–22

Robinson G M (1993c) The world food problem. *Geography Review* 6(4): 25–36

Robinson G M (1994a) *Conflict and Change in the Countryside: Rural Development in the Developed World*. John Wiley and Sons, Chichester and New York, revised edition

Robinson G M (1994b) Agriculture as a land use in Scotland. In: Fenton A, Gillmor D (eds) *Rural land use on the Atlantic periphery: Scotland and Ireland*. Royal Irish Academy/Royal Society of Edinburgh, Dublin and Edinburgh, pp. 71–96

Robinson G M (1994c) Environmentally Sensitive Areas and the greening of agricultural policy. In: Fodor I, Walker G P (eds) *Environmental Policy and Practice in Eastern and Western Europe*. Centre for Regional Studies, Hungarian Academy of Sciences, Pecs, pp. 189–200

Robinson G M (1994d) The greening of agricultural policy: Scotland's Environmentally Sensitive Areas (ESAs). *Journal of Environmental Planning and Management* 37: 215–25

Robinson G M (1995) New Zealand's trading policy in an age of globalisation: GATT, APEC and CER. *Pacific Viewpoint* 36(2): 129–41

Robinson G M (1996a) Globalisation and trading strategies in the South Pacific. In: Yeung Y (ed.) *Global Change and the Commonwealth*. Hong Kong Institute of Asia-Pacific Studies, The Chinese University of Hong Kong, pp. 323–38

Robinson G M (1996b) Globalisation and trading strategies in the South Pacific. *Australian Studies* 10: 70–85

Robinson G M (1997a) Greening and globalizing: agriculture in the new times. In: Ilbery B W, Chiotti Q, Rickard T (eds) *Agricultural restructuring and sustainability: a geographical perspective*. CAB International, Wallingford, pp. 41–54

Robinson G M (1997b) Farming without subsidies: lessons for Europe from New Zealand? *British Review of New Zealand Studies* 10: 89–104

Robinson G M (1998a) *Methods and techniques in human geography*. John Wiley and Sons, Chichester and New York

Robinson G M (1998b) Sustainable development and the search for sustainable agriculture. Unpublished paper presented to the Geographical Association Annual Conference, University of Leeds, April

Robinson G M (1998c) Agricultural land use in the Belize river valley. In: Barker D, Newby C, Morrisey M (eds) *A reader in Caribbean geography*. Ian Randle Publishers, Kingston, Jamaica, pp. 118–24

Robinson G M (1999) Countryside recreation management. In: Pacione M (ed.) *Applied Geography: principles and practice*. Routledge, London and New York, pp. 257–73

Robinson G M (2002a) Sustainable rural development. In: Bowler I R, Bryant C R, Cocklin C (eds) *The sustainability of rural systems: geographical interpretations*. Kluwer, Dordrecht, pp. 35–57

Robinson G M (2002b) Guest editorial: sustainable development – from Rio to Johannesburg. *Geography* 87: 185–8

Robinson G M (2003) Environmentally-friendly farming: an investigation of policies and impacts in Ontario, Canada. Unpublished report for Department of Trade and Foreign Affairs, Canada. School of Geography, Kingston University, Surrey

Robinson G M, Ghaffar A (1997) Agri-environmental policy and sustainability, *North-West Geographer*, third series, 1: 10–23

Robinson G M, Harris F A (2003) Food and agriculture. In: Harris F A (ed.) *Global environmental problems*. John Wiley and Sons, Chichester and New York (in press)

Robinson G M, Ilbery B W (1993) Reforming the CAP: beyond MacSharry. *Progress in Rural Policy and Planning* 3: 197–207

Robinson G M, Lind M (1999) Set-aside and environment: a case study in Southern England. *Tijdschrift voor Economische en Social Geografie* 90: 296–311

Robinson G M, Gray D A, Healey R G, Furley P A (1989) Developing a Geographical Information System (GIS) for agricultural development in Belize, Central America. *Applied Geography* 9: 81–94

Robinson G M, Tranter P, Loughran R J (2000) *Australia and New Zealand: economy, society and environment*. Edward Arnold, London

Roche M M (1996) Britain's farm to global seller: food regimes and New Zealand's changing links with the Commonwealth. In: Yeung Y (ed.) *Global change and the Commonwealth*. Chinese University of Hong Kong, Shatin, Hong Kong, Research Monograph no. 26, pp. 339–60

Roche M M (1999) International food regimes: New Zealand's place in the international frozen meat trade, 1870–1935. *Historical Geography* 27: 129–151

Roche M M, McKenna M, Le Heron R B (1999) Making fruitful comparisons: Southern Hemisphere producers and the global apple industry. *Tijdschrift voor Economische en Sociale Geografie*, 90, pp. 410–26

Roest K, Menghi A (2000) Reconsidering 'traditional' food: the case of Parmigiano Reggiano cheese. *Sociologia Ruralis* 40: 439–51

Rogers E M (1995) *Diffusion of innovations*. Free Press, New York, fourth edition

Roseberry W (1995) Introduction. In: Roseberry W, Gudmundson L, Samper M (eds) *Coffee, society and power in Los Angeles*. Johns Hopkins University Press, Baltimore, MD, pp. 3–24

Rosenau J W (1992) The relocation of authority in a shrinking world. *Comparative Politics* 24: 253–72

Rosenzweig C, Parry M L (1993) Potential impacts of climate change on world food supply: a summary of a recent international study. In: Kaiser H M, Drennen T E (eds) *Agricultural dimensions of global climate change*. St. Lucie Press, Delray Beach, FL, pp. 87–116

Rosenzweig C, Allen L, Harper L, Hollinger S, Jones J (eds) (1995) *Climate change and agriculture: analysis of potential international impacts*. American Society of Agronomy, Special Publications, no. 59

Rounce N V (1937) Individual native smallholdings. *East African Agricultural Journal*, September, 1–7

Rovira G (2000) *Women of maize: indigenous women and the Zapatista rebellion*. Latin America Bureau, London

Royer J S, Rogers R T (eds) (1998) *The industrialization of agriculture: vertical co-ordination in the US food system*. Ashgate, Aldershot

Russell N P, Fraser I M (1995) The potential impact of environmental cross-compliance on arable farming. *Journal of Agricultural Economics* 46: 70–79

Russwurm L H (1975) Urban fringe and urban shadow. In: Bryfogle R C, Kreuger R R (eds) *Urban problems*. Holt, Rinehart and Winston, Toronto, revised edition, pp. 148–64

Russwurm L H (1977) *The surroundings of our cities*. Community Planning Press, Ottawa

Ruthenberg H (1980) *Farming systems in the Tropics*. Clarendon Press, Oxford, 3rd edition

Ryle S (2002) Banana war leaves the Caribbean a casualty. *The Observer*, 24.11.2002, Business, p. 7

Salamon S (1992) *Prairie patrimony*. University of North Carolina Press, Chapel Hill

Saltiel J (1994) Controversy over CRP in Montana: implications for the future. *Journal of Soil and Water Conservation* 49: 284–8

Salzman P (1980) Is nomadism a useful concept? *Nomadic Peoples* 6: 1–7

Sanchez P A, Buiol S W (1975) Soils of the Tropics and the world food crisis. *Science* 188: 598–603

Sayer A (1995) *Radical political economy: a critique*. Blackwell, Oxford

Sayer A, Walker R (1992) *The new social economy: reworking the division of labour*. Blackwell, Oxford

Scambler A (1989) Farmers' attitudes towards forestry. *Scottish Geographical Magazine* 105: 47–9

Scargill D I (1994) Crisis in rural France. *Geography* 79: 168–72

Schaffer G (1980) Ensuring Man's food supplies by developing new land and preserving cultivated land. *Applied Geography and Development* 16: 7–24

Schedvin R B (1990) Staples and regions of Pax Britannica. *Economic History Review*, 2nd series, 43: 533–59

Schlosser E (2002) *Fast food nation: the dark side of the all-American meal.* HarperCollins, New York

Scoones I (ed.) (2001) *Diversity and dynamics: soil fertility and farming in Africa.* Earthscan, London

Scott J (1976) *The moral economy of the peasantry.* Yale University Press, New Haven, CT

Scott J (1999) *Seeing like a state: how certain schemes to improve the human condition have failed.* Yale University Press, New Haven, CT

Scott P (1957) Agricultural regions of Tasmania. *Economic Geography* 33: 109–21

Scottish Office (1995) *Rural Scotland: people, prosperity and partnership.* HMSO, Edinburgh, Command Paper 3041

Seaborne A (2001) Crop diversification in Canada's breadbasket: land use changes in Saskatchewan's agriculture. *Geography* 86: 151–8

Sears J (1990) *A research review of current information on the effects on birds of organic and low input farming.* RSPB, Sandy

Seeman J, Chirkov Y I, Lomas J, Primault B (eds) (1979) *Agrometeorology.* Springer-Verlag, Berlin

Selby J A (1994) Primary sector policies and rural development in Finland. *Progress in Rural Policy and Planning* 4: 157–76

Selby J A, Petajisto L (1994) Field afforestation in Finland in the 1990s. *Finnish Forest Research Institute Research Paper* 502

Select Committee for Agriculture, UK (2001) *Agriculture – Second Report.* House of Commons, London

Self P, Storing H J (1962) *The state and the farmer.* George Allen and Unwin, London

Sen A K (1981) *Poverty and famines: an essay on entitlement and deprivation.* Clarendon, Oxford

Sen A K (1990) Food, economics and entitlements. In: Dreze J, Sen A K (eds) *The political economy of hunger.* Clarendon, London, 2 vols, pp. 34–50

Serageldin I (1996) Sustainability and the wealth of nations: first steps in an ongoing journey, *Environment and Sustainable Development Studies, Monograph Series, The World Bank*, no. 5

Serageldin I (1999) New partnerships and new paradigms for the new century. *Current Science* 75: 501–6

Seymour J (1991) *The fat of the land.* Metanonia Press, New Ross, Ireland

Seymour J (1996) *The complete book of self-sufficiency.* Dorling Kindersley, London

Seymour S, Cox G (1992) Nitrates in water: the politics of pollution. In: Gilg A W (ed.) *Restructuring the countryside.* Avebury, Aldershot, pp. 18–202

SFSG (Scottish Food Strategy Group) (1993) *Scotland means quality.* SFSG, Edinburgh

Shackleton R (1998) Exploring corporate culture and strategy: Sainsbury at home and abroad during the early to mid 1990s. *Environment and Planning A* 30: 921–40

Shand H (1997) *Human nature: agricultural biodiversity and farm-based food security.* FAO, Rome

Shanin T (ed.) (1987) *Peasants and peasant societies: selected readings.* Basil Blackwell, Oxford

Shaw G, Williams A (1994) *Critical issues in tourism: a geographical perspective.* Blackwell, Oxford

Shiva V (1993) *Monocultures of the mind: perspectives on biodiversity and biotechnology.* Zed Books, London

Shiva V, Moser I (1995) *Biopolitics: a feminist and ecological reader on biotechnology.* Zed Books, London

Shoard M (1980) *The theft of the countryside.* Temple Smith, London

Short D (1996) Subsuming the family farm: from land use study to political economy in rural geography. *Scottish Geographical Magazine* 112: 51–3

Shortall S (1994) Farm women's groups: feminist or farming or community groups, or new social movements? *Sociology* 28: 279–91

Shortall S (2002) Gendered agricultural and rural restructuring: a case study of Northern Ireland, *Sociologia Ruralis* 42: 160–76

Shucksmith M (1993) Farm household behaviour and the transition to post-productivism. *Journal of Agricultural Economics* 44: 466–78

Shucksmith M, Herrmann V (2002) Future changes in British agriculture: projecting divergent farm household behaviour. *Journal of Agricultural Economics* 53: 37–50

Shucksmith M, Smith R (1991) Farm household strategies and pluriactivity in upland Scotland. *Journal of Agricultural Economics* 42: 340–53

Shucksmith M, Bryden J, Rosenthall P, Short C, Winter M (1989) Pluriactivity, farm structures and rural change. *Journal of Agricultural Economics* 40: 345–60

Shueb S, Atkins P J (1991) Crop area estimation: a comparison of remote sensing and census methods. *Geography* 76: 235–9

Sillitoe P (1998) Its all in the mound: shifting cultivation and indigenous soil management in Papua New Guinea. *Mountain Research and Development* 18: 123–34

Silson A (1995) The scenery of set aside. *Geographical Magazine* 67(3): 58–9

Simon D, Narman A (eds) (1999) *Development theory and practice: current perspectives on development and development co-operation.* Longman, Harlow

Simon D, Van Spengen W, Dixon C, Marman A (eds) (1995) *Structurally adjusted Africa: poverty, debt and basic needs.* Pluto Press, London

Simpson S, Robinson G M (2001) Agri-environment policies, post-productivism and sustainability: England's Countryside Stewardship Scheme. In: Molinero F, Baraja E, Alario M (eds) *Proceedings of the Second Anglo-Spanish Symposium on Rural Geography, University of Valladolid, July 2000.* University of Valladolid, Spain, Section 3–3, pp. 1–16

Simpson-Lewis W, Moore J, Pocock N, Taylor M, Swan H (1979) Canada's Special Resource Lands: National perspective of selected land uses. *Map Folios, Lands Directorate, Environment Canada,* no. 4

Singh J S (ed.) (1985) *Environmental regeneration in Himalaya.* Gyanodaya Prakashan, Nainital, India

Sittirak S (1998) *The daughters of development: women and the changing environment.* Zed Books, London

Sivertsen A, Lundberg A (1996) Farming practices and environmental problems in an arid landscape – a case study from the region of Lambayeque, Peru. *Geografiska Annaler* 78B: 147–61

Skerratt S J (1994) Itemised payment systems within a scheme – the case of Breadalbane. In: Whitby M (ed.) *Incentives for Countryside Management: the Case of Environmentally Sensitive Areas.* CAB International: Wallingford, pp. 105–34

Skerratt S J, Dent J B (1996) The challenge of agri-environment subsidies: the case of Breadalbane Environmentally Sensitive Area, Scotland. *Scottish Geographical Magazine* 112: 92–100

Skogstad G (1998) Ideas, paradigms and institutions: agricultural exceptionalism in the EU and the US. *Governance* 11: 463–90

Slatford R, Fishpool I (2001) India and the Green Revolution. *Geography Review* 15(2): 2–5

Slaymaker O (2001) Why so much concern about climate change and so little attention to land use change? *Canadian Geographer* 45: 71–8

Smil V (1999) China's agricultural land. *The China Quarterly* 158: 414–29

Smil V (2000) *Feeding the World: a challenge for the 21st century.* MIT Press, Cambridge, MA

Smit B (1994) Global change, climate, and geography. *The Canadian Geographer* 38: 81–9

Smit B E, Cocklin C (1981) Future urban growth and agricultural land: alternatives for Ontario. *Ontario Geographer* 18: 47–55

Smit B E, Burton I, Klein R J T, Wandel J (2000) An anatomy of adaptation to climate change and variability. *Climatic Change* 45: 223–51

Smith D F, Hill D M (1975) National agricultural ecosystems. *Journal of Environmental Quality* 4(2): 143–5

Smith D W (1998) Urban food systems and the poor in developing countries. *Transactions of* the Institute of British Geographers, new series, **23**: 207–20

Smithers J, Blay-Palmer A (2001) Technology innovation as a strategy for climatic adaptation in agriculture. *Applied Geography* **21**: 175–97

Smithers J, Smit B (1997) Agricultural systems response to environmental stress. In: Ilbery B W, Chiotti Q, Rickard T (eds) *Agricultural restructuring and sustainability: a geographical perspective*. CAB International, Wallingford, pp. 167–84

Sobhan R (1993) *Agrarian reform and social transformation: preconditions to development*. Zed Books, London

Soil Association (2000) *Organic facts and figures*. Soil Association, Bristol

Solow R M (1986) On the intergenerational allocation of natural resources. *Scandinavian Journal of Economics* **88**: 141–9

Soper M H R (1983) *Dairy farming and agriculture*. Association of Agriculture, London

Sparks T H, Mountford J O, Parish T (1994) Opportunities for conservation in the field boundaries of arable crops. *Aspects of Applied Biology* **40**: 461–8

Spellman G, Field K (2002) The changed fortunes of United Kingdom viticulture? *Geography* **87**: 324–30

Stamp L D (1940) Fertility, productivity and classification of land in Britain. *Geographical Journal* **96**: 389–412

Stamp L D (1948) *The land of Britain: its use and misuse*. Longman Green & Co, Geographical Publications, London

Stayner R, Reeve I (1990) *Uncoupling: relationships between agriculture and the local economies of rural areas in New South Wales*. Rural Development Centre, University of New England, Armidale, New South Wales

Steiner F R (1990) *Soil conservation in the United States: policy and planning*. Johns Hopkins University Press, Baltimore, MD

Stokes S (1989) *Saving America's countryside: a guide to rural conservation*. Johns Hopkins University Press, Baltimore, MD

Stone G D (2001) Theory of the square chicken: advances in agricultural intensification theory. *Asia Pacific Viewpoint* **42**: 163–80

Sturm L S, Smith F J (1993) Bolivian farmers and alternative crops: some insights into innovation adoption. *Journal of Rural Studies* **10**: 141–52

Suhariyanto K, Thirtle C (2001) Asian agricultural productivity and convergence. *Journal of Agricultural Economics* **52**: 96–110

Suli-Zakar I, Santha A, Tykkylainen M, Neil C (1998) Coping with socialist restructuring and the transition to a market economy in rural Hungary. In: Neil C, Tykkylainen M (eds) *Local economic development: a geographical comparison of rural community restructuring*. United Nations University Press, Tokyo, pp. 125–53

Sumberg J (1998) Mixed farming in Africa: the search for order, the search for sustainability. *Land Use Policy* **15**: 293–318

Susman P, O'Keefe P, Wisner B (1984) Global disasters: a radical interpretation. In: Hewitt K (ed.) *Interpretations of calamity*. Allen and Unwin, Boston, MA, pp. 264–83

Sutton K, Buang A (1995) A new role for Malaysia's FELDA: from land settlement agency to plantation company. *Geography* **80**: 125–38

Swinnen J F M (ed.) (1997) *Political economy of agrarian reform in Central and Eastern Europe*. Ashgate, Aldershot

Swinnen J F M (1999) The political economy of land reform choices in Central and Eastern Europe. *Economics of Transition* **7**: 637–64

Swinnen J F M, Mahtijs E (1997) Agricultural privatisation, land reform and farm restructuring in Central and Eastern Europe: a comparative analysis. In: Swinnen J F M, Buckwell A, Mathijs E (eds) *Agricultural privatisation, land reform and farm restructuring in Central and Eastern Europe*. Ashgate, Aldershot, pp. 333–73

Swinnen J F M, Tangermann S (1999) Conclusions and implications for food and agricultural policy in the process of accession to the EU. In: Tangermann S (ed.) *Agricultural implications of CEEC accession to the EU. Final report*. Institut fur Agrarokonomie, Georg-August-Universitat, Gottingen

Swinnen J F M, Buckwell A, Mathijs E (eds) (1997) *Agricultural privatisation, land reform and farm restructuring in Central and Eastern Europe.* Ashgate, Aldershot

Swyngedouw E (1997) Neither global nor local: 'glocalisation' and the politics of scale. In: Cox K (ed.) *Spaces of globalisation: reasserting the power of the local.* Guilford Press, New York, pp. 137–66

Syam T, Nishide H, Salam A K, Utomo M, Mahi A K, Lumbanraja J, Nugroho S G, Kimura M (1997) Land use and cover changes in a hilly area of south Sumatra, Indonesia (from 1970 to 1990). *Soil Science and Plant Nutrition* **43**: 587–99

Symes D (1991) Changing gender roles in productionist and post-productionist capitalist agriculture. *Journal of Rural Studies* **7**: 85–90

Symes D (1992) Agriculture, the state and rural society in Europe: trends and issues. *Sociologia Ruralis* **32**: 193–208

Symes D (1993) Agrarian reform and the restructuring of rural society in Hungary. *Journal of Rural Studies* **19**: 291–8

Symes D, Jansen A (eds) (1994) *Agricultural restructuring and rural change in Europe.* Studies in Sociology, Agricultural University, Wageningen

Symons L (1968) *Agricultural geography.* Bell, London

Tan S B-H (2000) Coffee frontiers in the Central Highlands of Vietnam: networks of connectivity. *Asia Pacific Viewpoint* **41**: 51–68

Tangermann S (ed.) (1999) *Agricultural implications of CEEC accession to the EU. Final report.* Institut fur Agrarokonomie, Georg-August-Universitat, Gottingen

Tansey G, Worsley A (1995) *The food system.* Earthscan, London

Tansley A E (1953) *The British Isles and their vegetation.* Cambridge University Press, Cambridge, 2 vols

Tarrant J R (1974) *Agricultural geography.* David & Charles, Newton Abbot

Tarrant J R (1980a) Agricultural trade within the European Community. *Area* **12**: 37–42

Tarrant J R (1980b) The geography of food aid. *Transactions of the Institute of British Geographers* new series **5**: 125–40

Tarrant J R (1980c) *Food policies.* Wiley, London

Tarrant J R (1985a) A review of international food trade. *Progress in Human Geography* **9**: 235–54

Tarrant J R (1985b) The significance of variability in Soviet cereal production. *Transactions of the Institute of British Geographers* new series **9**: 387–400

Tarrant J R (1992) Agriculture and the state. In: Bowler I R (ed.) *The geography of agriculture in developed market economies.* Longman, London, pp. 239–74

Teather E K (1996) Farm women in Canada, New Zealand and Australia redefine their rurality. *Journal of Rural Studies* **12**: 1–14

Theobald D M (2001) Land-use dynamics beyond the American urban fringe. *Geographical Review* **91**: 544–64

Thiesenhusen W C (1995) *Broken promises: agrarian reform and the Latin American campesino.* Westview Press, Boulder, CO

Thirsk I J (1997) *Alternative Agriculture. A history from the Black Death to the present day.* Oxford University Press, Oxford

Thomas R W, Huggett R J (1980) *Modelling in geography: a mathematical approach.* Harper & Row, London

Thompson S J, Cowan J J (2000) Globalizing agro-food systems in Asia. *World Development* **28**: 401–7

Thrift N J (1996) *Spatial formations.* Sage, London

Thrift N J (2000) Actor-network theory. In: Johnston R J, Gregory D, Pratt G, Watts M (eds) *The dictionary of human geography.* Blackwell, Oxford, 4th edition, pp. 5–6

Tickell A, Peck J A (1992) Accumulation, regulation and the geographies of post-Fordism: missing links in regulationist research. *Progress in Human Geography* **16**: 190–218

Tiffen M (1976) *The enterprising peasant: economic development in Gombe Emerite, North Eastern State, Nigeria, 1900–1968.* HMSO, London

Tilzey M (2000) Natural Areas, the whole countryside approach and sustainable agriculture. *Land Use Policy* **17**: 279–94

Tilzey M (2002) Conservation and sustainability. In: Bowler I R, Bryant C R, Cocklin C (eds) *The sustainability of rural systems: geo-*

graphical interpretations. Kluwer, Dordrecht, pp. 147–68

Tivy J (1987) Nitrogen cycling in agro-ecosystems. *Applied Geography* 7: 93–111

Tivy J (1990) *Agricultural ecology.* Longman, Harlow

Tiwari P C (1995) *Natural resources and sustainable development in Himalaya.* Shree Almora Book Depot, Almora, India

Tiwari P C (2000) Land-use changes in Himalaya and their impact on the plains ecosystem: need for sustainable land use. *Land Use Policy* 17: 101–111

Tokar M (2001) Monsanto: a profile of corporate arrogance. In: Goldsmith E, Mander J (eds) *The case against the global economy: And for a turn towards localization.* Earthscan, London, Ch. 7

Tolba M K (1992) *Saving our planet.* Chapman and Hall, London

Tordjman A (1995) *Trends in Europe. Consumer attitudes and the supermarket 1995.* Research Department, Food Marketing Institute, Washington, DC

Tovey H (2000) Agricultural development and environmental regulation in Ireland. In: Buller H, Hoggart K (eds) *Agricultural transformation, food and the environment.* Ashgate, Basingstoke, pp. 109–29

Townroe P M (1992) Skills and strengths in new rural businesses. Unpublished paper presented to the ESRC Small Business International Research Seminar, University of Warwick, June

Townsend A (1991) New forms of employment in rural areas: a national perspective. In: Champion A G, Watkins C (eds) *People in the countryside.* Paul Chapman Publishing, London, pp. 84–96

Troughton M J (1986) Farming systems in the modern world. In: Pacione M (ed.) *Progress in agricultural geography.* Croom Helm, London, pp. 93–123

Troughton M J (1993) Conflict or sustainability: contrasts and commonalities between global rural systems. *Geography Research Forum* 13: 1–11

Troughton M J (1997) Scale change, discontinuity and polarization in Canadian farm-based rural systems. In: Ilbery B W, Chiotti Q, Rickard T (eds) *Agricultural restructuring and sustainability: a geographical perspective.* CAB International, Wallingford, pp. 279–92

Troughton M J (2002) Enterprises and commodity chains. In: Bowler I R, Bryant C R, Cocklin C (eds) *The sustainability of rural systems: geographical interpretations.* Kluwer, Dordrecht, pp. 123–46

Turner B L, II, Brush S B (eds) (1987) *Comparative farming systems.* Guilford Press, New York

Turner M D (1999) Merging local and regional analyses of land-use change: the case of livestock in the Sahel. *Annals of the Association of American Geographers* 89: 191–219

Ufkes F (1993) Trade liberalisation, agro-food politics and the globalisation of agriculture. *Political Geography* 12: 215–31

UKCCIRG (United Kingdom Climate Change Impacts Review Group) (1991) *The potential effects of climate change in the United Kingdom.* HMSO, London

UNICEF (1998) *State of the world's children 1998.* UN Children's Fund/Oxford University Press, New York

United Nations (UN) Water Conference (1977) *Water for agriculture.* UN, New York

Unwin T (1987) Household characteristics and agrarian innovation adoption in north-west Portugal. *Transactions of the Institute of British Geographers*, new series, 12: 131–46

Unwin T (1988) The propagation of agricultural change in north-west Portugal. *Journal of Rural Studies* 4: 223–38

Unwin T (1994) Structural change in Estonian agriculture: from command economy to privatisation. *Geography* 79: 246–61

Unwin T (1997) Agricultural restructuring and integrated rural development in Estonia. *Journal of Rural Studies* 13: 93–112

USDA (United States Department of Agriculture) (2000) *National Resources Inventory: 1997 State of the Land update.* National Resources Conservation Service, USDA, Fort Worth, TX

USDC (United States Department of Commerce) (1993) *Census of Agriculture 1992.* Bureau of the Census, Washington, DC

Valdiya K S (1985) Accelerated erosion and landslide prone zones in the central Himalayan region. In: Singh J S (ed.) *Environmental regeneration in Himalaya*. Gyanodaya Prakashan, Nainital, India, pp. 12–38

Van den Ban A W, Hawkins H S (1988) *Agricultural extension*. Longman, New York

Vandergeest P (1988) Commercialisation and commodufication: a dialogue between perspectives. *Sociologia Ruralis* 28: 7–12

Van der Ploeg J (1992) The reconstitution of locality: technology and labour in modern agriculture, pp. 19–43 in Marsden *et al.* (eds) *Labour and locality*. David Fulton Publishers, London

Van der Ploeg J, Long A (eds) (1994) *Born from within: practice and perspectives of endogenous rural development*. Van Gorcum, Assen

Van der Ploeg J, Van Dijk G (eds) (1995) *Beyond modernization: the impact of endogenous rural development*. Van Gorcum, Assen

van Hemessen D, O'Grady L, Martin R (1994) *Report on landowner contact information for the Carolinian Canada, Niagara Escarpment, and Wetland Habitat Agreement Programs*. Ontario Heritage Foundation, Toronto

Van Huylenbroeck G, Whitby M (1999) *Countryside stewardship: policies, farmers and markets*. Pergamon, Oxford

Vanzetti D (1996) The next round: game theory and public choice perspectives. *Food Policy* 21: 461–77

Veeck G, Shaohua W (2000) Challenges to family farming in China. *Geographical Review* 90: 57–82

Veldman J (1984) Proposal for a theoretical basis for the human geography of rural areas. In: Clark G, Groenendijk J, Thissen F (eds) *The changing countryside*. Geo Books, Norwich, pp. 17–26

Village Aid (1996) Beyond PRA: a new approach to village-led development. Unpublished paper, Village Aid, London

Vinas C D (2000) The Spanish agrarian population on the threshold of the 21st century. Unpublished paper presented to the Anglo-Spanish Rural Geography Symposium, Universidade de Valladolid

Vine H (1953) Experiments on the maintenance of soil fertility at Ibadan, Nigeria, 1922–51. *Empire Journal of Experimental Agriculture* 21, supplement 82: 65–85

Visser S (1980) Technology change and the spatial structure of agriculture. *Economic Geography* 56: 311–19

Vogelmann J E, Sohl T, Howard S M (1998) Regional characterisation of land cover using multiple sources of data. *Photogrammetric Engineeriung and Remote Sensing* 64: 45–57

Von Hipple E (1988) *The sources of innovation*. Oxford University Press, Oxford

Wagstaff H (1987) Husbandry methods and farm systems in industrialised countries which use lower levels of external inputs: a review. *Agriculture, Ecosystems and Environment* 19: 1–27

Walford N S (2001) Patterns of development in tourist accommodation enterprises on farms in England and Wales. *Applied Geography* 21: 331–45

Walford N S (2002) Agricultural adjustment: adoption of and adaptation to policy reform measures by large-scale commercial farmers. *Land Use Policy* 19: 243–58

Walford N S, Burton R (2000) *The development of large-scale commercial farming in south-east England*. Report on an ESRC-funded project, School of Geography, Kingston University, Kingston-upon-Thames

Walker R A (1996) Another round of globalisation in San Francisco. *Urban Geography* 17: 60–94

Walker R A (1997) Fields of dreams, or the best game in town. In: Goodman D, Watts M J, (eds) *Globalising food: agrarian questions and global restructuring*. Routledge, London and New York, pp. 273–86

Walker R A, Moran E, Anselin L (2000) Deforestation and cattle ranching in the Brazilian Amazon: external capital and household processes. *World Development* 28: 683–99

Wallace I (1985) Towards a geography of agribusiness. *Progress in Human Geography* 9: 411–514

Wallerstein I (1991) *Geopolitics and geoculture: essays on the changing world system*. Cambridge University Press, Cambridge

Wang J Y (1972) *Agricultural meteorology.* Milieu Information Service, San Jose, CA, 3rd edition

Ward N (1990) A preliminary analysis of the UK food chain. *Food Policy* 15: 439–41

Ward N (1993) The agricultural treadmill and the rural environment in the post-productivist era. *Sociologia Ruralis* 33: 348–64

Ward N (1996) Surfers, sewage and the new politics of pollution. *Area* 28: 331–8

Ward N, Lowe P (1994) Shifting values in agriculture: the farm family and policy regulation. *Journal of Rural Studies* 11: 173–84

Ward N, McNicholas K (1998) Objective 5b of the Structural Funds and rural development in Britain. *Journal of Regional Studies* 32: 369–74

Ward N, Marsden T K, Munton R J C (1990) Farm landscape change. *Land Use Policy* 7: 291–302

Ward N, Lowe P, Seymour S, Clark J (1995) Rural restructuring and the regulation of farm pollution. *Environment and Planning A* 27: 1193–2111

Ward N, Clark J, Lowe P, Seymour S (1998) Keeping matters in its place: pollution regulation and the reconfiguration of farmers and farming. *Environment and Planning A* 30: 1165–78

Warren A, Batterbury S, Osbahr H (2001) Sustainability and Sahelian soils: evidence from Niger. *Geographical Journal* 167: 324–41

Warriner G K, Moul T M (1992) Kinship and personal communication network influences on the adoption of agricultural conservation technology. *Journal of Rural Studies* 8: 279–92

Watkins K (1997) Globalization and liberalization: implications for poverty, distribution and inequality. *UNDP Human Development Reports*, no. 32. UNDP, New York

Wathern P, Young S N, Brown I W, Roberts D A (1986) The EEC Less Favoured Areas Directive: implementation and impact on upland land use in the UK, *Land Use Policy*, 3: 205–12

Wathern P, Young S N, Brown I W, Roberts D A (1988) Recent upland use change and agricultural policy in Clwyd, North Wales. *Applied Geography* 8: 147–63

Watkins K (1995) *The Oxfam poverty report.* Oxfam, Oxford

Watts M J (1983) *Silent violence: food, famine and peasantry in northern Nigeria.* University of California Press, Berkeley, CA

Watts M J (1989) The agrarian crisis in Africa. *Progress in Human Geography* 13: 1–41

Watts M J (1996) Development III in the global agrofood system and the late twentieth-century development (or Kautsky reduxe). *Progress in Human Geography* 20: 230–45

Watts M J (2000) The great tablecloth. In: Clark G, Gertler M, Feldmann M (eds) *A handbook of economic geography.* Oxford University Press, Oxford, pp. 195–215

Watts M J (2001) Author's response: lost in space. Classics in human geography revisited. In Watts M (1983) Silent violence. *Progress in Human Geography* 25: 625–8

Watts M J, Bohle H G (1993) The space of vulnerability: the causal structure of hunger and famine. *Progress in Human Geography* 17: 43–67

Watts M J, Goodman D (1997) Agrarian questions. Global appetite. Local metabolism: nature, culture, and industry in fin-de-siecle agro-food systems. In: Goodman D, Watts M J (eds) *Globalising food: agrarian questions and global restructuring.* Routledge, London and New York, pp. 1–32

Weaver J C (1954a) Crop combinations in the Middle West. *Geographical Review* 44: 175–200

Weaver J C (1954b) Crop combination regions for 1919 and 1929 in the Middle West. *Geographical Review* 44: 560–72

Weaver J C (1954c) Changing patterns of cropland use in the Middle West. *Economic Geography* 30: 15–47

Weaver J C, Harg L P, Fenton B L (1956) Livestock units and combination regions in the mid-West. *Economic Geography* 32: 237–59

Webber P (1991) Agrarian change in Ghana. *Geography Review* 9(3): 25–30

Webster S, Felton M (1993) Targeting for nature conservation in agricultural policy. *Land Use Policy* 10: 67–82

Wells M (1996) *Strawberry fields: politics, class and work in California agriculture.* Cornell University Press, Ithaca, NY

Welsh Office (1996) *A working countryside for Wales.* HMSO, Cardiff, Command Paper 3180

Whatmore S J (1991a) *Farming women*. Macmillan, London

Whatmore S J (1991b) Life cycle or patriarchy? Gender divisions in family farming. *Journal of Rural Studies* 7: 71–6

Whatmore S J (1994) Global agro-food complexes and the refashioning of rural Europe. In: Amin A, Thrift N J (eds) *Globalization, institutions and regional development in Europe*. Oxford University Press, Oxford, pp. 46–67

Whatmore S J (1997) Dissecting the autonomous self: hybrid cartographies for a relational ethics. *Environment and Planning D, Society and Space* 15: 37–53

Whatmore S J (2002) From farming to agribusiness: the global agro-food system. In: Johnston R J, Taylor P J, Watts M (eds) *Geographies of global change: remapping the world in the late twentieth century*. Blackwell, Oxford, second edition, pp. 57–67

Whatmore S J, Thorne L (1997) Nourishing networks: alternative geographies of food. In: Goodman D, Watts M J (eds) *Globalising food: agrarian questions and global restructuring*. Routledge, London and New York, pp. 287–304

Whatmore S J, Munton R J C, Little J K, Marsden T K (1987a) Interpreting a relations typology of farm businesses in southern England. *Sociologia Ruralis* 27: 103–22

Whatmore S J, Munton R J C, Little J K, Marsden T K (1987b) Towards a typology of farm businesses in contemporary British agriculture. *Sociologia Ruralis* 27: 21–37

Whatmore S J, Munton R J C, Marsden T K, Little J K (1996) The trouble with subsumption and other rural tales. *Scottish Geographical Magazine* 112: 54–7

Whitaker D P (1984) The economy. In: Evans-Smith W (ed.) *Poland: a country study*. Department of Army, US Government, Washington, DC, pp. 163–222

Whitby M (ed.) (1994) *Incentives for Countryside Management: the Case of Environmentally Sensitive Areas*. CAB International: Wallingford

Whitby M (ed.) (1996) *The European environment and CAP reform: policies and prospects for conservation*. CAB International, Wallingford

Whitby M (2000) Reflections on the costs and benefits of agri-environment schemes. *Landscape Research* 25: 365–74

Whitby M, Ward N (1994) Of motherhood and apple pie. In: Whitby M, Ward N (eds) *The UK strategy for sustainable agriculture: a critical analysis*. Centre for Rural Economy, University of Newcastle-upon-Tyne, Newcastle-upon-Tyne, pp. 3–11

Whitehand J W R (1967) Fringe-belts: a neglected aspect of urban geography. *Transactions of the Institute of British Geographers*, **41**: 223–33

Whitehand J W R (1972) Building cycles and the spatial pattern of urban growth. *Transactions of the Institute of British Geographers* 56: 39–55

Whiteside M (1998) *Living farms: encouraging sustainable smallholders in South Africa*. Earthscan, London, KU Lib

Whittlesey D (1936) Agricultural regions of the world. *Annals of the Association of American Geographers* 26: 198–240

Wibberley J (1995) Cropping intensity and farming systems: integrity and intensity in international perspective. *Journal of the Royal Agricultural Society of England* 156: 43–55

Wickramasinghe A (1997) Women and minority groups in environmental management. *Sustainable Development* 5: 11–20

Wiggins S (2000) Interpreting changes from the 1970s to the 1990s in African agriculture through village studies. *World Development* 28: 631–62

Wiley J (1998) The banana industries of Costa Rica and Dominica in a time of change. *Tijdschrift voor Economische en Sociale Geografie* 89: 66–81

Williams D, Lloyd T, Warkins C (1994) Farmers not foresters. Constraints on the planting of new farm woodland. *Working Papers, Department of Geography, University of Nottingham*, no. 27

Willis R (2001) New Zealand in the 1990s: Farming. *Asia Pacific Viewpoint* 42: 55–66

Wilson G A (1994) German agri-environmental schemes I – a preliminary review. *Journal of Rural Studies* 10: 27–45

Wilson G A (1995) German agri-environmental schemes II – the MEKA Programme in Baden-Wurttemberg. *Journal of Rural Studies* **11**: 149–59

Wilson G A (1996) Farmer environmental attitudes and ESA participation. *Geoforum* **27**: 115–31

Wilson G A (1997a) Assessing the environmental impacts of the ESA scheme: a case for using farmers' environmental knowledge? *Landscape Research* **22**: 303–26

Wilson G A (1997b) Factors influencing farmer participation in the Environmentally Sensitive Areas scheme. *Journal of Environmental Management* **50**: 67–93

Wilson G A (1997c) Selective targeting in Environmentally Sensitive Areas: implications for farmers and the environment. *Journal of Environmental Planning and Management* **40**: 199–215

Wilson G A (2001) From productivism to post-productivism . . . and back again? Exploring the (un)changed natural and mental landscapes of European agriculture. *Transactions of the Institute of British Geographers*, new series, **26**: 103–20

Wilson G A, Hart K (2000) Financial imperative or conservation concern? EU farmers' motivations for participation in voluntary agri-environmental schemes. *Environment and Planning A* **32**: 2161–85

Wilson G A, Hart K (2001) Farmer participation in agri-environment schemes: towards conservation-oriented thinking? *Sociologia Ruralis* **41**: 254–74

Wilson G A, Wilson O J (2001) *German agriculture in transition: society, policies and environment in a changing Europe.* Macmillan, Basingstoke

Wilson G A, Lezzi M, Egli C (1996) Agri-environmental schemes in Switzerland. *European Urban and Regional Studies* **3**: 205–24

Wilson G A, Petersen J E, Holl A (1999) EU member state responses to agri-environment Regulation 2078/92/EEC: towards a conceptual framework? *Geoforum* **30**: 185–202

Wilson J, Fuller R (1992) Set-aside potential and management for wildlife conservation. *Ecos* **13**(3): 24–9

Wilson O J (1994) 'They changed the rules': farm family responses to agricultural deregulation in Southland, New Zealand. *New Zealand Geographer* **50**: 3–13

Wilson O J (1995) Farm household responses to agricultural deregulation: preliminary findings from a South Island case study. *Sociologia Ruralis* **335**: 756–66

Winchester H P M (1993) *Contemporary France.* Longman, Harlow

Winter M (1996a) Landwise or land foolish? Free conservation advice for farmers in the wider English countryside. *Landscape Research* **21**: 243–64

Winter M (1996b) *Rural politics: policies for agriculture, forestry and the environment.* Routledge, London

Winter M (2000) Strong or weak policy? The environmental impact of the 1992 reforms to the CAP arable regime in Great Britain. *Journal of Rural Studies* **16**: 47–60

Winter M (2002) The alternative food economy. Unpublished paper delivered to the Conference on Alternative Food Economies, Royal Geographical Society, March 2002

Winter M, Gaskell P (1998) *The effects of the 1992 reform of the Common Agricultural Policy on the countryside of Great Britain.* Countryside and Community Press, Countryside Commission, Cheltenham

Winter M, Gaskell P, Short C (1998) Upland landscapes in Britain and the 1992 CAP reforms. *Landscape Research* **23**: 273–88

Wisner B (1988) *Power and need in Africa.* Earthscan, London

Wolpert J (1964) The decision process in the spatial context. *Annals of the Association of American Geographers* **54**: 537–58

Wood L J (1994) Pyrethrum and essential oils: new cropping ventures in Tasmania. *Geography* **79**: 357–60

Woods M (1998) Researching rural conflicts: hunting, local politics and actor networks. *Journal of Rural Studies* **14**: 321–40

Wookey B (1987) *Rushall, the story of an organic farm.* Basil Blackwell, Oxford

World Bank (1991) *Gender and poverty in India.* World Bank, Washington DC

World Bank (1997) *World Development Report. The state in a changing world.* Oxford University Press, Oxford

World Bank (2000) *World development indicators.* World Bank, New York

Worthington B, Gant R L (1983) *Techniques in map analysis.* Macmillan Education, Basingstoke

Wrathall J E, Moore R (1996) Oilseed rape in Great Britain – the end of a 'revolution'. *Geography* 71: 351–5

WRI (World Resources Institute) (1996) *World resources 1996–7.* Oxford University Press, Oxford

Wright J (1993) Cultural geography and land trusts in Colorado and Utah. *Geographical Review* 83: 269–79

Wrigley N (1991) Is the 'golden age' of British grocery retailing at a watershed? *Environment and Planning A* 23: 1537–44

Wrigley N (1992) Anti-trust regulation and the restructuring of grocery retailing in Britain and the USA. *Environment and Planning A* 24: 727–49

Wrigley N (1994) After the store wars? Towards a new era of retail competition. *Journal of Retailing and Consumer Services* 1: 5–20

Wrigley N S (1999) Market rules and spatial outcomes: insights from the corporate restructuring of US food retailing. *Geographical Analysis* 31: 288–309

Wrigley N S (2002) Transforming the corporate landscape of US food retailing: market power, financial re-engineering and regulation. *Tijdschrift voor Economisce en Social Geografie* 93: 62–82

Wrigley N, Lowe M S (eds) (1996) *Retailing, consumption and capital: towards the new retail geography.* Longman, London

Wrigley N, Lowe M (2001) *Reading retail: a geographical perspective on retailing and consumption spaces.* Edward Arnold, London

Wynn G, Crabtree R, Potts J (2001) Modelling farmer entry into the ESA schemes in Scotland. *Journal of Agricultural Economics* 52: 65–82

Xu H (1992) Recent changes in land use in southeast Scotland: an approach with integration of remote sensing and GIS. Unpublished PhD thesis, Department of Geography, University of Edinburgh

Xu H, Young J A T (1989) Synergism of remotely-sensed and contextual data to monitor changes in land use. *Proceedings of IGARSS89, Vancouver* volume 1, 69–72

Xu H, Young J A T (1990) Monitoring changes in land use through integration of remote sensing and GIS. *Proceedings of IGARSS90, Washington, DC,* volume 1, 957–60

Yang H, Li X (2000) Cultivated land and food supply in China. *Land Use Policy* 17: 73–88

Yang H, Zhang X (1998) Can China feed itself? versus how should China feed itself? *China Newsletter* 136: 2–7

Yao A Y M (1981) Agricultural climatology. In: Landsberg H R (ed.) *World survey of climatology.* Elsevier, Amsterdam, pp. 189–298

Yapa L (1998) The poverty discourse and the poor in Sri Lanka. *Transactions of the Institute of British Geographers,* new series, 23: 95–115

Yarrington D (1997) *A coffee frontier: land, society and politics in Duaca, Venezuela.* University of Pittsburg Press, Pittsburg

Yarwood R (2002) *Countryside conflicts.* Geographical Association, Sheffield

Yarwood R, Evans N (1998) The changing geographies of domestic livestock animals. *Society and Animals* 6: 137–66

Yarwood R, Evans N (1999) The changing geography of rare livestock breeds in Britain. *Geography* 84: 80–7

Yarwood R, Evans N (2000) The geography of UK agriculture: mapping rare breeds. *Geography Review* 14(2): 18–22

Yeates M H (1985) Land in Canada's urban heartland. *Land Use in Canada Series, Lands Directorate, Environment Canada,* no. 27

Yeung Y (ed.) (1996) *Global change and the Commonwealth.* Chinese University of Hong Kong Research Monograph, no. 26

Young A (1998) *Land resources now and for the future.* Cambridge University Press, Cambridge

Young C (1993) A bitter harvest? Problems of restructuring East-Central European agriculture. *Geography* 78: 69–72

Young D, Bechtel A, Coupal R (1994) Comparing performance of the 1985 and the 1990

Conservation Reserve Programs in the West. *Journal of Soil and Water Conservation* **49**: 484–7

Young C, Morris C, Andrews C (1995) Agriculture and the environment in the UK: towards an understanding of the role of farming culture. *Greener Management International* **12**: 63–80

Young E M (1996) World hunger: a framework for analysis. *Geography* **81**: 97–110

Young E M (1997) *World hunger*. Routledge, London

Young W, Welford R (2002) *Ethical shopping: where to shop, what to buy and what to do to make a difference*. Vision Paperbacks, London

Zekeri A A (1992) Benefits from agricultural development projects: another lesson from Nigeria. *Journal of Rural Studies* **8**: 303–8

Zering K (1998) The changing US pork industry: an overview. In: Royer J S, Rogers R T (eds) *The industrialization of agriculture: vertical co-ordination in the US food system*. Ashgate, Aldershot, pp. 205–16

Index

Cairns Group 116
Calgene 55, 258
Cambodia 177
Cameroun 169
Canada 11, 20–1, 24, 98, 122–6, 143, 179, 199, 215–16, 256
    Alberta 21, 124, 126, 216
    British Columbia 215–16, 256
    Cowichan estuary 256
    Fraser delta 256
    Lake Superior 122
    Maritime provinces 215
    New Brunswick 215
    Niagara Peninsula 215
    Nova Scotia 215, 256
    Okanagan 215
    Ontario 21–2, 24, 124, 215, 237, 256
    Palliser's Triangle 20–1
    Peace River district 215
    Prairie provinces 122–5, 199, 256
    Prince Edward Island 256
    Quebec 215, 256
    Saskatchewan 122–4
    Southern Ontario 215
    Thunder Bay 122
    Toronto 216
    Western Canada 124, 215
Canada Land Use Monitoring Program (CLUMP) 215
Canadian Pacific Railway 122
Cape Verde 169
Capital/Capitalism 36–73, 131, 137–8, 140, 146–7, 162, 177, 228–9, 237
Carbon cycle 15
Carbon pathways 4–5
Cárdenas regime 188
Caribbean 57, 59, 166–70, 172
Carrefour 81
Carson, Rachel 222
Cash crops 150, 152, 155, 191, 193
Cassava 155, 195
Cattle, see Beef and Dairying
Central America 1, 57, 148–9, 166, 172, 203
Central Asia 9, 19
Central and East European Countries 107–15, 117
Central Europe 14, 107–115

Cereals/Grain 92, 96, 98–101, 105, 112, 121–3, 140, 172, 179, 181, 184–5, 190–1, 195, 199, 201, 220–1
Chiquita 166, 169
Chile 7, 159, 162
China 7, 14, 19, 58, 108, 110, 146–7, 159, 162–3, 172, 177, 186, 190, 195–6, 200–1
Citrus 168
Classification of agricultural systems 23–8, 30
Climate 3–9, 18–23
Climate change 18–23
Coca Cola 54
Cocoa 148, 168
Co-evolution 31
Coffee 148, 164–6, 168–9
Collectives/Co-operatives 110–15, 153, 186–7, 189, 193, 257
Colonialism 146–77, 180, 188, 229
Comecon 113–14
Commercialisation 148–66
Commodification/Commoditisation 65, 81–2
Commodity fetishism 77
Common Agricultural Policy (EU) 51, 58, 65, 91–118, 134, 166, 248, 251, 255, 257, 262
    Pillars 106–7, 116
    Reforms 92, 98–118, 134, 241
Commonwealth Scientific and Industrial Research Organisation 24
Communal Areas Management Programme For Indigenous Resources (CAMPFIRE) 190
Community labour co-operation 152, 196
ConAgra 59, 126
Concentration 62–3, 65, 67, 69
Congress (US) 118
Conservation 100, 118, 211, 214, 233, 242–3, 250
Conservation Reserve Program 100, 118
Consultative Group on International Agricultural Research (CGIAR) 191
Consumption/Consumers see Food consumption
Contract production 59, 131–2

Contrats Territoriaux d'Expoitation 107
Co-op (UK) 79, 170
Coordination Rurale 115
Corporate retailers see Supermarkets
Costa Rica 166–7
    Golfito 167
    Limon 167
Cost-price squeeze 66, 137
Cotonou Agreement 169
Cotton 9, 62, 121, 194, 258–60
Council for the Protection of Rural England 85
Countryside Agency 85
Countryside Premium Scheme 250
Countryside Commission 250, 252
Countryside Stewardship Scheme (CSS) 222, 242, 247–55
    Arable Option 242, 254
Cover crops 103
Credit 176–7, 195, 206
Crick, Francis 259
Crofting 218
Crop combinations 24
Crop rotation 15–16, 24, 150, 154, 220, 233, 241
    Norfolk four-course 15–16, 154
Crop yield 4–7, 20, 25, 63, 194–5
Cross-compliance 106
Cross-over pattern 160–2
Crow Rate 122
Cuba 148
Cultural ecology 151
Cultural turn 41–2, 242, 262
Culture economies 85
Czech Republic 108, 238

Dairying/Dairy cattle 91, 96, 99, 103–5, 112, 118, 124, 126, 140, 144, 222–3, 239, 252
De-agrarianisation 162
Decentralisation 210
Decoupling 117
Decree 207 186
Deer 223
Deforestation 153–4, 198
DeKalb 55
Del Monte 71, 168
Delocalisation 74, 76
Democratic Party 169
Denmark 79, 81, 91, 93, 169